Jennifer E. Van Eyk and Michael J. Dunn (Editors)

Proteomic and Genomic Analysis of Cardiovascular Disease

Related Titles from Wiley-VCH:

Stefan Lorkowski and Paul Cullen (Eds.)

Analysing Gene Expression
A Handbook of Methods. Possibilities and Pitfalls
(2 Volumes)

2002
ISBN 3-527-30488-6

Reiner Westermeier, Tom Naven

Proteomics in Practice
A Laboratory Manual of Proteome Analysis

2002
ISBN 3-527-30354-5

Proteomics
The premier international source for information in the field of proteomics
(Journal with 12 issues a year)

ISSN 1615-9853
http://www.wiley-vch.de/home/proteomics

Proteomic and Genomic Analysis of Cardiovascular Disease

Edited by:
Jennifer E. Van Eyk and Michael J. Dunn

WILEY-VCH

WILEY-VCH GmbH & Co. KGaA

Editors:

Jennifer E. Van Eyk
Department of Physiology
Queen's University
429 Botterell Hall
Kingston, Ontario K7L 2N6
Canada

Michael J. Dunn
Department of Neuroscience
King's College
De Crespigny Park
London, SE5 8AF
UK

Library of Congress Card No.: applied for

British Library Cataloguing-in-Publication Data
A catalogue record for this book is available from the British Library.

**Bibliographic information published
by Die Deutsche Bibliothek**
Die Deutsche Bibliothek lists this publication in the Deutsche Nationalbibliografie; detailed bibliographic data is available in the Internet at <http://dnb.ddb.de>

© 2003 WILEY-VCH Verlag GmbH & Co. KGaA, Weinheim

Printed in the Federal Republic of Germany
Printed on acid-free paper

Composition   K+V Fotosatz GmbH, Beerfelden
Printing   Druckhaus Darmstadt GmbH, Darmstadt
Bookbinding   Litges & Dopf Buchbinderei, Heppenheim

ISBN   3-527-30596-3

Preface

Over the last several years, the increased level of interest in and recognition of the power and potential of genomics and proteomics has been astonishing. The use of these approaches for the study of complex disease is accelerating and is especially applicable to cardiovascular disease, the number one cause of death in the world.

This is the first book in which leaders in the fields of genomics and proteomics address current issues about their applications to cardiovascular disease. By exploring the various strategies and technical aspects of both, using examples from cardiac or vascular biology, the limitations and the potential of these methods can be clearly seen. The book is divided into three sections: the first focuses on genomics, the second on proteomics, and the third provides an overview of the importance of these two scientific disciplines in drug and diagnostic discovery. The goal of this book is the transfer of their hard-earned lessons to the growing number of cardiovascular researchers using these techniques, with the hope of saving time and accelerating the pace of discovery.

The study of genomics most often entails the quantification of the transcriptome – the many thousands of messenger RNA's associated with the cellular response to stress and disease. Chapters by M. Schinke, L. Riggi, A.J. Butte and S. Izumo; C.C. Liew provide wonderful overviews of large-scale expression profiling in several chronic cardiac diseases. The strategic choices and decisions required to create a valid genomic study are described by M. Krenz, G. Schwartzbauer and J. Robbins – and emphasizes that physiology and genomics complement one another. Equally eloquent is the chapter by T.T. Tuomisto and S. Ylä-Herttuala in which they examine the use of genomics to study atherosclerosis and plaque formation – an area not yet influenced by proteomics. Although genomic analysis is often focused on chronic heart failure (e.g., S. Arab, M. Husain and P. Liu), it can also be applied to more acute settings, such as myocardial protection (E. Murphy, C.A. Afshari, J.G. Petranka and C. Steenbergen) and the stress response of the pulmonary vascular system (S.-F. Ma, S.Q. Ye and J.G.N. Garcia); these chapters provide insight into these distinct but clinically relevant issues.

Most genomic chapters focus on DNA microarrays; however, the use of subtractive hybridization in gene profiling is highlighted in the chapter by C. Depre. Throughout these powerful chapters on genomics and transcriptomics are the hard-learned lessons; one such cautionary and sobering tale is retold (A.C. Alpin and C.E. Murry) which also underscores the chapter entitled the skeptic's viewpoint (L. Adams, R.E. Bumgarner and S.M. Schwartz). Overall, the chapters balance the advantages and limitations of genomics, and outline the remaining challenges while acknowledging the enormous quantity of information that can be gleaned by this approach.

Proteomics, the study of the proteome, is still in its infancy. Its technical side is benefiting from new approaches and strategies based on previous studies. Technically, proteomics essentially consists of protein separation, identification and characterization (co- and post-translational modifications, function, cell localization, structure and binding partners). Although straightforward in theory, the technologies utilized in proteomics are challenging in practice. This is, in part, a reflection of the complexity of the proteome – a result of the complex chemical nature of proteins, each with its unique amino acid sequence, as well as their many (over 100) possible co- and post-translational modifications, modifications that change their chemical properties. Thus, separation and identification of these extremely complex protein mixtures require skill, fortitude, and patience (of both the investigator and the granting agency).

The classical, broad-based screening approach based on two-dimensional gel electrophoretic separation of proteins coupled with their identification by mass spectrometry is highlighted by one of the major founders of proteomics in cardiac disease (E. McGregor and M.J. Dunn). The underlying technologies of 2-DE and mass spectrometry are still the most commonly used methods for studying changes within the proteome; as such, mass spectrometry is vital. An overview of the evolution of mass spectrometry and its diverse roles in proteomics (H.L. Byers and M.A. Ward) is complemented by the chapter by M.E. Wright and R. Aebersold writing on ICAT, the first robust method for differential quantification of mass spectrometry data, and that by R.E. Jenkins and S.R. Pennington on the application of new protein chip/mass spectrometry techniques. This is followed by a series of chapters focusing on subproteomics – the detailed study of specific groups of proteins within cardiac and vascular cells. Subproteomic analysis can be based on: investigation of discrete cellular organelles (R.A. Gottlieb and H. He); dynamic signaling cascades (J.A. MacDonald and T.A.J. Haystead; T.M. Vondriska, J. Zhang and P. Ping); functional groups either within cells (J.A. Simpson, S. Iscoe and J.E. Van Eyk) or external to cells, as with the study of secreted proteins and peptides (Z.-G., Jin, D.-F. Liao, C. Yan and B.C. Berk). Collectively, these chapters describe strategies that allow detailed insights into the dynamics of the cell.

Translating the identification of a gene or protein change in disease to an approved commercial product – either a diagnostic marker or therapeutic drug – is a long and difficult process. The final two chapters by A.J. Marion and M.H. Gollob and by S.T. Rapundalo explore this process and the promise of genomics and proteomics, respectively, with regard to drug discovery. Both provide an interesting and insightful perspective on the use of these scientific disciplines in the broader scheme of clinical and commercial application. The detailed knowledge that can be obtained from proteomics and genomics should provide a greater probability of success.

Genomics and proteomics are two sides of the same coin. The link between these two disciplines, however, has yet to be fully appreciated and exploited; this should occur in the next few years.

The chapters in this book collectively provide an overview of the current status of genomics and proteomics in the cardiovascular field. When looked at in its entirety, one obtains tantalizing glimpses into the incredible complexity of cellular function – the complexity that drives the cell's adaptation to stress and its environment, and how disease alters its normal physiological response, whether compensatory or pathological. New and exciting discoveries based on genomic and proteomic strategies provide insight into cardiovascular disease, information that may aid in development of new diagnostic markers and therapeutic approaches for its control and treatment. While editors regularly express the sentiment that "the future holds unlimited potential," the examples provided herein are a testament to the fact that this potential is already being realized. What remains to be determined is the magnitude of their influence in shaping future scientific discovery and how this will impact on clinical treatment of cardiovascular disease.

Jennifer E. Van Eyk and Michael J. Dunn
Kingston (Canada) and London (UK)
December 2002

Contents

List of Contributors

LAWRENCE ADAMS
Department of Pathology
University of Washington
Box 357335, 1959 NE Pacific Street
Seattle, WA 98195-7335
USA

RUEDI AEBERSOLD
Institute for Systems Biology
1441 North 34th Street
Seattle, WA 98103-8904
USA
E-mail: raebersold@systemsbiology.org
http://www.systemsbiology.org

CYNTHIA A. AFSHARI
Microarray Group
National Institute of Environmental
Health Sciences, NIH
111 Alexander Drive
Research Triangle Park, NC 27709
USA
E-mail: afshari@niehs.nih.gov
http://div.niehs.nih.gov/microarray/
home.htm

ALFRED C. APLIN
Department of Pathology
University of Washington
Seattle, WA 98195-7335
USA

SARA ARAB
Heart & Stroke
Richard Lewar Center of Excellence
EN 12-324
Toronto General Research Institute
200 Elizabeth Street
Toronto, ON M5G 2C4
Canada

BRADFORD C. BERK
Aab Institute of Biomedical Sciences
University of Rochester
601 Elmwood Avenue
Rochester, NY, 14642
USA
E-mail:
bradford_berk@urmc.rochester.edu

ROGER E. BUMGARNER
Departments of Microbiology
University of Washington
Seattle, WA 98195
USA
E-mail: rogerb@u.washington.edu

ATUL J. BUTTE
Children's Hospital Informatics
Program
320 Longwood Avenue
Boston, MA 02115
E-mail: atul_butte@harvard.edu
http://www.chip.org

Helen L. Byers
Proteome Sciences plc
Institute of Psychiatry
Kings College, De Crespigny Park
London, SE5 8AF
UK

Christophe Depre
Cardiovascular Research Institute
University of Medicine and Dentistry
New Jersey Medical School
185 South Orange Avenue
Newark, NJ 07103
USA
E-mail: deprech@umdnj.edu

Michael J. Dunn
Department of Neurosciences
Institute of Psychiatry
Kings College, De Crespigny Park
London, SE5 8AF
UK
E-mail: m.dunn@iop.kcl.ac.uk

Joe G. N. Garcia
Division of Pulmonary and Critical
Care Medicine
Johns Hopkins University School
of Medicine
5501 Hopkins Bayview Circle
Baltimore, MD 21224
USA
E-mail: drgarcia@jhmi.edu

Michael H. Gollob
Section of Cardiology
Baylor College of Medicine
One Baylor Plaza, 506D
Houston, TX, 77030
USA

Roberta A. Gottlieb
Department of Molecular
& Experimental Medicine
The Scripps Research Institute MEM220
10550 N. Torrey Pines Road
La Jolla, CA 92037
USA
E-mail: robbieg@scripps.edu

Timothy A. J. Haystead
Department of Pharmacology
and Cancer Biology
Center for Chemical Biology
Duke University Medical Center
Durham, NC 27710
USA
E-mail: hayst001@mc.duke.edu
http://pharmacology.mc.duke.edu

Huaping He
Department of Molecular
& Experimental Medicine
The Scripps Research Institute
MitoKor, 11494 Sorrento Valley Road
San Diego, CA 92121
USA
E-mail: hej@mitokor.com

Mansoor Husain
Heart & Stroke
Richard Lewar Center of Excellence
EN 12-324
Toronto General Research Institute
200 Elizabeth Street
Toronto, ON M5G 2C4
Canada

Steven D. Iscoe
Department of Physiology
Queen's University
Kingston ON, K7L 3N6
Canada
E-mail: iscoes@post.queensu.ca

SEIGO IZUMO
Beth Israel Deaconess Medical Center
330 Longwood Avenue
Boston, MA 02215
USA
E-mail: sizumo@caregroup.harvard.edu

ROSALIND E. JENKINS
Department of Human
Anatomy & Cell Biology
University of Liverpool
Ashton Street
Liverpool, L69 3GE
UK
E-mail: r.jenkins@liverpool.ac.uk

ZHENG-GEN JIN
Center for Cardiovascular Research
University of Rochester
601 Elmwood Avenue
Rochester, NY 14642
USA
E-mail:
zheng-gen_jin@urwc.rochester.edu

MAIKE KRENZ
Division of Molecular Cardiovascular
Biology, MLC 7020
Children's Hospital Research Foundation
3333 Burnet Avenue
Cincinnati, OH 45229-3039
USA

DUAN-FANG LIAO
Center for Cardiovascular Research
University of Rochester
601 Elmwood Avenue
Rochester, NY 14642
USA

CHOONG CHIN LIEW
Cardiovascular Genome Unit
Brigham and Women's Hospital
Harvard Medical School
75 Francis Street Thorn 1334
Boston, MA 02115
E-mail: Cliew@rics.bwh.harvard.edu

PETER LIU
Heart & Stroke
Richard Lewar Center of Excellence
EN 12-324
Toronto General Research Institute
200 Elizabeth Street
Toronto, ON M5G 2C4
Canada
E-mail: peter.liu@utoronto.ca

SHWU-FAN MA
Division of Pulmonary and Critical
Care Medicine
Johns Hopkins University School
of Medicine
5501 Hopkins Bayview Circle
Baltimore, MD 21224
USA

JUSTIN A. MacDONALD
Department of Pharmacology
and Cancer Biology
Center for Chemical Biology
Duke University Medical Center
Durham, NC 27710
USA

ALI J. MARIAN
Section of Cardiology
Baylor College of Medicine
One Baylor Plaza, 506D
Houston, TX 77030
USA
E-mail: amarian@bcm.tmc.edu
http://public.bcm.tmc.edu

EMMA McGREGOR
Proteome Sciences plc
Institute of Psychiatry
Kings College, De Crespigny Park
London, SE5 8AF
UK

ELIZABETH MURPHY
Laboratory of Signal Transduction
National Institute of Environmental
Health Sciences, NIEHS
111 Alexander Drive
Research Triangle Park, NC 27709
USA
E-mail: murphy1@niehs.nih.gov
http://dir.niehs.nih.gov/dirlst/home.htm

CHARLES E. MURRY
Department of Pathology
Box 357470
University of Washington
Seattle, WA 98195-7335
USA
E-mail: murry@u.washington.edu

STEPHAN R. PENNINGTON
Department of Human
Anatomy & Cell Biology
University of Liverpool
Ashton Street
Liverpool, L69 3GE
UK
E-mail: srpenn@liv.ac.uk
http://www.liv.ac.uk/hacb

JOHN G. PETRANKA
Laboratory of Signal Transduction
National Institute of Environmental
Health Sciences, NIH
111 Alexander Drive
Research Triangle Park, NC 27709
USA
E-mail: petranka@niehs.nih.gov

PEIPEI PING
Cardiology Research, Department
of Physiology & Biophysics, Suite 122
Baxter Biomedical Research Building
570 South Preston St.
Louisville, KY 40202-1783
USA
E-mail: ping@ntr.net
http://Louisville.edu/medschool/
physiology

STEPHEN T. RAPUNDALO
Department of Cardiovascular
Molecular Sciences
Pfizer Global R & D
Ann Arbor Laboratories
2800 Plymouth Road
Ann Arbor, MI 48105
USA
E-mail: stephen.rapundalo@pfizer.com

LAUREN RIGGI
Beth Israel Deaconess Medical Center
Harvard Medical School
300 Longwood Avenue
Boston, MA 02115
USA
E-mail: lriggi@caregroup.harvard.edu

JEFFREY ROBBINS
Division of Molecular Cardiovascular
Biology, MLC 7020
Children's Hospital Research Foundation
3333 Burnet Avenue
Cincinnati, OH 45229-3039
USA
E-mail: jeff.robbins@chmcc.org
http://www.cincinnatichildrens.org

MARTINA SCHINKE
Beth Israel Deaconess Medical Center
Cardiovascular Division SL 215
330 Brookline Ave.
Boston, MA 02215
USA
E-mail:
mschinke@caregroup.harvard.edu
http://www.cardiogenomics.org

STEPHEN M. SCHWARTZ
Department of Pathology
University of Washington
Box 357335, 1959 NE Pacific Street
Seattle, WA 98195-7335
USA
E-mail: steves@u.washington.edu

GARY SCHWARTZBAUER
Division of Molecular Cardiovascular
Biology, MLC 7020
Children's Hospital Research Foundation
3333 Burnet Avenue
Cincinnati, OH 45229-3039
USA

JEREMY A. SIMPSON
Department of Physiology
Queen's University
Kingston ON, K7L 3N6
Canada

CHARLES STEENBERGEN
Department of Pathology
Duke University Medical Center
Durham, NC 27710
USA
E-mail: Steen001@mc.duke.edu

TIINA T. TUOMISTO
A.I. Virtanen Institute of Molecular
Sciences
University of Kuopio
P.O. Box 1627
FIN-70211 Kuopio
Finland
E-mail: Tiina.Tuomisto@uku.fi

JENNIFER E. VAN EYK
Department of Physiology
Queen's University
Kingston ON, K7L 3N6
Canada
E-mail: jve1@post.queensu.ca
http://meds.queensu.ca/medicine/
physiol/about.html

THOMAS M. VONDRISKA
Departments of Physiology
& Biophysics and Medicine
Division of Cardiology
University of Louisville
Louisville, KY 40202
USA
E-mail: tvondriska@hotmail.com

MALCOLM A. WARD
Proteome Sciences plc
Institute of Psychiatry
Kings College, De Crespigny Park
Box P045, South Wing Laboratory
London, SE5 8AF
UK
E-mail: m.ward@iop.kcl.ac.uk

MICHAEL E. WRIGHT
Institute for Systems Biology
1441 North 34th Street
Seattle, WA 98103
USA
E-mail: mwright@systemsbiology.org

CHEN YAN
Center for Cardiovascular Research
University of Rochester
601 Elmwood Avenue
Rochester, NY 14642
USA
E-mail: chen_yan@urmc.rochester.edu

SHUI QING YE
Division of Pulmonary and Critical
Care Medicine
Johns Hopkins University School of
Medicine
5501 Hopkins Bayview Circle
Baltimore, MD 21224
USA

SEPPO YLÄ-HERTTUALA
A.I. Virtanen Institute of Molecular
Sciences
University of Kuopio
P.O. Box 1627
FIN-70211 Kuopio
Finland
E-mail: Seppo.Ylaherttuala@uku.fi
http:/www.uku.fi/laitokset/aivi/index.htm

JUN ZHANG
Departments of Physiology
& Biophysics and Medicine
Division of Cardiology
University of Louisville
Louisville, KY 40202
USA
E-mail: J0Zeng1@louisville.edu

Abbreviations

2DE	two-dimensional polyacrylamide-gel electrophoresis
ACAT	acyl coenzyme A-cholesterol acyltransferase
ADME	absorption, distribution, metabolism, elimination/excretion
Ang II	angiotensin II
Apo	apolipoprotein
ApoE	apolipoprotein E
ASMC	arterial smooth muscle cells
Avg Diff	average difference
BAD	bipolar affective disorder
Bcl-2	B-cell leukemia/lymphoma 2
CALI	chromophore-assisted laser inactivation
CBB	colloidal coomassie blue
cDNA	complementary DNA
CI	chemical ionisation
CID	collision induced dissociation
CIEF	capillary isoelectric focusing
CNL	constant neutral loss
CNS	central nervous system
CS	cyclic stretch
COPD	chronic obstructive pulmonary disease
CSF	colony stimulating factor
CuZnSOD	cytosolic copper/ zinc-containing superoxide dismutase
CV	cardiovascular
CZE	capillary zone electrophoresis
DCM	dilated cardiomyopathy
DIGE	differential gel electrophoresis
DTT	dithiothreitol
Egr-1	early growth response 1
EH	enoyl-CoA-hydratase
EI	electron impact
ERK	extracellular signal regulated kinase
ESI	electro spray ionization
EST	expressed sequence tag

FALI	fluorophore assisted light inactivation
FDA	Federal Drug Administration
FGF	fibroblast growth factor
FHC	familial hypertrophic cardiomyopathy
FTICR	fourier transform ion cyclotron resonance mass spectrometry
FPP	farnesyl pyrophosphate
GABA	γ-aminobutyric acid
GIST	global internal standard technology
GOLD	genomics online database
GPX	glutathione peroxidase
HAEC	human arterial endothelial cells
HCM	hypertrophic cardiomyopathy
HF	heart failure
HGP	human genome project
HIV	human immunodeficiency virus
HMG Co A	hydroxy methylglutaryl coenzyme A synthase
µHPLC-ESI-MS/MS	microcapillary reversed-phase high performance liquid chromatography electrospray ionization tandem mass spectrometry
HTS	high throughput screening
HUVEC	human umbilical vein endothelial cells
ICAM	intracellular adhesion molecule
ICAT™	isotope-coded affinity tags
IEF	isoelectric focusing
IFN-γ	interferon-γ
IGF-1	insulin like growth factor-1
IL	interleukin
IPG	immobilized pH gradient
IPP	isopentenyl-diphosphate δ-isomerase
LC	liquid chromatography
LDL	low density lipoprotein
LSS	laminar shear stress
LV	left ventricle
MAS	Affymetrix® Microarray Suite
MDLC	multidimensional liquid chromatography
MGED	microarray gene expression database
MHC	myosin heavy chain
MM	mismatch
MMAC1	*m*utated in *m*ultiple *a*dvanced *c*ancers-1
MMP	matrix metalloproteinases
MnSOD	mitochondrial manganese-containing superoxide dismutase
mRNA	messenger RNA
MS	mass spectrometry
MudPIT	multidimensional protein identification technology
NF-κB	nuclear factor-κB
NO	nitric oxide

NSAID	non-steroidal anti-inflammatory drugs
PCR	polymerase chain reaction
PDGF	platelet-derived growth factor
PECAM	platelet-endothelial cell adhesion molecule
pI	isoelectric point
PIP2	phosphatidylinositol 3,4-biphosphate
PIP3	phosphatidylinositol 3,4,5-triphosphate
PK	pharmacokinetic
PKCε	protein kinase C-epsilon
PM	perfect match
PMA	phorbol 12-myristate 13-acetate
PMF	peptide mass fingerprinting
PPAR	peroxisome proliferator-activated receptors
PPIase	peptidyl-prolyl cis-trans isomerase
PSD	post-source decay
PTEN	*phosphatase and tensin* homologue on chromosome *ten*
PTM	post-translational modification
Q-TOF	quadrupole time of flight
ROS	reactive oxygen species
R & D	Research & Development
RP-HPLC	reversed-phase high performance liquid chromatography
RT-PCR	real time polymerase chain reaction
RT-PCR	reverse transcriptase PCR
SAGE	serial analysis of gene expression
SAR	structure activity relationship
SCA	synthetic carrier ampholytes
SDS-PAGE	sodium dodecyl sulfate polyacrylamide gel electrophoresis
SELDI	surface enhanced laser desorption/ionization
SMC	smooth muscle cell
SNP	single nucleotide polymorphism
SOM	self-organizing map
SOXF	secreted oxidative stress-induced factors
SQS	squalene synthase
TAP	tandem affinity purification
TEP1	*TG*F-regulated, *e*pithelial cell *e*nriched *p*hosphatase
THP-1	human monocyte/macrophage cell line
TIMP	tissue inhibitor of metalloproteinase
TNF*a*	tumor necrosis factor *a*
TSS	turbulent shear stress
UCH	ubiquitin carboxyl-terminal hydrolase
UV	ultra violet
VCAM	vascular cell adhesion molecule
VEGF	vascular endothelial growth factor
VLDL	very low density lipoprotein
WHHL	Watanabe heritable hyperlipidemic

Section 1
Genomics

1

Large Scale Expression Profiling
in Cardiovascular Disease Using Microarrays:
Prospects and Pitfalls

Martina Schinke, Lauren Riggi, Atul J. Butte, and Seigo Izumo

1.1
DNA Microarray Technologies

Microarray technologies have revolutionized the way we identify gene expression changes in biological events [1] and complex diseases [2]. Instead of looking at a handful of genes, recent technology advances allow us to scan up to 12,000 transcripts at once. As the draft sequencing of the human genome has been completed and it becomes clear that it merely contains 33,000 genes [3], it is hoped that global expression profiling integrated with other types of genomic information will fully describe the regulation of gene-expression networks and how they malfunction in disease.

In this chapter we will provide an overview of current microarray technologies, describe how to design meaningful microarray experiments, and address the type of answers it can be expected to provide. We also consider the pitfalls and shortcomings of the technology, and different approaches of data analysis developed to deal with the data avalanche. Finally, we review current literature to see how microarray studies have helped us to decipher the transcriptional regulatory network underlying cardiovascular diseases.

1.1.1
cDNA Microarrays or Oligonucleotide Arrays?

The construction of gene expression databases requires technologies that can accurately and reproducibly measure changes in global mRNA expression levels. Ideally, these technologies should be able to screen all gene transcripts, be applicable across a wide range of cell and tissue types, require minimal amounts of biological material, and be capable of processing large number of samples. Although several different technology platforms have been developed [4–9], each comes with its own set of advantages and limitations in meeting these stringent requirements (Fig. 1.2). The two most commonly used platforms are complementary DNA (cDNA) microarrays and high-density oligonucleotide arrays.

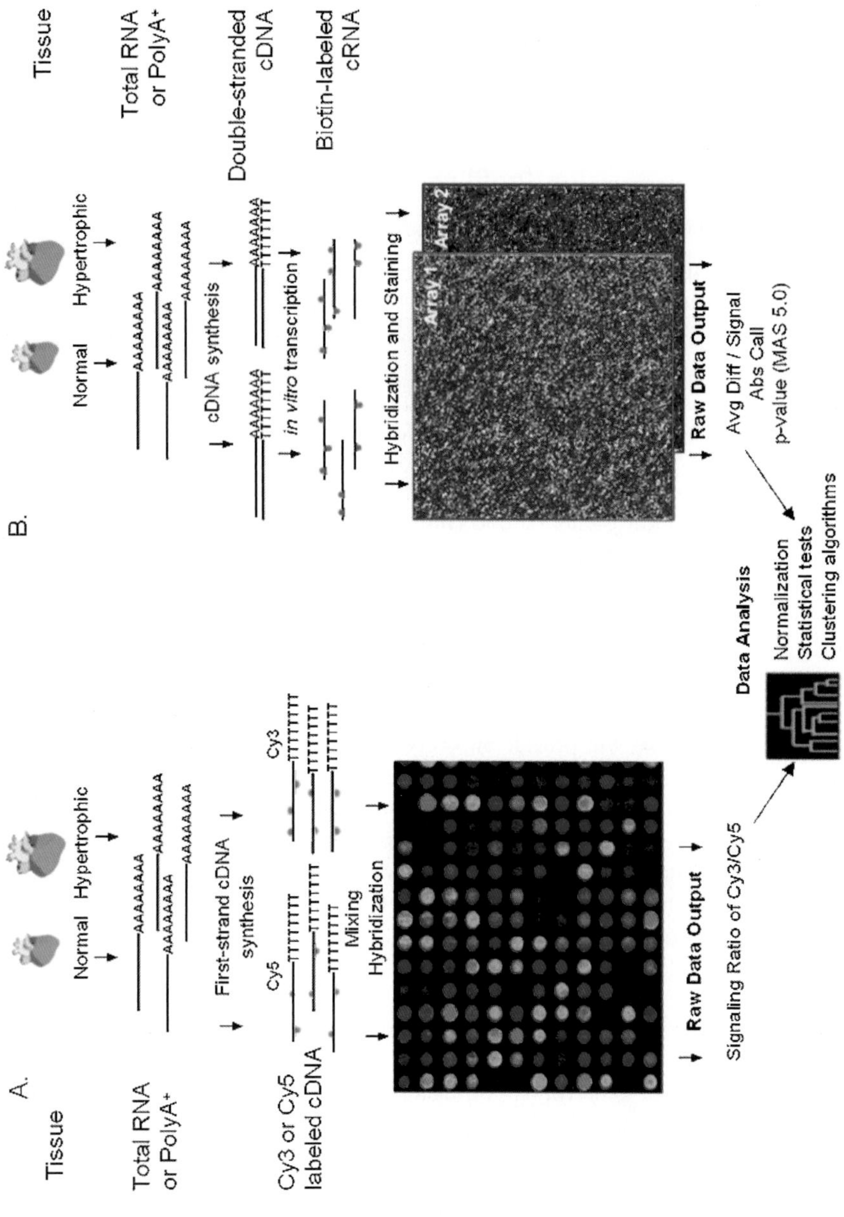
A.
Tissue
Normal
Hypertrophic
Total RNA or PolyA+
AAAAAAAA
First-strand cDNA synthesis
Cy5
Cy3
TTTTTTTT
Cy3 or Cy5 labeled cDNA
Mixing
Hybridization
Raw Data Output
Signaling Ratio of Cy3/Cy5
B.
Tissue
Normal
Hypertrophic
Total RNA or PolyA+
cDNA synthesis
Double-stranded cDNA
in vitro transcription
Biotin-labeled cRNA
Hybridization and Staining
Array 1
Array 2
Raw Data Output
Avg Diff / Signal
Abs Call
p-value (MAS 5.0)
Data Analysis
Normalization
Statistical tests
Clustering algorithms

cDNA microarrays have been developed by several academic groups as well as commercial suppliers. In general, cDNA libraries or clone collections are amplified by polymerase chain reaction (PCR) at an average product size of 1,000 nucleotides in length. These PCR products are printed in a two-dimensional grid onto glass slides or nylon membranes as spots at defined locations. Spots are typically 100–300 μm in size and are spaced about the same distance apart. Using this technique, arrays consisting of more than 30,000 cDNAs can be fitted onto the surface of a conventional microscope slide (Fig. 1.2).

The process of spotting is generally not accurate enough to allow direct comparison between different arrays. Therefore, the two RNA samples from tissues to be compared are typically used to generate first-strand cDNA targets labeled with two different fluorescent dyes, for example Cy3 and Cy5. These are then purified, pooled and hybridized to the same array, resulting in competitive binding of the differentially labeled cDNAs to the arrayed sequences (Fig. 1.1). After hybridization and washing, the slide is scanned in a high-resolution confocal fluorescent scanner using two different wavelengths corresponding to the dyes used, and the intensity of the same spot in both channels is compared. From these measurements, a ratio of transcript levels for each gene represented on the array can be calculated. In order to be able to compare results from different arrays to each other, a reference RNA, e.g. a mixture of all the samples of one experiment or a commercially available standard, is used to normalize the Cy3 and Cy5 intensities (see 2.2 Comparing expression data).

Protocols for total RNA isolation and nucleic acid labeling are available on the internet from a variety of academic laboratories (Tab. 1.1). To achieve the most linear relationship between starting material and labeled probe, incorporation of fluorescent-labeled nucleotides during first strand cDNA synthesis has been the method of choice. However, this requires up to 100 μg of total RNA as starting material, which excludes studies using primary cells or human tissues of limited availability.

Fig. 1.1 Schematic overview of cDNA and oligonucleotide microarray sample preparation. Total RNA is isolated from tissues of the experimental sample (e.g. hypertrophic heart), and from the control sample (e.g. normal heart). **A** For cDNA microarrays, RNA from both samples is reverse transcribed into single-stranded cDNA in the presence of two fluorescent dyes (such as Cy3 and Cy5). Both samples are mixed in the hybridization buffer and hybridized to the array, usually under a coverslip. After washing, array slides are scanned by a high-resolution confocal fluorescence scanner with two wavelengths corresponding to the two fluorescent dyes used, and independent images for the control and experimental channels are generated. Signal intensities for both dyes are used to calculate the ratios of mRNA abundance. **B** Total RNA is isolated from the experimental and control tissue sample, and cDNA is synthesized using an Oligo-dT primer that carries the T7 promotor sequence. T7 DNA polymerase is then used for in vitro transcription with biotin-labeled nucleotides that become incorporated into the cRNA. Each labeled cRNA is then hybridized to an array in a rotating hybridization oven. After hybridization, the array is washed and stained using a fluorescent dye coupled to streptavidin. Arrays are scanned and signal intensities are converted into Average Difference (AvgDiff) values or Signals. Modified from [64].

Tab. 1.1 Links to Microarray Resources. This table provides a selection of commercial and academic facilities and resources related to microarray technologies

---

Microarray Core Facilities

http://sequence-www.stanford.edu/	Stanford DNA Sequencing and Technology Center
http://www.cgr.harvard.edu	Bauer Center for Genomics Research
http://www.nhgri.nih.gov/DIR/Microarray/main.html	National Human Genome Research Institute
http://microarrays.com	Microarrays Inc., Nashville, TN

Public Microarray Databases

http://pga.lbl.gov/PGA/PGA_inventory.html	NHLBI Programs for Genomic Applications Inventory Site
http://www.cardiogenomics.org	Raw data and normalized data related to cardiomyopathies in mouse models and humans using Affymetrix GeneChips^{TM}
http://www.dnachip.org	Microarray data storage, image files, data retrieval, analysis, and visualization
http://www.ncbi.nlm.nih.gov/geo/	Gene expression and hybridization array data repository, and online resource for the retrieval of gene expression data from any organism or artificial source
http://www.ebi.ac.uk/microarray/ArrayExpress/	Public database for microarray based gene expression data

Protocols and Software

https://www.affymetrix.com/	Affymetrix online resource center for GeneChip arrays
http://biosun1.harvard.edu/complab/dchip/	dCHIP, freely available analysis tool for Affymetrix GeneChip arrays
http://cmgm.stanford.edu/pbrown/protocols/index.html	A guide to cDNA microarrays
http://www.microarrays.org	Public source for microarray protocols and software
http://www.tigr.org/software/	Software tools freely available to the scientific community
http://rana.lbl.gov/EisenSoftware.htm	Software tools for image analysis, and cluster analysis and visualization
http://www.biodiscovery.com/	Commercial software products for microarray research and analysis
http://ep.ebi.ac.uk/	Set of tools for clustering, analysis and visualization of gene expression data

Development of standards for array experiment annotation and data representation

http://www.mged.org	Microarray Gene Expression Database Group

---

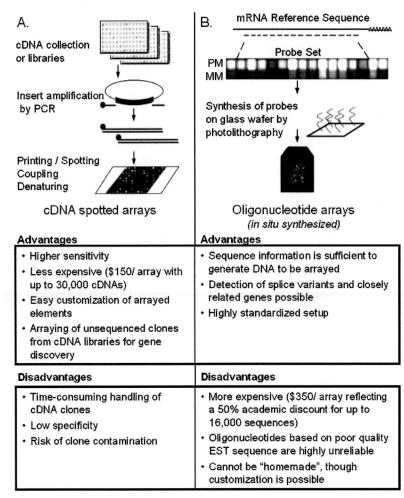

A.

cDNA collection or libraries

Insert amplification by PCR

Printing / Spotting
Coupling
Denaturing

cDNA spotted arrays

B.

mRNA Reference Sequence

Probe Set

PM
MM

Synthesis of probes on glass wafer by photolithography

Oligonucleotide arrays
(in situ synthesized)

Advantages	Advantages
• Higher sensitivity • Less expensive ($150/ array with up to 30,000 cDNAs) • Easy customization of arrayed elements • Arraying of unsequenced clones from cDNA libraries for gene discovery	• Sequence information is sufficient to generate DNA to be arrayed • Detection of splice variants and closely related genes possible • Highly standardized setup
Disadvantages	**Disadvantages**
• Time-consuming handling of cDNA clones • Low specificity • Risk of clone contamination	• More expensive ($350/ array reflecting a 50% academic discount for up to 16,000 sequences) • Oligonucleotides based on poor quality EST sequence are highly unreliable • Cannot be "homemade", though customization is possible

Fig. 1.2 Schematic overview of the manufacturing process, and a summary of advantages and disadvantages of cDNA versus oligonucleotide arrays. **A** Inserts from cDNA libraries or clone collections are amplified by PCR. PCR products are spotted or printed by an arrayer or through ink-jet technology onto glass slides or nylon membranes at specified intervals, for example by using chemical linkers that covalently bind the DNA to the glass slide. **B** On Affymetrix GeneChips™, each gene is represented by a probe set of 16–20 short oligonucleotides (25mers) that are chosen from different regions of the gene. Each perfect match (PM) oligonucleotide is paired with an oligonucleotide that has a mismatch (MM) base in the central position, making up a probe pair. Oligonucleotides are synthesized in-situ to a glass wafer by a light-directed chemical synthesis process called photolithography. Multiple probe arrays are synthesized simultaneously on a large glass wafer, which is then diced into individual probe arrays. Prices for cDNA microarrays and Affymetrix GeneChips™ are estimates as of 3/2002.

In contrast to high-density oligonucleotide arrays, cDNA are "home-made" by several academic laboratories. Detailed information about the manufacture of cDNA microarrays is available from Pat Brown's laboratory in Stanford (Tab. 1.1). The advantage of cDNA microarrays is the great degree of flexibility in the choice of arrayed elements, allowing for the preparation of customized and tissue-specific arrays for specific investigations. In addition, spotting clones from unsequenced libraries makes a great tool for new gene discovery. These advantages, together with the comparatively low costs, have made cDNA microarrays the most frequently used platform in academic institutions (Fig. 1.2). However, cDNA microarrays require the time-consuming handling and amplification of cDNA libraries or clone collections (I.M.A.G.E., Riken), which carry a high risk of clone contamination. Using oligonucleotides (50mers and up) instead of cDNAs might provide a solution to this problem, but on the expense of cost efficiency and detection sensitivity [10]. Spotted or printed cDNA microarrays and related technology are available from Agilent, Ambion, Clontech, Packard Bioscience/Perkin Elmer, and others, as well as through Core Facilities that have been established at diverse academic institutions (Tab. 1.1).

High density oligonucleotide microarrays consist of short 20–25mer oligonucleotides that are either printed onto glass slides or synthesized in situ by ink-jet technology (Agilent Technologies) or by photolithography onto silicon wafers (Affymetrix). Affymetrix high-density oligonucleotide GeneChips™ contain up to 1,000,000 unique oligonucleotide features covering more than 39,000 transcript variants (represented on two microarrays HG-U133A and HG-U133B). Each gene is represented by at least one set of 11–20 different "probe pairs" (Fig. 1.2). A probe pair consists of a 25-base-pair (bp) "perfect-match" (PM) oligonucleotide probe and a 25-bp "mismatch" probe (MM), in which the 13th position is designed not to match the target (cellular) sequence. The information across all 20 paired PM and MM probes (the "probe set") is integrated by proprietary Affymetrix Microarray Suite (MAS) software.

For hybridization of eukaryotic samples to Affymetrix GeneChips™, double-stranded cDNA is synthesized from 5 µg total RNA (or a minimum of 0.2 µg purified poly(A)+ messenger RNA) isolated from tissue or cells. The first strand synthesis is primed using a T7-(dT)$_{24}$ primer. After second strand synthesis, an in vitro transcription (IVT) reaction is done to produce biotin-labeled cRNA from the cDNA template. The cRNA is fragmented before hybridization. In contrast to cDNA microarrays, each mRNA preparation is hybridized to a separate oligonucleotide array. Hybridization, washing, and staining of Affymetrix GeneChips™ takes place in a highly standardized setup, which greatly reduces the variability of microarray processing. After scanning of the arrays, various analytical methods are applied to distinguish specific signals from noise, calculate background, and to scale and normalize the signals (see 2. Computational analysis of microarray data) before expression levels can be compared across arrays.

If the amount of starting material is not sufficient to generate the minimum amount of RNA needed for the standard protocols, for example if human biopsy samples or microdissected single cells are to be analyzed, alternative labeling protocols have to be used. Starting amounts of less than 100 ng of total RNA require two rounds of amplification, which generally results in a loss of signals from low abundance RNAs. Optimized protocols have been developed that allow to faithfully amplify as little as 2 ng total RNA [11, 12]. However, it is also reported that the quality of the amplified product, as measured by the specific signal intensity after hybridization, drops significantly when less than 10–20 ng of starting material is used [12].

Advantages and disadvantages of both microarray technology platforms are summarized in Fig. 1.2. However, the crucial difference between cDNA and oligonucleotide microarrays is that cDNA microarrays return the amount of each transcript *relative to another sample*, whereas oligonucleotide microarrays theoretically return an *absolute amount* of each transcript. This implies a major difference in the ability to group and universally compare microarrays, which is discussed below under analysis techniques. Regardless of the technology platform chosen, microarray experiments yield far more information than we are used to process. Each experiment typically results in hundreds of genes that are differentially expressed across the time points, conditions, or phenotypes that are being compared. It is extremely important to design meaningful experiments in order to avoid "drowning" in a glut of uninterpretable data.

1.1.2
Designing Meaningful Experiments

The number of review articles on gene expression technologies probably exceeds the number of primary research publications in this field [13]. This is not the result of any paucity of primary data since Microarray Core Facilities have been established in nearly every academic institution, each processing hundreds of microarrays each year, and numerous laboratories manufacture and run their own cDNA microarrays, resulting in millions of independent gene expression measurements. One might argue that the limited availability of efficient publicly available tools for data processing, functional annotation, and literature-searching make it hard to filter through the vast amount of gene expression data, find meaningful results, and integrate it with existing knowledge. The more likely reason is that most scientists have been trained to look for an individual gene or signaling pathway, the expression change of which is responsible for a biological phenomenon or disease state. Facing lists of hundreds of gene expression changes leaves a formidable challenge in identifying changes that are important in establishing the phenotype from changes that are secondary phenomena. Browsing through the lists of genes in order to decide which gene represents a key regulatory molecule or a potential new therapeutic target, we will most likely chose candidate genes that are known to us and conform with pre-existing hypotheses rather than taking the risk to potentially waste time on unknown genes. However, genome-wide expression profiling should lead to the generation of new hypotheses rather than confirm existing pathways. How can we enrich the gene lists for positive candidates and get more meaningful results?

Firstly, performing independent replicates is essential to identify the biological and experimental variability in an experiment and will substantially decrease the number of false positive gene expression changes. In our experience, analyzing independent triplicate samples has been sufficient in most cases. However, high variability or limited availability of starting material, as for human biopsy tissues, might make replications impossible.

Secondly, although this might seem obvious, designing meaningful experiments is critical. In general, cell lines have the advantage of cell type homogeneity and will give relatively little biological variability. When looking for causes of phenotypic changes, for example the induction of differentiation, it is important to look for early events. Comparing undifferentiated to differentiated cells may not give you clues on the initial trigger, e.g. the underlying key regulatory molecules or signaling cascades, but reflect the consequences of the induction of differentiation. Similarly, comparing a transgenic mouse line with its transgene-negative littermate will give one long gene list but it is very difficult, if not impossible, to distinguish between primary and secondary changes in gene expression without additional effort placed into experimental design. You run the risk of identifying differences in expression that are secondary to the continuous overexpression of the transgene rather than being directly connected to the gene of interest.

These problems might be addressed by using inducible systems, for example inducible promoters as the Tet-on and Tet-off systems [14] to regulate transgene expression, together with carefully designed time courses. Experimental animal models for cardiovascular diseases should be analyzed at various time points to account for the onset, progression, and chronic phase of the disease. Integrating the gene expression profiles of different models with related phenotypic pathology, for example comparing mouse models of cardiac hypertrophy induced by transgenic overexpression of hypertrophic stimuli with models of pressure-overload induced hypertrophy, will help to further discriminate between primary causes and secondary changes associated with the disease that is being investigated. Currently, cross-species and cross-platform comparisons are hindered by the fact that individual microarray platforms interrogate different sets of genes, and by the lack of standardized RNA processing and labeling procedures, appropriate software and annotation tools, and the use of a common reference RNA that would allow for cross-experimental normalization (see 2.4. Data Sharing). Currently, only experiments performed by individual laboratories on the same microarray platform can easily be integrated, but only few laboratories have the funding needed to perform such large scale expression profiling studies. To address this problem, the National Heart, Lung, and Blood Institute (NHLBI) launched the "Programs of Genomic Applications (PGA)" initiative. Eleven PGA's were funded to develop genomic resources including microarray and SNP data, animal models, clone collections, and software tools for the scientific community in order to advance scientific research related to heart, lung, blood, and sleep disorders. PGA resources are accessible through the NHLBI and on the CardioGenomics website (Tab. 1.1).

1.2
Computational Analysis of Microarray Data

1.2.1
Raw Data Analysis

Most commercially available scanner manufacturers provide software to process the scanned raw image and to transform the fluorescent intensity pixels into a measure for gene expression, and there are additional image-processing software tools available for cDNA microarrays (Tab. 1.1). In contrast, Affymetrix Microarray Suite (MAS) is the only software to analyze high-density oligonucleotide arrays, with the exception of dCHIP [15], an analysis tool developed by the Wong laboratory at the Harvard School of Public Health that has not found widespread acceptance yet. MAS calculates the expression value based on the intensity of the PM/MM probe pairs (see 1.1 and Fig. 1.2), using the MM oligonucleotide hybridization to access background noise within the PM signal. This concept poses specific problems and will therefore be discussed in greater detail in this section.

MAS calculates a numerical value called the average difference (AvgDiff) for expression intensities of the transcripts as

$$\mathrm{AvgDiff} = \sum(\mathrm{PM\text{-}MM})/(\mathrm{Pairs\ in\ Avg}) \ .$$

For this purpose, the intensity of the MM oligonucleotide is subtracted from the intensity of the PM oligonucleotide and averaged across the "Pairs in Avg", which is the number of probe pairs for which the intensity differences (PM-MM) are within the range of three standard deviations (by default). Thus, in cases where the MM probe has a higher intensity value than the PM, the AvgDiff has a negative value. However, in a biological sense, transcripts can only be absent and have a value of zero or be present and have a positive value. Negative AvgDiff values are frequently observed for transcripts of low abundance.

The design intent behind the MM feature is to quantify background (scanner noise, etc.) and non-specific interaction resulting from cross-hybridization within the PM signal, and thus to provide a more reliable measure of the signal attributable to specific probe-target interactions. However, it has been shown that 66–80% of the MM signal is derived from probe-target interactions rather than from random binding [16]. Therefore, the AvgDiff value will under-represent the actual mRNA concentration.

In the recently released MAS 5.0 software, Affymetrix has changed the algorithm from empirical to statistical and has adapted its terminology to fit more standard terms. The AvgDiff that had been used for empirical expression analysis was changed to the "Signal". The signal is calculated using the One-Step Tukey's Biweight Estimate, which yields a robust weighted mean that is relatively insensitive to outliers. The Tukey's Biweight method gives an estimate of the amount of variation in the data, exactly as standard deviation measures the amount of variation for an average. Still, MAS 5.0 subtracts a "stray signal" estimate from the PM signal that is based on the intensity of the MM signal. However, in cases where the MM signal outweighs the PM signal, an adjusted value is used. While these adjustments will eliminate negative values, they still rely on the MM signal as an indicator for background noise, and will subtract a significant portion of the PM signal that is derived from specific target-hybridization. In our experience, using the Signal value from MAS 5.0 as the raw value for downstream comparison analysis significantly changes the numbers of genes that are differentially expressed, as compared to using the Avg-Diff value from MAS 4.0. It remains to be seen which analysis software provides a more reliable assessment of gene expression. As a side note, both software applications cannot be run on the same workstation. MAS also provides the option for comparison analysis, however, it only allows the user to compare two chips at a time and is therefore not useful as an analysis tool for multiple chip experiments.

Data Quality. It is important to visually inspect the scanned image in order to evaluate the quality of your sample and of the hybridization, and to detect potential flaws related to chip manufacturing. Some examples of flawed arrays include uneven or high background, small white speckles appearing throughout the array, dark circular areas where no hybridization has taken place, holes or cracks (uncommon), dark scratches running through large sets of probes, but may be as subtle as a gradation of intensity due to light leakage into the scanner. Small flaws can be manually masked, but for large flaws running of the labeled probe on a new chip is advisable.

In addition to the visual inspection, spiked in controls are frequently used to evaluate the hybridization efficiency. For Affymetrix arrays, oligonucleotide B2 creates a bright border around the array that is used for automatic placement of the grid. Also, each eukaryotic probe array contains probe sets for several prokaryotic genes that can be labeled and serve as controls for labeling and hybridization efficiencies. The integrity of the RNA, e.g. the efficiency of first strand synthesis (converting mRNA into cDNA), is evaluated by several internal controls, such as GAPDH and β-actin, which are represented by 3 probe pairs corresponding to the 5′ end, 3′ end or the middle part of the transcript. The 5′ value should be at least half of the 3′ value (or more). Other quality measures are RawQ values, which describe the degree of pixel-to-pixel variation among the probe cells used to calculate the background. The level of noise is used as a criterion to determine the significance of differences between PM and MM probe cells. Finally, the scaling factor (SF) is used to normalize the signal across the entire chip to an arbitrary target intensity (see 2.2. Comparing expression data – Normalization). The SF for each given experiment should be within a 2–3 fold range.

Potential Pitfalls. One potential problem for frequent Affymetrix users is the constant change of array types and the inability to compare data from old arrays to the new ones. As an example, the first human chip, HuGeneFL, contained probe sets for 5,600 full-length genes. The next generation of chips covered ~12,000 genes and 50,000 EST clusters that were spread over 5 chips, HG-U95 A-E. Although the majority of genes that were represented on HuGeneFL were preserved, the probe sets for these genes were not identical due to the constant up-dating of public databases and improvements in probe selection. In many cases, genes were represented by completely new probe sets. At this time, the actual oli-gonucleotide sequence was Affymetrix proprietary information, and so it hap-pened that Affymetrix designed probe sets based on GenBank entries that were entered in the incorrect sequence orientation. This led to 25–35% of the probe sets on the mouse (MG-U74v.1) and human (HG-U95Av.1) arrays being defective. Replacement arrays had only been around for about one year when new human HG-U133 arrays were released. This set of two arrays contains over 1,000,000 unique oligonucleotide features, which represent greater than 33,000 of the best characterized human genes. With the completion of the human genome se-quence, Affymetrix now uses genomic sequence information to verify their se-quence selection, orientation, and quality. Also, the actual oligonucleotide se-quences are now accessible on their website (Tab. 1.1). Another potential problem in comparing array data that have been collected over a longer period of time are instrument upgrades. Previous scanner versions used a photon multiplier tube (PMT) gain of 100%, which allowed for fluorescent intensities to become easily sa-turated. New scanner settings use a PMT of 10%, which increases the range of in-tensity values that are detected. This is problematic when your scanner settings have been changed in the middle of an ongoing experiment. Values from arrays that have been scanned at different settings can no longer be directly compared.

1.2.2
Comparing Expression Data

Normalization.   Before expression data from different cDNA or high-density oligo-nucleotide microarrays can be compared to each other, the data need to be nor-malized. Normalization attempts to identify the biological information by remov-ing the impact of non-biological influence on the data, and by correcting for sys-tematic bias in expression data. Systematic bias can be caused by differences in la-beling efficiencies, scanner malfunction, differences in the initial quantity of mRNA, different concentrations of DNA on the arrays (reporter bias), printing and tip problems and other microarray batch bias, uneven hybridization, as well as experimenter-related issues.

Every normalization procedure is likely to remove or even distort some of the biological information. Therefore, it is a good idea to address the problems lead-ing to systematic bias in order to keep normalization to a minimum. Misaligned lasers can easily be fixed, and reciprocal labeling with swapped color dyes will al-low correction for differences in labeling efficiencies in cDNA microarray experi-ments. Sensible cDNA microarray design can help to distinguish reporter bias caused by different DNA concentrations from "biological" effects caused by the systematic arrangement of reporters on the array. Uneven hybridization can be caused by insufficient amounts of labeled probe that fail to saturate the target spots. However, the experimenters themselves can be one of the largest sources of systematic variability. Considering the many steps necessary to perform a microarray experiment, it doesn't come as a surprise that experiments done by the same experimenter have been shown to cluster more tightly and have less variability than experiments done by several experimenters. Taken together, sensible design of arrays and experiments, systematic error checking, the use of reference samples, replicates, consistent methods, and good quality control can significantly enhance data quality and minimize the need for data normalization.

There are several techniques that are widely used to normalize gene-expression data (reviewed in [17]), such as total intensity normalization, linear regression techniques [18], ratio statistics [19], and LOWESS (LOcally WEighted Scatterplot Smoothing) or LOESS (LOcally wEighted regreSSion) correction [20]. Every normalization strategy relies on a set of assumptions. It is important to understand your data to know whether these assumptions are appropriate to use on your data set or not. In general, all of the strategies assume that the average gene does not change, either looking at the entire data set or at a user-defined subset of genes.

Total intensity normalization relies on the assumptions that equal amounts of labeled RNA sample have been hybridized to the arrays. Furthermore, when used with cDNA microarrays, this technique assumes that an equal number of genes are upregulated and downregulated, so that the overall intensity of all the elements on an array is the same for all RNA samples (control and experimental). Under these assumptions, a normalization factor can be calculated to re-scale the overall intensity value of the arrays. Affymetrix MAS uses a scaling factor to bring all the arrays in an experiment to a preset arbitrary target intensity value.

Linear regression techniques rely on the assumption that a significant fraction of genes are expressed at the same level when RNA populations from closely related samples are compared. Based on this assumption, plotting the intensity values of all genes from one sample against the intensity values of all genes from the other sample should result in genes that are expressed at equal levels clustering along a straight line. Regression techniques are then used to adjust the slope of this line to one. However, it has been shown for both, cDNA and high-density oligonucleotide experiments, that the signal intensities are nonlinear [21]. In these cases, a robust local regression technique such as LOWESS correction is more suitable [22]. Some techniques rely on a sufficient number of non-differentially expressed genes, such as "housekeeping" genes or exogenous control genes that have been spiked into the RNA before labeling. However, if the number of pre-determined "housekeeping" genes is small or their intensities do not cover a range of different intensity levels, this approach is a bad choice to fit normalization curves. Also, many of the so-called housekeeping genes do exhibit a natural variability in their expression level. Spiked in controls can span a broad range of ratio and intensity levels and may be useful to detect systematic bias, but cannot account for differences in the initial amount of RNA.

Comparison analysis. After normalization, the data for each gene are typically reported as an " expression ratio" (or its logarithm) of the normalized value of the expression level for the experimental sample divided by the normalized value of the control sample if cDNA arrays have been used. For oligonucleotide arrays,

Average Difference (Affymetrix GeneChip Analysis Suite) or Signals (Affymetrix MAS 5.0) are used as measure for absolute gene expression levels.

For experiments involving a pair of conditions, the next step is to identify genes that are differentially expressed. Various techniques have been proposed for the

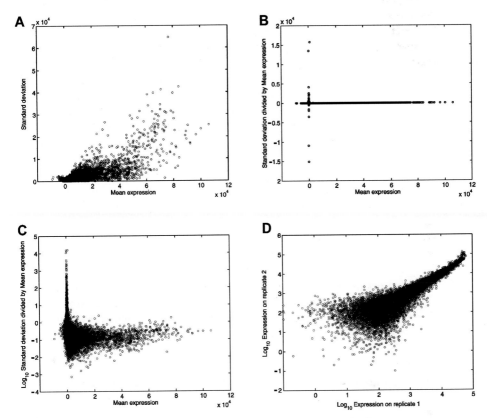

Fig. 1.3 Different views on the reproducibility of microarray measurements. In an Affymetrix microarray experiment involving three types of mice (CSX −/−, +/− and +/+), with three independent replicates for each type, **Panel A** shows the standard deviation of each gene in each type of mouse (y-axis) plotted against the mean expression level (x-axis) for that gene for that group (there are 13,179 gene measurements and 3 types of mice, or 39,537 points). In this view, there appears to be a higher standard deviation (and thus higher irreproducibility) in genes measured at higher expression levels. **Panel B** shows the same points, with the y-axis now showing standard deviation divided by the mean expression level. For most genes, the normalized standard deviation now appears to be the same across the higher mean gene expression level, except for a few genes at a very low mean expression level with large standard deviation. **Panel C** shows the same graph, with the y-axis showing the logarithm of the standard deviation divided by the mean expression level. Here, most genes' standard deviation is around one-tenth of their mean expression, except for several genes with low expression levels. **Panel D** shows all genes from two of the replicate experiments, with each axis representing the expression measurement on each chip. Ideally, this plot should represent a line with slope 1. Instead, one sees a typical "fishtail" diagram, with lower expression levels seemingly having less reproducibility.

selection of differentially expressed genes. Earlier studies have used arbitrary cut-off values such as twofold increase or decrease post-normalization without providing the theoretical background for choosing this level as significant. The inherent problem with this simple technique lies in the fact that the experimental and biological variability is far greater for genes that are expressed at low levels than for genes that are expressed at high levels, and that variability is different across different experiments (Fig. 1.3). Therefore, selecting significant genes based on an arbitrary fold change across the entire range of experimental data tends towards preferentially selecting genes that are expressed at low levels.

New data analysis techniques are continuously being developed, driven in part by the obvious need to move beyond setting arbitrary fold-change cut-off values, and because none of the existing techniques has found widespread acceptance in the community so far. Several approaches apply widely-used parametric statistical tests such as Student's *t*-Test [23] and ANOVA (ANalysis Of VAriance) [24], or non-parametric tests such as Mann-Whitney U test [25] or Kruskal-Wallis test (www.cardiogenomics.org) for every individual gene. However, due to the costs of microarray experiments, the number of replicates is usually low and thereby can lead to inaccurate estimates of variance.

The true power of microarray experiments does not come from the analysis of single experiments attempting to identify single gene expression changes or signaling pathways, but from analyzing many experiments that survey a variety of time points, phenotypes, or experimental conditions in order to identify global regulatory networks. Successful examples include genome-scale experiments identifying genes in the yeast mitotic cell cycle [1], or tumor classification [2]. In order to identify common patterns of gene expression from multiple hybridizations, more sophisticated clustering tools have to be used.

1.2.3
Clustering Algorithms

Identifying patterns of gene expression and grouping genes into expression classes provides much greater insight into their potential biological relevance than simple lists of up- and downregulated genes. Mathematical algorithms and computational tools to cluster the data are rapidly evolving, but no method has been described that seems universally suited to reveal biologically meaningful patterns in all data sets. Indeed, it is becoming increasingly clear that every analysis method might reveal a different aspect of the data set. Consequently, asking the right questions and choosing the appropriate algorithms to answer these questions is a crucial element of a meaningful experimental design. Within the scope of this chapter, we can only briefly describe some of the most commonly used techniques. This overview can by no means be comprehensive, because new tools and algorithms are continuously being developed.

Current methodologies in functional genomics that use larger RNA expression data sets for clustering can be roughly divided into two categories: *supervised learning* (analysis to determine ways to accurately split into or predict groups of samples or diseases), and *un-supervised learning* (analysis looking for characterization of the components of a data set, without a priori input on cases or genes). In supervised learning, algorithms attempt to learn a concept from labeled training examples, to predict the labels of the test set correctly. Within supervised learning, there are two classes of techniques, including (1) single feature or sample determination (finding genes or samples that match a particular pattern, using nearest-neighbor [26] or t-tests), (2) multiple feature determination (finding combinations of genes that match a particular a priori pattern, using decision trees [27], neural networks [28], or support vector machines [29–31]).

In unsupervised learning, algorithms are written to find patterns within a dataset, instead of trying to determine how best to predict a "correct answer." Within unsupervised learning, there are three classes of techniques, including (1) feature determination (determine genes with interesting properties, without specifically looking for a particular pattern determined a priori, such as using principal component analysis [32–36]), (2) cluster determination (determine groups of genes or samples with similar patterns of gene expression) using nearest neighbor clustering [26, 37], self-organizing maps [38, 39], k-means clustering, or one and two-dimensional dendrograms [40, 41]), and (3) network determination (determine graphs representing gene-gene or gene-phenotype interactions, using Boolean networks [42-44], Bayesian networks [45], and relevance networks [46–48]).

Cluster determination algorithms try to find genes that have similar expression levels and group them together into clusters. To approach this problem mathematically, an expression vector is defined that represents each gene as a multidimensional point in "expression space". In this view, each experiment represents a distinct axis, and the expression level of the gene gives its geometric coordinate. With this definition in place, there are several ways we can then define a "distance", ranging from the straight-forward Euclidean distance and correlation coefficient, to mutual-information. Mathematical descriptions of various distance metrics are available as supplementary material to [17].

Before we cluster the data, it might be useful to re-scale the data in order to enhance specific aspects of the genes in an experiment. In a process called "mean centering", each vector is re-scaled to set the average expression of each gene to 0. Consequently, we can easily identify up- or downregulation of each gene in respect to its average expression. If one agrees with the assumption that most genes are not changing between control and experimental samples, then mean centering is particularly useful. It is also useful in time-series experiments, when we are more interested in following the variations from the average expression rather than in the absolute expression values at each time point.

Hierarchical clustering algorithms have become the most popular tools for analyzing gene-expression data [40]. First, distances between points are calculated, with Euclidean distance measures being the most commonly used [49]. Related genes are thought to be closer to each other, therefore, the clustering process follows the simple principle:

0. Initial clusters = isolated data points
1. Update clusters by merging the two "nearest" clusters
2. Go to Step 1 until only one cluster is left.

The resulting cluster can be visualized as a single hierarchical tree that resembles a phylogenetic classification with a number of nested subsets. Hierarchical clustering assumes that the nature of the data structure is fundamentally hierarchical, as in evolutionary studies. However, it is questionable whether this is also true for complex microarray data. Another potential problem of methods that use Euclidean distances is the difficulty in finding genes that are negatively related to each other [48].

Examples of non-hierarchical clustering include k-means clustering, self-organizing maps (SOMs), and relevance networks. In *k-means clustering*, genes are partitioned into a pre-defined number of different clusters, without trying to specify the relationship between individual genes [50]. Therefore, it requires an advanced prior knowledge about the number of clusters that are represented within the data set. Some groups use other tools, such as hierarchical tree algorithms, to first identify the optimal number of clusters before applying the k-means algorithms [51]. *SOMs* have initially been developed to model complex data structures in biological neural networks [52], such as the topological relationship of neurons in the cerebral cortex of the brain. SOMs have been found to be well-suited to cluster and visualize large high-dimensional data sets, and have therefore found widespread application in telecommunications, artificial speech, and speech recognition, before being used for microarray data analysis [53]. Similar to k-means clustering, SOMs partition the genes into clusters based on their similarity in gene expression, with the additional constraint that the cluster centers are restricted to lie in a predefined topology. Reference vectors are defined for each partition that are adjusted to best fit to the expression vectors of the assigned genes. *Relevance networks* are based on comprehensive pairwise comparisons that can be used to find correlations between disparate biological measures, such as RNA expression and susceptibility to pharmaceuticals [48].

Algorithms like the ones cited above are referred to as "unsupervised" methods, because they don't rely on any prior assumptions regarding the functions of genes to be clustered. On the other hand, "supervised" methods, such as decision trees, nearest neighbor, and support vector machines (*SVMs*), use existing biological information of gene function as a "training data set" [30]. SVMs first learn to distinguish between different classes of genes within the training data set. Having learned the features of these classes, SVMs can then identify and classify unknown genes with similar features from gene expression data.

Though there are now many clustering techniques available for the functional genomics researcher, it is still crucial to have a question or hypothesis in mind before selecting a technique. Hypotheses such as "What uncategorized genes have an expression pattern similar to these genes that are well characterized?", "How different is the pattern of expression of gene X from other genes?", "What category of function might gene X belong to?", and "What are all the gene-gene interactions present among these tissue samples?" implicitly guide the choice of the appropriate type of algorithm (supervised or unsupervised) as well as the specific selection.

Although cluster analysis is a powerful tool, no cluster analysis gives absolute answers. Selecting different normalization strategies and distance metrices, or using different algorithms can lead to the identification of completely different clusters, and many of them might not be biological meaningful. Therefore, using our biological understanding about the system under study is essential in order to decide whether an analysis gave us valid data.

1.2.4
Data Sharing

Most microarray experiments interrogate thousands of genes on tens to hundreds of individual samples; each generating large lists of hundreds of differentially expressed genes. These data can be extremely complex. We have seen in our data sets that different biological insights may be uncovered by different approaches to analyses on the same data set. In most scientific publications of microarray data, the authors provide their own, sometimes subjective, interpretation on a subset of genes that they believe encompass the important aspects of the study. However, the identity of genes whose expression levels do not change is often as important as those that do. In addition, many researchers are only interested in one or a few genes; their interests are ill served if these genes do not make it into these "selected" gene lists. To make the most efficient use of microarray data, it is therefore important that the raw unprocessed data are available to others, as each researcher brings in a different perspective and different analytical methods that will help to extract insights beyond those identified by the original set of authors. Also, those data often serve as training data sets for the development of new cluster algorithms and other analytical tools [48].

Only the largest microarray laboratories have established their own databases (Tab. 1.1) [54], whereas microarray data accompanying publications are typically reported on the author's websites, if at all. Public microarray gene expression databases are being developed by the National Center for Biotechnology Information (NCBI), the Gene Expression Omnibus GEO, and by the European Bioinformatics Institute (EBI) (see Tab. 1.1 for links).

However, the formats and annotations for data exchange that would allow the interested scientist to easily download the data, understand the experiment, and evaluate the data quality, have not been established yet. In fact, most data come without any annotations at all, making them inaccessible for the general scientific community. On the first Microarray Gene Expression Database (MGED) meeting that took place in November 1999 in Cambridge, UK, a working group was established under the same name to facilitate the adoption of standards for microarray experiments (Tab. 1.1). This group addresses the most critical issues related to the exchange of microarray data: 1) The formulation of the minimum information about a microarray experiment required to interpret and verify the results (MIAME); 2) The establishment of a data exchange format (MAGE-ML for MicroArray Gene Expression markup Language) and object model (MAGE-OM); 3) The development of ontologies for microarray experiment description and biological material annotation; and 4) The development of recommendations regarding experimental controls and data normalization methods. MGED has submitted their proposal for MAGE-ML to the Object Management Group (OMG), an organization that develops industry guidelines and defines standards for software application development. Once these standards have been established, data exchange between laboratories will depend mainly on the willingness of individual researchers to share their expression data with the research community, or on scientific journals to enforce the release of the primary data upon publication of a microarray study.

1.3
Potential Use of This Technology in Understanding Complex Heart Disease

Large-scale expression analysis is a potentially powerful tool to characterize complex disease traits. Therefore, many scientists have used microarrays [24, 25, 51, 55–61] or other expression profiling technologies, such as subtraction [23, 62] or GeneCalling [63] to elucidate the transcriptional profiles in experimental models and human patient samples related to cardiac hypertrophy and heart failure (summarized in Tab. 1.2).

Hypertrophy is a universal adaptive response to increased cardiac workload, stress, and injury that can be induced by many different stimuli. Microarray expression profiling studies have been performed on models for pressure-overload induced hypertrophy [62, 63]; hypertrophy induced by Isoproterenol and AngiotensinII [58]; hypertrophy induced by transgenic overexpression of Calsequestrin, Calcineurin, $G\alpha q$, and $\Psi \varepsilon$Rack [51], as well as on models of myocardial infarction [25, 55–57]. Ideally, comparison of these (in some aspects) phenotypically similar, but pathophysiological dissimilar models would enable us to differentiate among the multiple processes that cause hypertrophy (and heart failure), and to identify patterns of gene expression that are either common for these diseases or specific for a certain subset of conditions.

Unfortunately, direct comparison of results between these studies is nearly impossible due to substantial differences in the technology platforms being used, the number of transcripts being interrogated or sequenced, differences in experimental conditions and data analysis, and tissue-types that are being compared (Tab. 1.2). Given these differences, there is generally little overlap in the number and identity of genes that are differentially expressed between these studies. Still, some consensus can be found, like an induction of expression of the atrial and brain natriuretic peptide genes ANP and BNP, and extracellular matrix genes in cardiac hypertrophy, or an induction of genes involved in stress response, inflammation and wound healing in different models of myocardial infarction. Most of these changes confirm findings that have already been reported previously.

However, individual studies identify new sets of differentially expressed genes that have not yet been implicated in the disease under study, such as a decrease

Tab. 1.2 Publications on cardiovascular diseases using large scale expression profiling technologies

Model	Species	Tissue	Method	Time points	n	Transcripts on the array/changed*	Findings	Ref.
Transient ischemia	Pig	LV ischemic and control area	Subtraction	1	1	481	Upregulation of genes involved in survival mechanism	[23]
Experimental MI	Mouse	Total ventricle	cDNA array (Clontech)	1	1	588/78	Upregulation of genes associated with heart muscle development and stress response	[55]
Experimental MI	Rat	LV free wall versus IVS	Heart cDNA array	5	4	4,258/230	Changes in expression of genes that are implicated in cytoskeletal architecture, ECM, contractility, and metabolism	[57]
ACE inhibition after experimental MI	Rat	LV	Oligonucleotide array (Affymetrix)	1	6	8,000/37	Upregulation of natriuretic genes, ECM genes, genes involved in inflammation/wound healing, and muscle proteins	[25]
CHF after experimental MI	Rat	LV	Heart cDNA array	6	4–7	1,075/68	Upregulation of natriuretic petides ANF and BNP, matrix genes, and genes that are involved in wound healing, infiltrating mononuclear phagocytes, and vascular reactivity	[56]
Pressure-overload induced hypertrophy	Rat	Whole heart	GeneCalling	1	1	18,000/74 fragments (23 genes)	Elevated expression of ECM genes and secreted factors	[63]
Pressure-overload hypertrophy	Mouse	Total heart	Subtraction	1	1	100	Upregulation of genes involved in transcription, translation; proteins that form or regulate the cytoskeleton; and genes involved in signaling	[62]
Hypertrophy, AII and ISO-induced	Mouse	LV	cDNA array	14 (AII), 1 7 (ISO)	1	4,000/359	Induction of secreted factors, structural genes, genes related to energy metabolism, and different signaling pathways	[58]

Tab. 1.2 (continued)

Model	Species	Tissue	Method	Time points	n	Transcripts on the array/ changed*	Findings	Ref.
Hypertrophy in 4 transgenic models	Mouse	LV	cDNA array (Incyte)	1	1	8,800/276	No genes are found to be regulated in common among the 4 models. A set of apoptotic genes is uniquely upregulated in the Gαq model	[51]
Exposure to IGF-1	Rat	Neonatal ventricular cardiomyocytes	1.2 cDNA array (Clontech)	4	6	1,193/68	Upregulation of genes involved in intracellular signaling, cell cycle, transcription/translation, cellular respiration and mitochondrial function, cell survival, ion channels and calcium signaling, and humoral factors	[59]
End-stage hypertrophic cardio-myopathy	Human	Not specified	Heart cDNA array	1	1	10,368/38	Identification of several genes that show differential expression in heart failure	[60]
End-stage ICM and DCM	Human	LV free wall	Oligonucleotide array (Affymetrix)	1	2	6,800/19	Altered expression of cytoskeletal and myo-fibrillar genes, genes involved in protein degradation, stress response, and metabolism	[24]
Myocarditis	Mouse	Whole heart	Rat heart cDNA array (Incyte)	3	1	4,200/169	Upregulation of genes related to viral replication, host defense, metabolic changes, and ECM proteins	[61]

* The number of transcripts listed in the table might include redundant transcripts; the actual number of interrogated genes might be lower. AII angiotensin II; DCM dilated cardiomyopathy; ECM extracellular matrix, ICM ischemic cardiomyopathy, ISO isoproterenol, IVS intraventricular septum; LV left ventricle; MI myocardial infarction; n number of replicates.

in SLIM1 (striated muscle LIM protein-1) expression and an increase in gelsolin expression in failing human hearts [24]. The degree of confidence in the data quality and analysis depends on whether replicate experiments were performed, dye-swaps were used to adjust for differences in labeling efficiencies (for cDNA microarrays), how cut-off values were determined in order to identify differentially expressed genes, and whether or not differential gene expression was confirmed by alternative methods, such as Northern blotting or RT-PCR. Also, were the tissues and time points chosen appropriate for the questions asked? For example, in cases where the whole left ventricle was used for analysis, it is questionable whether the sensitivity was sufficient to survey genes whose differential expression is confined to a specific region of the heart, such as the infarcted area. Additional confidence and potentially new insight would be gained if the complete data set, the raw data, and an explicit description of the experimental conditions would be available for other laboratories to reproduce the results. Although data supplements of extended gene lists and additional method descriptions were available in some cases [57], none of the microarray studies cited above provided access to their raw data. Finally, every new hypothesis has to be further tested in biological systems, such as through transgenic overexpression, dominant-negative, antisense, or gene targeting strategies, in order to establish a causal relationship of the change in gene expression with the disease – one gene at a time.

Although it becomes clear that microarrays hold much promise for the analysis of cardiovascular diseases, many of these studies illustrate the experimental and technological hurdles imposed by current technologies. This includes the fact that genome-wide chips are still unavailable for most species. Currently available arrays cover only a fraction of genes in the genome, setting a limit to the number of differentially expressed genes that can be detected. Some laboratories tried to increase the number of genes that are potentially involved in the cardiac disease by developing heart-specific cDNA arrays [57, 60, 61] or, even more selective, by performing subtractive hybridization experiments in order to pre-select differentially expressed clones, which were then used to generate cDNA microarrays [56]. Once these limitations and problems have been overcome, large scale expression monitoring should enhance the development of therapeutic strategies for treatment of cardiac diseases.

1.4
Acknowledgements

We thank Pascal Braun, Sekwon Kong, and Jeffrey Brown for critical reading of the manuscript, and Ashish Nimgaonkar for help with Fig. 1.3. The work is supported by the NHLBI Programs for Genomic Applications, U01 HL66582.

1.5
References

1 CHO RJ, CAMPBELL MJ, WINZELER EA, STEINMETZ L, et al. A genome-wide transcriptional analysis of the mitotic cell cycle. *Mol. Cell.* **1998**; *2*(1):65–73.

2 SHIPP MA, ROSS KN, TAMAYO P, WENG AP, et al. Diffuse large B-cell lymphoma outcome prediction by gene-expression profiling and supervised machine learning. *Nat. Med.* **2002**; *8*(1):68–74.

3 LANDER ES, LINTON LM, BIRREN B, NUSBAUM C, et al. Initial sequencing and analysis of the human genome. *Nature* **2001**; *409*(6822):860–921.

4 ADAMS MD, KELLEY JM, GOCAYNE JD, DUBNICK M, et al. Complementary DNA sequencing: expressed sequence tags and human genome project. *Science* **1991**; *252*(5013):1651–1656.

5 SCHENA M, SHALON D, DAVIS RW, BROWN PO. Quantitative monitoring of gene expression patterns with a complementary DNA microarray. *Science* **1995**; *270*(5235):467–470.

6 VELCULESCU VE, ZHANG L, VOGELSTEIN B, KINZLER KW. Serial analysis of gene expression. *Science* **1995**; *270*(5235):484–487.

7 LOCKHART DJ, DONG H, BYRNE MC, FOLLETTIE MT, et al. Expression monitoring by hybridization to high-density oligonucleotide arrays. *Nat. Biotechnol.* **1996**; *14*(13):1675–1680.

8 BRENNER S, JOHNSON M, BRIDGHAM J, GOLDA G, et al. Gene expression analysis by massively parallel signature sequencing (MPSS) on microbead arrays. *Nat. Biotechnol.* **2000**; *18*(6):630–634.

9 SUTCLIFFE JG, FOYE PE, ERLANDER MG, HILBUSH BS, et al. TOGA: an automated parsing technology for analyzing expression of nearly all genes. *Proc. Natl. Acad. Sci. USA.* **2000**; *97*(5):1976–1981.

10 KANE MD, JATKOE TA, STUMPF CR, LU J, et al. Assessment of the sensitivity and specificity of oligonucleotide (50mer) microarrays. *Nucleic Acids Res.* **2000**; *28*(22):4552–4557.

11 OHYAMA H, ZHANG X, KOHNO Y, ALEVIZOS I, et al. Laser capture microdissection-generated target sample for high-density oligonucleotide array hybridization. *Biotechniques* **2000**; *29*(3):530–536.

12 BAUGH LR, HILL AA, BROWN EL, HUNTER CP. Quantitative analysis of mRNA amplification by in vitro transcription. *Nucleic Acids Res.* **2001**; *29*(5):E29.

13 SOLOVIEV M. EuroBiochips: spot the difference! *Drug Discov. Today* **2001**; *6*(15):775–777.

14 GOSSEN M, BUJARD H. Tight control of gene expression in mammalian cells by tetracycline-responsive promoters. *Proc. Natl. Acad. Sci. USA.* **1992**; *89*(12):5547–5551.

15 LI C, WONG WH. Model-based analysis of oligonucleotide arrays: expression index computation and outlier detection. *Proc. Natl. Acad. Sci. USA.* **2001**; *98*(1):31–36.

16 CHUDIN E, WALKER R, KOSAKA A, WU SX, et al. Assessment of the relationship between signal intensities and transcript concentration for Affymetrix Gene-Chip(R) arrays. *Genome Biol.* **2002**; *3*(1): Research 0005.1–0005.10.

17 QUACKENBUSH J. Computational analysis of microarray data. *Nat. Rev. Genet.* **2001**; *2*(6):418–427.

18 HEDENFALK I, DUGGAN D, CHEN Y, RADMACHER M, et al. Gene-expression profiles in hereditary breast cancer. *N. Engl. J. Med.* **2001**; *344*(8):539–548.

19 CHEN Y, DOUGHERTY ER, BITTNER ML. Ratio-based decisions and the quantitative analysis of cDNA microarray images. *J. Biomed. Opt.* **1997**; *2*:364–374.

20 CLEVELAND WS, DEVLIN SJ. Locally weighted regression: an approach to regression analysis by local fitting. *J. Am. Stat. Assoc.* **1988**; *83*:596–610.

21 RAMDAS L, COOMBES KR, BAGGERLY K, ABRUZZO L, et al. Sources of nonlinearity in cDNA microarray expression measurements. *Genome Biol.* **2001**; *2*(11).

22 TSENG GC, OH MK, ROHLIN L, LIAO JC, et al. Issues in cDNA microarray analysis: quality filtering, channel normalization, models of variations and assessment of gene effects. *Nucleic Acids Res.* **2001**; *29*(12):2549–2557.

23 DEPRE C, TOMLINSON JE, KUDEJ RK, GAUSSIN V, et al. Gene program for cardiac cell survival induced by transient ischemia in conscious pigs. *Proc. Natl. Acad. Sci. USA.* **2001**; *98*(16):9336–9341.

24 YANG J, MORAVEC CS, SUSSMAN MA, DI-PAOLA NR, et al. Decreased SLIM1 expression and increased gelsolin expression in failing human hearts measured by high-density oligonucleotide arrays. *Circulation* **2000**;*1 02*(25):3046–3052.

25 JIN H, YANG R, AWAD TA, WANG F, et al. Effects of early angiotensin-converting enzyme inhibition on cardiac gene expression after acute myocardial infarction. *Circulation* **2001**; *103*(5):736–742.

26 GOLUB TR, SLONIM DK, TAMAYO P, HUARD C, et al. Molecular classification of cancer: class discovery and class prediction by gene expression monitoring. *Science* **1999**; *286*(5439):531–537.

27 QUINLAN J. C4.5: programs for machine learning. San Mateo, Calif.: Morgan Kaufmann; **1992**.

28 RUMELHART D, MCCLELLAND J, University of California SDPRG. Parallel distributed processing : explorations in the microstructure of cognition. Cambridge, Mass.: MIT Press; **1986**.

29 FUREY TS, CRISTIANINI N, DUFFY N, BEDNARSKI DW, et al. Support vector machine classification and validation of cancer tissue samples using microarray expression data. *Bioinformatics* **2000**; *16*(10):906–914.

30 BROWN MP, GRUNDY WN, LIN D, CRISTIANINI N, et al. Knowledge-based analysis of microarray gene expression data by using support vector machines. *Proc. Natl. Acad. Sci. USA.* **2000**; *97*(1):262–267.

31 CHOW ML, MOLER EJ, MIAN IS. Identifying marker genes in transcription profiling data using a mixture of feature relevance experts. *Physiol. Genomics* **2001**; *5*(2):99–111.

32 ALTER O, BROWN PO, BOTSTEIN D. Singular value decomposition for genome-wide expression data processing and modeling. *Proc. Natl. Acad. Sci. USA.* **2000**; *97*(18):10101–10106.

33 RAYCHAUDHURI S, STUART JM, ALTMAN RB. Principal components analysis to summarize microarray experiments: application to sporulation time series. *Pac. Symp. Biocomput.* **2000**, 455–466.

34 FIEHN O, KOPKA J, DORMANN P, ALTMANN T, et al. Metabolite profiling for plant functional genomics. *Nat. Biotechnol.* **2000**; *18*(11):1157–1161.

35 WEN X, FUHRMAN S, MICHAELS GS, CARR DB, et al. Large-scale temporal gene expression mapping of central nervous system development. *Proc. Natl. Acad. Sci. USA.* **1998**; *95*(1):334–339.

36 HILSENBECK SG, FRIEDRICHS WE, SCHIFF R, O'CONNELL P, et al. Statistical analysis of array expression data as applied to the problem of tamoxifen resistance. *J. Natl. Cancer Inst.* **1999**; *91*(5):453–459.

37 BEN-DOR A, BRUHN L, FRIEDMAN N, NACHMAN I, et al. Tissue classification with gene expression profiles. *J. Comput. Biol.* **2000**; *7*(3–4):559–583.

38 TAMAYO P, SLONIM D, MESIROV J, ZHU Q, et al. Interpreting patterns of gene expression with self-organizing maps: Methods and application to hematopoietic differentiation. *Proc. Natl. Acad. Sci. USA.* **1999**; *96*(6):2907–2912.

39 TORONEN P, KOLEHMAINEN M, WONG G, CASTREN E. Analysis of gene expression data using self-organizing maps. *FEBS Lett.* **1999**; *451*(2):142–146.

40 EISEN MB, SPELLMAN PT, BROWN PO, BOTSTEIN D. Cluster analysis and display of genome-wide expression patterns. *Proc. Natl. Acad. Sci. USA.* **1998**; *95*(25):14863–14868.

41 ROSS DT, SCHERF U, EISEN MB, PEROU CM, et al. Systematic variation in gene expression patterns in human cancer cell lines. *Nat. Genet.* **2000**; *24*(3):227–235.

42 LIANG S, FUHRMAN S, SOMOGYI R. Reveal, a general reverse engineering algorithm for inference of genetic network architectures. *Pac. Symp. Biocomput.* **1998**, 18–29.

43 WUENSCHE A. Genomic regulation modeled as a network with basins of attraction. *Pac. Symp. Biocomput.* **1998**, 89–102.

44 SZALLASI Z, LIANG S. Modeling the normal and neoplastic cell cycle with "realistic Boolean genetic networks": their application for understanding carcinogen-

esis and assessing therapeutic strategies. *Pac. Symp. Biocomput.* **1998**, 66–76.

45 FRIEDMAN N, LINIAL M, NACHMAN I, PE'ER D. Using Bayesian networks to analyze expression data. *J. Comput. Biol.* **2000**; 7(3–4):601–620.

46 BUTTE A, KOHANE I. Unsupervised Knowledge Discovery in Medical Databases Using Relevance Networks. In: LORENZI N, editor. Fall Symposium, American Medical Informatics Association; 1999; Washington, DC: Hanley and Belfus; **1999**, p. 711–715.

47 BUTTE AJ, KOHANE IS. Mutual information relevance networks: functional genomic clustering using pairwise entropy measurements. *Pac. Symp. Biocomput.* **2000**, 418–429.

48 BUTTE AJ, TAMAYO P, SLONIM D, GOLUB TR, et al. Discovering functional relationships between RNA expression and chemotherapeutic susceptibility using relevance networks. *Proc. Natl. Acad. Sci. USA.* **2000**; 97(22):12182–12186.

49 LELE S, RICHTSMEIER JT. Euclidean distance matrix analysis: confidence intervals for form and growth differences. *Am. J. Phys. Anthropol.* **1995**; 98(1):73–86.

50 HARTIGAN JA, WONG MA. A K-Means Clustering Algorithm. *Applied Statistics* **1979**; 28:100–108.

51 ARONOW BJ, TOYOKAWA T, CANNING A, HAGHIGHI K, et al. Divergent transcriptional responses to independent genetic causes of cardiac hypertrophy. *Physiol. Genomics* **2001**; 6(1):19–28.

52 KOHONEN T. Physiolocigal interpretation of the self-organizing map algorithm. *Neural Networks* **1993**; 6(7):895–905.

53 TAMAYO P, SLONIM D, MESIROV J, ZHU Q, et al. Interpreting patterns of gene expression with self-organizing maps: methods and application to hematopoietic differentiation. *Proc. Natl. Acad. Sci. USA.* **1999**; 96(6):2907–2912.

54 SHERLOCK G, HERNANDEZ-BOUSSARD T, KASARSKIS A, BINKLEY G, et al. The Stanford Microarray Database. *Nucleic Acids Res.* **2001**; 29(1):152–155.

55 LYN D, LIU X, BENNETT NA, EMMETT NL. Gene expression profile in mouse myocardium after ischemia. *Physiol. Genomics* **2000**; 2(3):93–100.

56 SEHL PD, TAI JT, HILLAN KJ, BROWN LA, et al. Application of cDNA microarrays in determining molecular phenotype in cardiac growth, development, and response to injury. *Circulation* **2000**; 101(16):1990–1999.

57 STANTON LW, GARRARD LJ, DAMM D, GARRICK BL, et al. Altered patterns of gene expression in response to myocardial infarction. *Circ. Res.* **2000**; 86(9):939–945.

58 FRIDDLE CJ, KOGA T, RUBIN EM, BRISTOW J. Expression profiling reveals distinct sets of genes altered during induction and regression of cardiac hypertrophy. *Proc. Natl. Acad. Sci. USA.* **2000**; 97(12):6745–6750.

59 LIU T, LAI H, WU W, CHINN S, WANG PH. Developing a strategy to define the effects of insulin-like growth factor-1 on gene expression profile in cardiomyocytes. *Circ. Res.* **2001**; 88(12):1231–1238.

60 BARRANS JD, STAMATIOU D, LIEW C. Construction of a human cardiovascular cDNA microarray: portrait of the failing heart. *Biochem. Biophys. Res. Commun.* **2001**; 280(4):964–969.

61 TAYLOR LA, CARTHY CM, YANG D, SAAD K, et al. Host gene regulation during coxsackievirus B3 infection in mice: assessment by microarrays. *Circ. Res.* **2000**; 87(4):328–334.

62 JOHNATTY SE, DYCK JR, MICHAEL LH, OLSON EN, ABDELLATIF M. Identification of genes regulated during mechanical load-induced cardiac hypertrophy. *J. Mol. Cell. Cardiol.* **2000**; 32(5):805–815.

63 SHIMKETS RA, LOWE DG, TAI JT, SEHL P, et al. Gene expression analysis by transcript profiling coupled to a gene database query. *Nat. Biotechnol.* **1999**; 17(8):798–803.

64 SCHULZE A, DOWNWARD J. Navigating gene expression using microarrays – a technology review. *Nat. Cell. Biol.* **2001**; 3(8):E190–195.

2

Global Genomic Analyses of Cardiovascular Disease:
A Potential Map or Blind Alley?

MAIKE KRENZ, GARY SCHWARTZBAUER, and JEFFREY ROBBINS

2.1
Blindly Searching for Structure-Function

Some of the first successes in the new age of molecular biomedicine involved identifying the gene or genes responsible for a particular disease. These initial studies
relied heavily upon extensive information about the gene product, or what is now
called today biological annotation. Thus, a wealth of information about the pro-
tein(s) or biochemical pathways underlying such diseases as sickle cell anemia,
Tay-Sachs disease and phenylketonuria existed, and these largely biochemical data
drove the nucleic acid studies, in which specific mRNAs were isolated and used
as probes to isolate the respective genomic sequences. This general approach,
known as *functional cloning*, presupposes and depends upon extensive data at the
protein level, often laboriously gathered over a long period of time. Starting approximately 25 years ago, this methodology was complemented by and then supplanted
using the so-called "reverse genetics" or what is more accurately termed *positional
cloning*, in which pedigrees are constructed, linkage analyses carried out and the investigator then hones in on the (presumably) affected locus. Importantly for the purposes of the present topic, unlike functional cloning, positional cloning presupposes
no information about the gene product, and the genes for Duchenne muscular dystrophy [1], Fragile X Syndrome [2] and Huntington's disease [3] are only a few of the
many disease-causing genes identified using this technology.

Identification of the candidate gene does not necessarily translate into an understanding of the pathological processes that actually underlie disease development.
Thus, investigators have had to struggle for 15 years to understand how dystrophin actually causes the pathology of Duchenne Muscular Dystrophy, or how the
protein huntingtin leads to Huntington's disease. For the cardiovascular system,
these issues are illustrated by focusing on the disease, familial hypertrophic cardiomyopathy (FHC). The genetic etiology and molecular pathology of FHC has been
the subject of intense investigation since the seminal observation by Seidman and
her colleagues, linking mutations in the major cardiac contractile protein, the β
myosin heavy chain, with the development of a genetically-based hypertrophic cardiomyopathy [4]. It has become clear over the ensuing years that a variety of mutations in multiple sarcomeric proteins can lead to cardiac disease and failure [5].

FHC is characterized by unexplained cardiac hypertrophy in the absence of increased cardiac load or other systemic abnormalities. Clinically, the autosomal dominant diseases show variable penetrance with hypertrophy occurring in either ventricle; almost always (>95%), there is involvement of the intraventricular septum [6]. FHC is a major cause of sudden death in otherwise healthy appearing young adults [7]. However, even within a family in which the disease is due to a single genetic defect, the severity, onset and penetrance of the pathology is highly variable, presumably due to the existence of modifier loci, although this has not yet been formally demonstrated. Early on, it was recognized that FHC mapped to multiple loci and soon after the first reports linking the disease to *β-MyHC*, gene linkage/positional cloning approaches zeroed in on other disease loci and identified mutations in the other sarcomeric protein genes. These studies and "candidate gene" studies, in which mutations in closely related proteins are looked for, identified mutations in cardiac myosin binding protein-C, components of the troponin complex, actin, titin, *α*-tropomyosin and others. The association of FHC with mutations in this set of functionally related proteins led to the syndrome being called "a disease of the sarcomere [8]."

These studies illustrate the logical progression of a positional cloning approach and its strengths and weaknesses. The strengths include the relatively rapid progress that can be made once the pedigrees are established, the locus or loci mapped, and the linkage validated by statistical analyses. Identification of a gene or group of genes gives one the necessary biological annotation to begin "blind" candidate gene studies in which structurally related or functionally related sequences are scanned for mutations. Collecting linkage data for these candidates is time consuming and the data are often incomplete, precluding the use of linkage analysis. Thus, the candidate gene approach, in which closely related genes (such as actin or troponin or myosin light chain in the case of FHC), or genes that can be logically placed as playing a possible role in the characterized pathology are chosen, has been quite useful. However, often the resultant biological validation, that is showing that the particular mutation is causative, is lacking or the results ambiguous [9, 10].

Unfortunately, identifying the primary genetic etiology has led to only limited insight into understanding the resultant pathology, which is the result of and reflects the downstream changes in the transcriptome and proteome. The progression from functional cloning → positional cloning → candidate gene is all intellectually founded in a reductionist approach: understanding the cause(s) of the disease by focusing on the primary genetic lesion. For cardiovascular disease, even those that are monogenically based, we now realize that reductionism has its limitations, particularly when it comes to understanding a disease pathology in terms of choosing potential therapeutic targets. In a paradigm shift it is now appreciated that, in order to understand the molecular pathologies at work, the overall regulatory networks and interplay of the different organ systems, cell types and even subcellular compartments and organelles need to be described.

It is this explicit recognition of the value of a global approach for understanding cardiovascular disease that has generated the excitement associated with exploring the transcriptome. The advent of multiple methodologies for defining a cell's total transcriptional output and how it might change during disease progression offers heretofore unrealized potential for truly understanding the pathology of cardiovascular disease. The different methods currently in use for profiling gene expression include differential display [11], expression sequence tag (EST) identification and sequencing [12, 13], serial analysis of gene expression (SAGE) [14] and microarray analyses with either cDNA or oligonucleotide based chips. However, the commercialization of the microarray technology through Affymetrix's™ development of comprehensive chip sets for the human, mouse and other model organisms, has brought global analysis out of the boutique into the general cardiovascular research community in a surprisingly short period of time. A large number of cardiovascular laboratories have now adopted microarrays for analyzing the genetic output of both normal and diseased cardiovascular tissue at different developmental times. Instead of measuring the output of only a few genes, we can now quantitate and compare tens of thousands of transcripts in a more or less reproducible manner across different conditions.

2.2
The Starting Line: Garbage In – Garbage Out?

The field of cardiovascular biology appears, at first glance, to be particularly well suited for the types of comprehensive analyses needed for the technology's productive application to discovery-based research. After all, there are now close to 200 animal models of human cardiovascular disease, which have been produced through the use of genetic manipulation using either transgenesis or gene ablation. The use of these mouse-based genetic models where the primary etiology is known and the disease course can be closely monitored should, at first glance, easily provide the basic reagent – a relatively well defined disease in an organism with a well-defined genetic background, in which the disease's progression can be monitored closely. Thus, using large groups of animals, the longitudinal progression of disease development becomes a timeline along which points can be taken for global transcriptional analyses and indeed, large data sets are beginning to be gathered across a number of these disease models (http://www.nhlbi.nih.gov/resources/pga/resource.htm). However, considering the dependence on these animal models, it is worthwhile to take a closer look at how they are made and what the phenotypes might mean in terms of genetic output.

Our ability to precisely manipulate the mammalian genome of the mouse has given us access to exciting new approaches for studying gene function in a physiological context. Selective changes in a protein or the protein profile of the heart can be accomplished in order to study both gain-of-function disorders (where an active protein is overexpressed or a mutant protein with altered function is expressed) and loss-of-function disorders (an endogenous gene is inactivated or a mutant protein can serve as a dominant-negative mutation). Even the most abundant cardiomyocyte proteins, the components of the contractile apparatus, can now be altered or replaced in a defined manner [15–17].

For models of cardiovascular disease it is highly desirable that expression be restricted to the heart and preferably, to a specific cell type, e.g. the cardiomyocyte, within the heart. Systemic or ectopic expression would certainly raise the "biological noise" of the transcriptome analyses as the heart responded to ectopic stimuli. For transgene expression the ideal promoter should drive high levels of gene activity in a cardiac-specific manner at appropriate times during development and should also display a minimum of position-dependent effects. The most successful and widely used elements are derived from contractile genes themselves [18] with the a-myosin heavy chain promoter most often chosen due to its high levels of expression in both the adult atria and ventricles, its cardiac specificity, and the presence of sequences necessary for copy number-dependent and position-independent expression [19, 20].

Creation of a transgenic animal is not genomically benign. Using pronuclear microinjection techniques, the construct is introduced into fertilized eggs, which are then transferred to pseudopregnant females. Inside the pronucleus, the transgene will either randomly integrate into the mouse genome or be degraded by exonuclease activity. Usually several copies of the transgene are found in a tandem head-to-tail array. As transgene insertion occurs early during development, most or all of the cells of the resulting mouse will carry the genetic modification. Occasionally, the transgene is integrated at more than one site in the genome and offspring of such founders may inherit one or the other copy of the transgene, resulting in multiple lines with different copy numbers and potentially distinct phenotypes from a single founder [21]. This could, if undetected, have a significant detrimental effect on any subsequent transcriptome-based analysis, as it is more than likely that the different transgene numbers would result in phenotypes of varying severity.

In cardiovascular disease models, the expression level of the transgene is often positively correlated with the copy number and therefore the severity of the phenotype, but the relationship is not a simple one. The site of integration within the genome, the proximity of the transgene to transcriptional activators and silencers can both alter the level and tissue selectivity of protein expression. Moreover, transgene insertion can also influence the expression of neighboring genes, since it might disrupt a protein coding sequence, a promoter element, or other regulatory regions [22–24]. Fortunately, the consequences of such insertional mutagenic effects are recessive in many cases and therefore do not alter the phenotype unless the transgene is bred to homozygosity.

In contrast to the random insertion of the construct into the genome with the transgenic approach, gene targeting replaces a DNA sequence with the exogenous sequence at a specific site within the genome using homologous recombination [25]. Deletions, point mutations, or replacements can be used to achieve gene inactivation or mutation. The construct is introduced into embryonic stem cells via electroporation: the rare homologous recombination can be selected for and those cells are then microinjected into blastocysts, which are then implanted into pseudopregnant females. Mice heterozygous for the targeted allele can then be intercrossed to generate homozygous mice. If the introduced genetic alteration is, for example, a deletion of a critical region of the protein of interest and results in a null allele, the gene will be completely ablated, producing a "gene knockout" in the homozygous mice.

Gene targeting can also be used to introduce subtle mutations into a gene of interest [26], or as part of a strategy to generate a tissue-specific or inducible knockout [27]. Since this strategy directs integration of the construct to a particular site, the site-specific effects and copy-number dependency that can confound the transgenic approach can be avoided [28]. Moreover, the targeted protein will be expressed under the control of the endogenous promoter, resulting in endogenous expression levels that reflect the correct temporal and spatial distribution of the wild-type protein. Although more precise, gene targeting is often not the approach chosen since it is much more time-consuming than transgenesis. Moreover, if the gene of interest plays an important role in early development, either high prenatal mortality or distortion of the phenotype due to developmental defects can occur.

For transcriptome-based analyses, more precise manipulation of altered gene expression would be a significant advantage, as one could lower the biological noise generated either by systemic expression of the altered gene (as is the case in most gene targeting experiments, with rare exceptions [29]), or by transgene expression throughout development. Tissue-specificity of a gene ablation can be achieved using the *E. coli* bacteriophage P1 enzyme Cre recombinase. Cre recombinase binds to a 34 bp DNA sequence called a *loxP* site, which contains two 13 bp repeated sequences in opposite orientation flanking an 8 bp spacer. Removal of the DNA fragment flanked by *loxP* (a 'floxed' allele) occurs when Cre is expressed. Mice homozygous for a floxed gene are bred to a mouse line expressing Cre recombinase driven by a promoter specific for the tissue of interest: the offspring will carry a deletion of the floxed gene only in that tissue. The Cre/*lox* system has been successfully used by a number of groups to create cardiac-specific knockouts [30] and can also be combined with an inducible system to add temporal control [27, 31, 32].

In order to accomplish temporal control (including expression of a transgene in a reversible manner), a variety of inducible systems have been developed and potentially offer the ability to create a well-defined start point for transgene expression of gene ablation in the adult, freeing the transcriptome of the noise resulting from the consequences of transgene expression during other developmental times. All systems share a common theme, exemplified by the tetracycline regulatory system. Mice with tissue-specific expression of the tetracycline transactivator (tTA) are generated. tTA is a transcription factor that regulates the expression of any gene downstream of its cognate promoter sequence (tetOp). Therefore, a second mouse line containing the desired transgene downstream of tetOp is bred to the first mouse line, creating a so-called 'tet-off' system, in which the interaction of tTA and tetOp and thereby the expression of the gene of interest is inhibited in the presence of tetracycline. If a mutant form of tTA, reverse tTA (rtTA), which is activated (i.e., will bind to the operator DNA) rather than repressed by tetracycline, is used, tetracycline can be administered to the mice only for the period of time when expression of the transgene is desired ('tet-on') [33]. The inducible system can be combined with the Cre/*lox* in order to provide precise control of *cre* expression and has been successfully used to control gene ablation and transgene expression in mouse hearts [27, 31, 32]. The current technology, while somewhat flawed in terms of its reproducibility and "leakiness" nevertheless offers the potential of improving dramatically our ability to precisely affect and control the transcriptome, decreasing the inherent biological noise of globally based RNA analyses.

2.3
Is The Mouse A Valid Model for Human-Based Disease Transcriptome Studies?

When considering the validity of *in vitro* models for studies of human disease mechanisms, the limitations are clear. *In vitro* models are designed to reduce a complex situation to the supposedly most relevant components, which can then be tightly controlled under the chosen experimental conditions. This simplification is also the major shortcoming of all *in vitro* approaches and determines the validity for a particular model. However, the question of validity cannot be answered as easily when genetic manipulations are discussed. With increasingly sophisticated technologies, we can now study the consequences of alterations in the genome in the whole animal. The genomic alteration is clearly defined and controlled – but we study the consequences in a complex system, in which many potentially interfering variables are unknown. Consequently, the only criterion of validity is the generation of the 'correct' phenotype. Hence, the conclusions depend to a large extent on the thoroughness and quality of the phenotypic characterization. As far as the new miniaturized technologies allow us to detect, mice appear to be a valid model system for a number of human disorders [34]. However, the mouse heart clearly differs from the human: the basic motor protein complement, calcium handling and the electrophysiology are all distinctly different. Another source of difficulty in interpreting murine data are the dramatic strain-dependent phenotypes. Depending on the genetic background, both penetrance and the characteristics of a particular mutation can vary considerably [35, 36]. This strain dependency is thought to be due to modifier loci [37, 38] but little is known about these elements. Does the lack of a phenotype after genetic manipulation demonstrate that the altered gene does not play a critical role for the disease of interest? Or does it merely indicate that the response to the genetic alteration can only be detected in an alternate mouse strain? Similarly, the positive result of a particular study alone does not necessarily prove its general validity, as the strain used could be predisposed to a pathology that normally would not present. The different strains have, under baseline conditions, different transcriptome profiles, which add further to the difficulties in comparing data sets between labs, even when supposedly identical strains are being used. Therefore, every individual mouse model has to be carefully tested by designing appropriate internal and external controls.

2.4
Arrays and Cardiovascular Disease

The Good: Why carry out transcriptome analysis? Altered gene expression has been a focus of biomedicine ever since it became apparent that a cell responds to external and internal stimuli by altering its genetic output. In the past, experiments were often designed to monitor the appearance/disappearance of a particular transcript or set of transcripts in response to a well-defined stimulus. The large-scale interrogation of cardiac mRNAs offers a relatively easy way of carrying out a first pass at understanding gene function in the heart. Instead of being restricted to the parallel processing of 1–20 gene transcripts *via* traditional hybridization techniques such as Northern analysis or dot blotting, the experimentalist can assess the expression profile of an entire genome in one experiment.

The Bad: However, the inherent limitations of the transcriptome-based studies are daunting. First and most importantly, the transcriptome is not representative of the proteome, as many transcripts are not being actively translated. Second, contextual information is lost, as the phenotype of the cell is highly dependent upon the protein-protein interactions: information that is not present in the transcriptome data. Third, in terms of cell function, a protein's value, like that of real estate is often dependent upon location and its state of being. There are ample examples of protein activity and concomitant biological function being highly or even completely dependent upon post-translational modifications, which either modify protein-protein interactions or are responsible for shuttling the protein from one part of the cell to another [39]. Thus, the fact that a transcript is present or upregulated does not necessarily translate into altered biological activity at the protein level and conversely, the lack of any observed regulation, when the transcript(s) in question is compared to an unstimulated control, does not mean that the cognate protein's activity has not been significantly modified, either by a change in its subcellular location, or modification of its inherent catalytic activity. Finally, it is now abundantly clear that the "1 gene-1 protein" dogma deduced from studies in simple procaryotic systems no longer holds. Rather, "1 gene" can give rise to many polypeptides of radically different function, through alternative splicing of the primary transcript and/or post-translational processing [40]. Thus, interrogation of a transcript is often confounded by questions of whether a particular exon is represented in the array or if the probe sequences can distinguish between closely related, but functionally distinct isoforms.

The Ugly: The multiple experimental designs of different experimentalists, while perhaps logical for each group's purposes, make it impossible to meaningfully compare results between them. Thus, widely disparate numbers of up- or down-regulated genes in various disease models of the heart exist, with only marginal overlap apparent between the data sets. This is to be expected during the early phases of a technology's development, as the laboratories struggle to assimilate the technique into their own experimental programs. Nevertheless, a certain degree of rigor has been lacking as laboratories leap headfirst into the "bag of chips," with experiments not taking into account both the inherent noise of the technique, which dictates multiple, independent sample preparations and experimental duplication, and biological noise, which leads to a significant amount of scatter in the data. Both sources of error can be significant but solvable, through the use of multiple experiments carried out along a well-defined experimental timeline as the phenotype develops. Thus, rather than depending upon a snapshot of data at a single time point, a collection of points during a period of time will be much more valuable in actually determining whether a particular transcript's up- or down-regulation is associated with an aspect of a pathological state.

An obvious application of the technology is to uncover the changes that precede, accompany and result from the expression of a primary genetic insult. Numerous catalogues now exist, both in real [41, 42] and virtual space (http://www.nhlbi.nih.gov/resources/pga/). The first point to consider is that even at this early stage of cardiovascular disease-based global genomics, we are drowning in data. At the authors' institution (http://www.cincinnatichildrens.org/Research/Research_Cores/Pediatric_Informatics/BioSoftware/default.htm), approximately 10 gigabytes of data are collected and archived every week. Second, because of the different methodologies, the different investigators and filters that are being applied both before and after data acquisition, there is depressingly little agreement concerning which genes are up and down regulated. As alluded to above, the lack of detail in published experimental protocols and post-data acquisition algorithms make it impossible to actually compare cardiovascular data sets that are posted in the literature or in virtual space. Third, very few, if any of the analyses are truly complete. Rather, most consist of a very limited number of points in the disease model and the interrogation is rarely comprehensive. Fourth, validation is often lacking or incomplete, consisting, for the most part, of confirming the general up- or down-regulation of a gene either by *in situ* hybridization or Northern analysis with the cognate cDNA or oligonucleotide probe. Fifth, gene annotation is also incomplete, with the data presented either as parsed, column based lists with an abbreviated one phrase description, or placed into functional groups on the basis of rather ill-defined criteria. Only very rarely are comprehensive analyses presented and the data filtering rigorously annotated, so that the overall quality of the categorization can be objectively determined.

In short, we are for the most part at the level of preparing catalogues, with the biological values waiting to be assigned in the future as the initial filters identify candidates for further validation. A valid analogy is that we know the names in the phone book, but certainly cannot describe the society that results from the cumulative actions of these individuals, in time, at different locations with different combinations of the others. Consider the hallmark of a good reporter. When s/he wants to fully describe a story so that the reader can place the information in context, the following questions must be asked and answered: *who* or *what, why, where* and *how*? In terms of global analyses, most experiments are simply asking *who* or *what*. To answer the other questions, *why, where* and *how*, which is necessary to fully comprehend the processes involved, we must understand how the transcriptome translates into the proteome and be able to place a product into its corresponding subcellular or extracellular location, as well as identify the relevant biological partners. Identifying the "why" and "how" aspects will require an exceptionally rich amount of biological annotation, placing the action(s) of the gene product in their spatial and temporal contexts within the biological system. The recent successes in computational simulations of selected networks [43, 44] provides proof-of-principle for the effectiveness of using these algorithms on the cardiovascular databases.

2.5
Filtering The Transcriptome: Enhancing The Value

Gene arrays as applied to cardiovascular disease are inherently noisy experiments as even in a seemingly genetically homogenous population the severity and penetrance of the disease is not necessarily uniform. In order to raise the signal-to-noise ratio so that interpretable data are obtained, some form of filtering is essential. This can occur at either the "back-end" using bioinformatics, or at the "front-end" by careful selection of the sample population (http://www.jax.org/research/churchill/research/expression/in dex.html) [45] or by sub selection of only the most relevant portion of the transcriptome (see below). Additionally, the overall experimental design is critical and details are often overlooked or not reported so that it becomes problematic for another experimentalist to exactly repeat the procedure. Seemingly trivial matters can significantly affect a particular experiment. For example, how are the animals sacrificed? If an animal smells the blood of another in the procedure room, a stress response can be initiated. Similarly, the circadian rhythm can have a significant impact on the hormonal status of an animal, which could impact cardiovascular status. These considerations are rarely acted upon but could, if carefully considered, minimize the inherent biological noise of the experiment.

The Back-end – Very few, if any studies have documented the fulfillment of the initial catalogue's promise. For example, hypertrophy is a common adaptive response of the heart to increased workload, injury and stress, and has been characterized in many models as being defined by the activation of a common set of genes, which are normally only expressed during early cardiac development. While a number of different models of hypertrophy [46] or myocardial infarction [47] have been catalogued at the transcriptional level, the difficulty lies in translating what are essentially thousands upon thousands of observations into a coherent, testable model that can be biologically validated either through gain or loss-of function studies. However, one of the more carefully considered cardiovascular studies illustrates the value of multiple comparisons and rigorous mining of the resultant data. Aronow et al. set out to examine the "monolithic" hypothesis, that is, a gene program common to different hypertropies exists, by comparing the transcriptomes of four genetic hypertrophy models that showed varying degrees of the hypertrophic response [48]. DNA microarrays were used to compare approximately 9000 mRNAs in the four transgenic mouse models. Although the total number of regulated genes (defined as a certain – fold up- or down-regulated) varied between the models with the numbers corresponding to the relative severity of the phenotypic response, no commonality between the four models could be defined. However, by applying a modest amount of analysis using hierarchical-tree and K-means clustering to the data sets, patterns involving 276 genes that were regulated among the four models could be discerned, including a subset associated with the activation of apoptosis. Northern analyses were subsequently used to confirm the microarray patterns and the biological annotation initiated using the existing databases.

The Front-end – In order to enhance the signal-to-noise ratio of an experiment, its design can also apply filters at the "front-end". For example, rather than interrogating the entire RNA complement of the cell, one can attempt to restrict it only to those transcripts that will be processed into the proteome [49]. We recently carried out such a study in order to explore the efficacy of the "front-filter" approach [50]. The *β*-agonist, isoproterenol, when administered over a 10–14 day period is a simple and well-characterized protocol that results in a characteristic 20–30% cardiac hypertrophy as measured by the heart to body weight ratios [51]. However, instead of interrogating the entire RNA complements of the treated and untreated animals, only the actively translated RNA was studied. Polysomes derived from the animals' ventricles were loaded on sucrose gradients, size fractionated using velocity density centrifugation and the RNA from these fractions used to select for transcripts that were loaded onto polysomes in response to isoproterenol. Four Clontech Atlas 1.2 microarray filters were simultaneously hybridized to radiolabeled cDNA probes derived from either vehicle-treated (control) free or polysome bound RNA, or from isoproterenol-treated free or polysome bound RNA. A numerical value for the shift to polysomes was calculated by taking the ratio of isoproterenol bound/free signal divided by the ratio of vehicle-treated bound/free signal. Signals were normalized to a median filter value to correct for differences in probe specific activities with increases of a particular transcript in the polysome fraction due to chronic isoproterenol infusion resulting in a value of >1.0. Thus, while a particular transcript's steady state level might not be increased during the treatment, if its translational efficiency was affected, it would be detected. This high-throughput screen, designed to identify only the transcripts that are actively translated during cardiac hypertrophy or whose translation is down-regulated during the physiological response, identified a number of genes with established links to hypertrophy, including Sp3, c-jun, annexin II, cathepsin B, and HB-EGF [52–56], confirming the screen's accuracy. However, in order to test the usefulness of the screen, we decided to focus on a candidate transcript that *had not* been previously linked to hypertrophy and found that protein levels of the tumor suppressor PTEN (phosphatase and *tensin* homologue on chromosome *ten*) were increased in the absence of increased messenger RNA levels (Fig. 2.1). While overall, the mRNA levels of PTEN were not increased as result of isoproterenol treatment, the movement of the existing transcripts into the heavy portion of the polysomes was. Quantitative western blot analyses showed that PTEN protein expression is, in fact, induced in isoproterenol-treated mouse hearts relative to vehicle-treated hearts [50], in agreement with the polysome-derived data. Taken to-

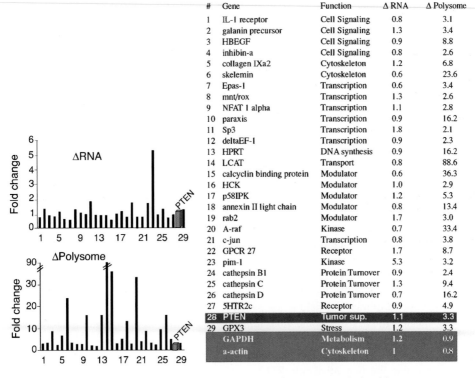

#	Gene	Function	Δ RNA	Δ Polysome
1	IL-1 receptor	Cell Signaling	0.8	3.1
2	galanin precursor	Cell Signaling	1.3	3.4
3	HBEGF	Cell Signaling	0.9	8.8
4	inhibin-a	Cell Signaling	0.8	2.6
5	collagen IXa2	Cytoskeleton	1.2	6.8
6	skelemin	Cytoskeleton	0.6	23.6
7	Epas-1	Transcription	0.6	3.4
8	mnt/rox	Transcription	1.3	2.6
9	NFAT 1 alpha	Transcription	1.1	2.8
10	paraxis	Transcription	0.9	16.2
11	Sp3	Transcription	1.8	2.1
12	deltaEF-1	Transcription	0.9	2.3
13	HPRT	DNA synthesis	0.9	16.2
14	LCAT	Transport	0.8	88.6
15	calcyclin binding protein	Modulator	0.6	36.3
16	HCK	Modulator	1.0	2.9
17	p58IPK	Modulator	1.2	5.3
18	annexin II light chain	Modulator	0.8	13.4
19	rab2	Modulator	1.7	3.0
20	A-raf	Kinase	0.7	33.4
21	c-jun	Transcription	0.8	3.8
22	GPCR 27	Receptor	1.7	8.7
23	pim-1	Kinase	5.3	3.2
24	cathepsin B1	Protein Turnover	0.9	2.4
25	cathepsin C	Protein Turnover	1.3	9.4
26	cathepsin D	Protein Turnover	0.7	16.2
27	5HTR2c	Receptor	0.9	4.9
28	PTEN	Tumor sup.	1.1	3.3
29	GPX3	Stress	1.2	3.3
	GAPDH	Metabolism	1.2	0.9
	a-actin	Cytoskeleton	1	0.8

Fig. 2.1 Polysome-derived RNA levels change although total RNA amounts remain stable. For each array position that had a signal greater than 0.5X the median filter value as well as a signal on all membranes, the ratio (TI/UI)/(TS/US), where: US, untranslated sham (vehicle solvent only); TS, translated sham; UI, untranslated isoproterenol treated; TI, translated isoproterenol treated. A ratio >1 indicates a shift toward polysomes in the isoproterenol treated hearts and <1 indicates a shift away from polysomes. Only candidates that had a ratio of >2 were selected. ΔRNA.

Total RNA changes were calculated by summing signals from all four array membranes. With the exception of the serine/threonine kinase, pim-1 (gene #23), the candidate genes exhibited minimal RNA fluctuation. ΔPolysomes. Significant changes occurred in the degree of polysome loading (movement to the heavy fraction). PTEN was selected on the basis of the subsequent, biological annotation. Note that the constitutive markers, GAPDH and alpha actin, remained relatively unchanged.

gether with the shift of PTEN mRNA into the heavy polysome fractions during hypertrophy and the minimal change of total PTEN mRNA, this finding is consistent with regulation of PTEN expression by increased translational initiation.

PTEN was originally identified as a human tumor-suppressor gene and is also called MMAC1/TEP1 (MMAC1, *m*utated in *m*ultiple *a*dvanced *c*ancers-1; TEP1, *T*GF-*β* regulated, *e*pithelial cell *e*nriched *p*hosphatase). The gene is either deleted or inactivated in a high percentage of breast, endometrial, brain and prostate cancers [57–59]. A potent tumor-suppressor function has been confirmed by performing an *in vivo* loss-of-function via gene ablation studies in mice. Mice with only one functional copy of the gene are more likely to develop tumors of multiple origins, while loss of both alleles leads to embryonic lethality [60–62].

PTEN is a dual-specificity phosphatase with homology to the focal adhesion-associated protein tensin [63]. *In vitro*, PTEN can dephosphorylate acidic polypeptides, focal adhesion kinase (FAK), and the adaptor protein, Shc. However, the major *in vivo* substrate for PTEN appears to be phosphatidylinositol 3, 4, 5-triphosphate (PIP3), as embryonic fibroblasts taken from PTEN null mouse strains have abnormally high levels of PIP3 and are resistant to apoptosis [61]. The PTEN$^{-/-}$ fibroblasts have very high levels of activated Akt, a serine/threonine kinase that is regulated by PIP3 and phosphatidylinositol 3, 4-biphosphate (PIP2). Intriguingly, Akt is an important regulator of both cell survival and growth [64], and PTEN has been defined genetically and biochemically to act as a negative regulator of Akt in opposition to the evolutionarily conserved IGF-1/PI3K/Akt signaling pathway [65–67]. Thus, the biological annotation for this candidate is extraordinarily rich, although the data are from systems other than the heart. Nevertheless, the overall functional annotation of PTEN was of sufficient value as to warrant further exploration.

High throughput functional screens are critical for assigning biological value to candidates identified by genome or proteome wide screens. A general limitation of the cardiovascular field is the paucity of accurate screens for determining a candidate protein's functional role, unless one assumes *in vitro* binding assays with putative partners (such as defined transcriptional factor activation domains, etc.) are truly accurate representations of a candidate's biological role. Therefore, to determine the (potential) role PTEN might play during cardiac hypertrophy, we restricted our approach to either cell culture or transgenic animals using complementary gain- and loss-of-function approaches whenever possible.

To explore the possibility of a biological function for PTEN in the hypertrophic response, adenovirus was used to overexpress the protein in primary cultures of neonatal rat cardiomyocytes. Overexpression resulted in fewer viable cells as a result of apoptotic pathway activation. More interestingly, expression of a catalytically inactive form of PTEN [63] we termed H123YPTEN, led to cardiomyocyte hypertrophy with a well-ordered sarcomeric structure being conserved in the cardiomyocytes as shown by α-actinin staining [50]. Molecular markers, cell volume and shape, as well as protein synthetic rates were all consistent with the inactive form mediating a robust hypertrophy in the cultured cardiomyocytes [50].

There are several possible mechanisms by which H123YPTEN might act as a dominant suppressor of endogenous PTEN, including the sequestration of PTEN binding partners needed for full activity. For example, PTEN can bind to focal adhesion kinase directly, leading to its dephosphorylation and inactivation. Consistent with the ability of H123YPTEN to sequester FAK from endogenous PTEN, FAK tyrosine phosphorylation in AdH123Y infected cells was increased. We are currently analyzing whether the H123Y mutation stabilizes the interaction with FAK in cardiomyocytes.

Although these data are intriguing they do not provide enough information about the way PTEN acts within the cellular networks to make any firm conclusions about its role in a physiologically relevant hypertrophy. Rather, the data illustrate that cell culture experiments represent yet another biological filter against which to test putative candidates isolated from the screens. The data do, however, justify a more extensive exploration of PTEN's activity within the whole organ and whole animal contexts using drug-inducible, cardiac-specific transgenesis.

2.6
Concluding Remarks

When used correctly, large-scale transcriptome analyses can and will provide critical insights into the biology that underlies cardiovascular function, but only when coupled with richly annotated databases. This has already been realized in other mammalian-based systems such as in the profiling of different tumors [68, 69], as well as in simpler systems such as yeast [70]. With the human, mouse and rat sequence information that is now available, the list of sparsely annotated, or completely unknown genes continues to grow: perhaps the best use of the array technology in the short term for cardiovascular discovery will be to identify new candidates from this list and generate the new hypotheses that will determine the experiments necessary to define the biology of the genes and their products.

The recent advances in transgenic and gene targeting approaches have significantly increased our insights into the causes and development of a number of cardiovascular diseases. Although these animal models have their limitations, the further development of inducible and conditional systems will allow us to even more precisely control single genes. However, the pathogenesis of many human disorders depends on the interaction of either a single or several genes with environmental factors and can be influenced by modifiers or interacting genes, thereby contributing to the heterogeneity of human disease. Interestingly, variable penetrance consistent with the human phenotype has also been observed in some mouse models [35, 71]. Despite the differences between mice and man, studying the interplay of multiple factors in genetically altered mice will give us a basis for unraveling modulating mechanisms in humans, and much of the information needed for validation of the mouse models will come from comparing the total human and murine transcriptomes during development of the various cardiovascular pathologies.

Transcriptional profiling, as represented in the current literature when applied to cardiovascular disease, suffers from a certain narrowness of view. Most studies have focused on the response of the heart to a single surgical or pharmacological treatment or genetic mutation. The power of applying massive, parallel analyses to the problem of transcript profiling is currently best illustrated in yeast, where Hughes and colleagues first constructed a reference database or "compendium" of expression profiles, corresponding to 300 mutations/chemical treatments of *S. cerevisiae*. This massive database showed that a compendium could serve as an important reference tool, against which any yeast profile could be compared, and the similarities and differences then used to determine which pathways were actually perturbed in the new, unknown transcriptome by simply comparing the data to the compendium. Thus, after the initial annotation was complete, data from the unknown could be quickly analyzed, similarities between single genes, subgroups and groups of genes identified and a "fingerprint" of the new transcriptional profile identified.

Such a task, while daunting for the heart, is obtainable; indeed elements of this are some of the goals of a current funding initiative by the National Institutes of Health (http://www.nhlbi.nih.gov/resources/pga/resource.htm) and research consortia (http://www.cellularsignaling.org/). However, the yeast data, when compared to that emanating from the cardiovascular research community, underscores the immense scope of a truly comprehensive and global analysis. A large consortium of experimentalists and bioinformatics experts were needed to plan the experiments, carry them out, filter the data and choose the appropriate mining algorithms for the data's true value to emerge. Wrapping the profiling data within the biological phenomenology, the so-called "phenome" in which the model under study is completely characterized in terms of the changes at the cellular, biochemical, organ and system levels needs to be done by the cardiovascular community in order to enhance the data's value. As we begin to develop more efficient ways of understanding and perturbing entire networks and the resultant functional consequences, the reductionist approaches that have served us so well in the last half of the twentieth century will assume less importance. Nothing of this scope has been attempted in mammalian species, but it is these approaches that will provide data mines of immense value for the formulation of novel and testable hypotheses that will lead to unexpected and important insights into the biogenesis of cardiovascular disease.

2.7
References

1 HOFFMAN, E.P., BROWN, R.H., JR., KUNKEL, L.M. Dystrophin: the protein product of the Duchenne muscular dystrophy locus. *Cell.* **1987**, *51*, 919–928.

2 SPARKMAN, D.R. Recombinant DNA approach to X-linked mental retardation. *J. Neurogenet.* **1984**, *1*, 199–211.

3 HOOGEVEEN, A.T., WILLEMSEN, R., MEYER, N., DE ROOIJ, K.E., ROOS, R.A., et al. Characterization and localization of the Huntington disease gene product. *Hum. Mol. Genet.* **1993**, *2*, 2069–2073.

4 GEISTERFER-LOWRANCE, A.A., KASS, S., TANIGAWA, G., VOSBERG, H.P., McKENNA, W., et al. A molecular basis for familial hypertrophic cardiomyopathy: a beta cardiac myosin heavy chain gene missense mutation. *Cell.* **1990**, *62*, 999–1006.

5 SEIDMAN, C.E., SEIDMAN, J.G. Molecular genetic studies of familial hypertrophic cardiomyopathy. *Basic Res. Cardiol.* **1998**, *93*, 13–16.

6 DAVIES, M.J., McKENNA, W.J. Hypertrophic cardiomyopathy–pathology and pathogenesis. *Histopathology.* **1995**, *26*, 493–500.

7 MARON, B.J., ROBERTS, W.C., McALLISTER, H.A., ROSING, D.R., EPSTEIN, S.E. Sudden death in young athletes. *Circulation.* **1980**, *62*, 218–229.

8 THIERFELDER, L., WATKINS, H., MacRAE, C., LAMAS, R., McKENNA, W., et al. Alpha-tropomyosin and cardiac troponin T mutations cause familial hypertrophic cardiomyopathy: a disease of the sarcomere. *Cell.* **1994**, *77*, 701–712.

9 POETTER, K., JIANG, H., HASSANZADEH, S., MASTER, S. R., CHANG, A., et al. Mutations in either the essential or regulatory light chains of myosin are associated with a rare myopathy in human heart and skeletal muscle. *Nat. Genet.* **1996**, *13*, 63–69.

10 SANBE, A., NELSON, D., GULICK, J., SETSER, E., OSINSKA, H., et al. In vivo analysis of an essential myosin light chain mutation linked to familial hypertrophic cardiomyopathy. *Circ. Res.* **2000**, *87*, 296–302.

11 LIANG, P., PARDEE, A. B. Differential display of eukaryotic messenger RNA by means of the polymerase chain reaction. *Science.* **1992**, *257*, 967–971.

12 DEMPSEY, A. A., DZAU, V. J., LIEW, C. C. Cardiovascular genomics: estimating the total number of genes expressed in the human cardiovascular system. *J. Mol. Cell. Cardiol.* **2001**, *33*, 1879–1886.

13 LIEW, C. C., HWANG, D. M., WANG, R. X., NG, S. H., DEMPSEY, A., et al. Construction of a human heart cDNA library and identification of cardiovascular based genes (CVBest). *Mol. Cell. Biochem.* **1997**, *172*, 81–87.

14 VELCULESCU, V. E., ZHANG, L., VOGELSTEIN, B., KINZLER, K. W. Serial analysis of gene expression. *Science.* **1995**, *270*, 484–487.

15 JAMES, J., OSINSKA, H., HEWETT, T. E., KIMBALL, T., KLEVITSKY, R., et al. Transgenic over-expression of a motor protein at high levels results in severe cardiac pathology. *Transgenic Res.* **1999**, *8*, 9–22.

16 SEIDMAN, J. G., SEIDMAN, C. The genetic basis for cardiomyopathy: from mutation identification to mechanistic paradigms. *Cell.* **2001**, *104*, 557–567.

17 IZUMO, S., SHIOI, T. Cardiac transgenic and gene-targeted mice as models of cardiac hypertrophy and failure: a problem of (new) riches. *J. Card. Fail.* **1998**, *4*, 263–270.

18 JAMES, J., ROBBINS, J. Molecular remodeling of cardiac contractile function. *Am. J. Physiol.* **1997**, *273*, H2105–2118.

19 GULICK, J., HEWETT, T. E., KLEVITSKY, R., BUCK, S. H., MOSS, R. L., et al. Transgenic remodeling of the regulatory myosin light chains in the mammalian heart. *Circ. Res.* **1997**, *80*, 655–664.

20 SUBRAMANIAM, A., GULICK, J., NEUMANN, J., KNOTTS, S., ROBBINS, J. Transgenic analysis of the thyroid-responsive elements in the alpha-cardiac myosin heavy chain gene promoter. *J. Biol. Chem.* **1993**, *268*, 4331–4336.

21 TSIEN, J. Z., CHEN, D. F., GERBER, D., TOM, C., MERCER, E. H., et al. Subregion- and cell type-restricted gene knockout in mouse brain. *Cell.* **1996**, *87*, 1317–1326.

22 COSTANTINI, F., RADICE, G., LEE, J. L., CHADA, K. K., PERRY, W., et al. Insertional mutations in transgenic mice. *Prog. Nucleic Acids Res. Mol. Biol.* **1989**, *36*, 159–169.

23 KRULEWSKI, T. F., NEUMANN, P. E., GORDON, J. W. Insertional mutation in a transgenic mouse allelic with Purkinje cell degeneration. *Proc. Natl. Acad. Sci. USA.* **1989**, *86*, 3709–3712.

24 SINGH, G., SUPP, D. M., SCHREINER, C., MCNEISH, J., MERKER, H. J., et al. *legless* insertional mutation: morphological, molecular, and genetic characterization. *Genes Dev.* **1991**, *5*, 2245–2255.

25 DOETSCHMAN, T., GREGG, R. G., MAEDA, N., HOOPER, M. L., MELTON, D. W., et al. Targetted correction of a mutant HPRT gene in mouse embryonic stem cells. *Nature.* **1987**, *330*, 576–578.

26 VALANCIUS, V., SMITHIES, O. Testing an "in-out" targeting procedure for making subtle genomic modifications in mouse embryonic stem cells. *Mol. Cell. Biol.* **1991**, *11*, 1402–1408.

27 MINAMINO, T., GAUSSIN, V., DEMAYO, F. J., SCHNEIDER, M. D. Pre-selection of integration sites imparts repeatable transgene expression. *Circ. Res.* **2001**, *88*, 587–592.

28 WALLACE, H., ANSELL, R., CLARK, J., MCWHIR, J. Pre-selection of integration sites imparts repeatable transgene expression. *Nucleic Acids Res.* **2000**, *28*, 1455–1464.

29 KUO, H. C., CHENG, C. F., CLARK, R. B., LIN, J. J. C., LIN, J. L. C., et al. A defect in the Kv channel-interacting protein 2 (KChIP2) gene leads to a complete loss of I(to) and confers susceptibility to ventricular tachycardia. *Cell.* **2001**, *107*, 801–813.

30 FEINER, L., WEBBER, A. L., BROWN, C. B., LU, M. M., JIA, L., et al. Targeted disrup-

tion of semaphorin 3C leads to persistent truncus arteriosus and aortic arch interruption. *Development*. **2001**, *128*, 3061–3070.

31 FEIL, R., BROCARD, J., MASCREZ, B., LE-MEUR, M., METZGER, D., et al. Ligand-activated site-specific recombination in mice. *Proc. Natl. Acad. Sci. USA*. **1996**, *93*, 10887–10890.

32 SOHAL, D.S., NGHIEM, M., CRACKOWER, M.A., WITT, S.A., KIMBALL, T.R., et al. Temporally regulated and tissue-specific gene manipulations in the adult and embryonic heart using a tamoxifen-inducible Cre protein. *Circ. Res*. **2001**, *89*, 20–25.

33 GOSSEN, M., FREUNDLIEB, S., BENDER, G., MULLER, G., HILLEN, W., et al. Transcriptional activation by tetracyclines in mammalian cells *Science*. **1995**, *268*, 1766–1769.

34 FAZIO, S., LINTON, M.F. Mouse models of hyperlipidemia and atherosclerosis. *Front. Biosci*. **2001**, *6*, D515–525.

35 TADDEI, I., MORISHIMA, M., HUYNH, T., LINDSAY, E.A. Genetic factors are major determinants of phenotypic variability in a mouse model of the DiGeorge/del22q11 syndromes. *Proc. Natl. Acad. Sci. USA*. **2001**, *98*, 11428–11431.

36 NERBONNE, J.M., NICHOLS, C.G., SCHWARZ, T.L., ESCANDE, D. Genetic manipulation of cardiac K(+) channel function in mice: what have we learned, and where do we go from here? *Circ. Res*. **2001**, *89*, 944–956.

37 VALENZA-SCHAERLY, P., PICKARD, B., WALTER, J., JUNG, M., POURCEL, L., et al. A dominant modifier of transgene methylation is mapped by QTL analysis to mouse chromosome 13. *Genome Res*. **2001**, *11*, 382–388.

38 MONTAGUTELLI, X. Effect of the genetic background on the phenotype of mouse mutations. *J. Am. Soc. Nephrol*. **2000**, *11* Suppl 16, S101–105.

39 MOLKENTIN, J.D., LU, J.R., ANTOS, C.L., MARKHAM, B., RICHARDSON, J., et al. A calcineurin-dependent transcriptional pathway for cardiac hypertrophy. *Cell*. **1998**, *93*, 215–228.

40 PERIASAMY, M., STREHLER, E.E., GARFINKEL, L.I., GUBITS, R.M., RUIZ-OPAZO, N., et al. Fast skeletal muscle myosin light chains 1 and 3 are produced from a single gene by a combined process of differential RNA transcription and splicing. *J. Biol. Chem*. **1984**, *259*, 13595–13604.

41 YANG, J., MORAVEC, C.S., SUSSMAN, M.A., DIPAOLA, N.R., FU, D., et al. Decreased SLIM1 expression and increased gelsolin expression in failing human hearts measured by high-density oligonucleotide arrays. *Circulation*. **2000**, *102*, 3046–3052.

42 PAONI, N.F., LOWE, D.G. Expression profiling techniques for cardiac molecular phenotyping. *Trends Cardiovasc. Med*. **2001**, *11*, 218–221.

43 STREICHER, J., DONAT, M.A., STRAUSS, B., SPORLE, R., SCHUGHART, K., et al. Computer-based three-dimensional visualization of developmental gene expression. *Nat. Genet*. **2000**, *25*, 147–152.

44 ARNONE, M.I., DAVIDSON, E.H. The hardwiring of development: organization and function of genomic regulatory systems. *Development*. **1997**, *124*, 1851–1864.

45 NEWTON, M.A., KENDZIORSKI, C.M., RICHMOND, C.S., BLATTNER, F.R., TSUI, K.W. On differential variability of expression ratios: improving statistical inference about gene expression changes from microarray data. *J. Comput. Biol*. **2001**, *8*, 37–52.

46 HWANG, J.J., DZAU, V.J., LIEW, C.C. Genomics and the pathophysiology of heart failure. *Curr. Cardiol. Rep*. **2001**, *3*, 198–207.

47 STANTON, L.W., GARRARD, L.J., DAMM, D., GARRICK, B.L., LAM, A., et al. Altered patterns of gene expression in response to myocardial infarction. *Circ. Res*. **2000**, *86*, 939–945.

48 ARONOW, B.J., TOYOKAWA, T., CANNING, A., HAGHIGHI, K., DELLING, U., et al. Divergent transcriptional responses to independent genetic causes of cardiac hypertrophy. *Physiol. Genomics*. **2001**, *6*, 19–28.

49 ZONG, Q., SCHUMMER, M., HOOD, L., MORRIS, D.R. Messenger RNA translation state: the second dimension of high-throughput expression screening. *Proc. Natl. Acad. Sci. USA*. **1999**, *96*, 10632–10636.

50 SCHWARTZBAUER, G., ROBBINS, J. The tumor suppressor gene PTEN can regulate

51 KUDEJ, R.K., IWASE, M., UECHI, M., VAT-
NER, D.E., OKA, N., et al. Effects of
chronic beta-adrenergic receptor stimulation in mice. *J. Mol. Cell. Cardiol.* **1997**,
29, 2735–2746.

52 PERRELLA, M.A., MAKI, T., PRASAD, S., PI-
MENTAL, D., SINGH, K., et al. Regulation
of heparin-binding epidermal growth fac-
tor-like growth factor mRNA levels by hypertrophic stimuli in neonatal and adult
rat cardiac myocytes. *J. Biol. Chem.* **1994**,
269, 27045–27050.

53 TAKEMOTO, Y., YOSHIYAMA, M., TAKEUCHI,
K., OMURA, T., KOMATSU, R., et al. Increased JNK, AP-1 and NF-kappa B DNA
binding activities in isoproterenol-in-
duced cardiac remodeling. *J Mol. Cell.
Cardiol.* **1999**, *31*, 2017–2030.

54 MORARU, II, SYRBU, S., MICHAELS, K.,
KIM, D.H., MALCHOFF, D., et al. Pressure overload induces overexpression of
annexins II and V in aortic-banded rats.
Ann. N. Y. Acad. Sci. **1998**, *853*, 329–332.

55 SACK, M.N., DISCH, D.L., ROCKMAN,
H.A., KELLY, D.P. A role for Sp and nuclear receptor transcription factors in a
cardiac hypertrophic growth program.
Proc. Natl. Acad. Sci. USA. **1997**, *94*,
6438–6443.

56 CRIE, J.S., MORTON, P., WILDENTHAL, K.
Changes in cardiac cathepsin B activity
in response to interventions that alter
heart size or protein metabolism: comparison with cathepsin D. *J. Mol. Cell.
Cardiol.* **1983**, *15*, 487–494.

57 LI, J., YEN, C., LIAW, D., PODSYPANINA,
K., BOSE, S., et al. PTEN, a putative protein tyrosine phosphatase gene mutated
in human brain, breast, and prostate cancer . *Science.* **1997**, *275*, 1943–1947.

58 LI, D.M., SUN, H. TEP1, encoded by a
candidate tumor suppressor locus, is a
novel protein tyrosine phosphatase regulated by transforming growth factor beta.
Cancer Res. **1997**, *57*, 2124–2129.

59 STECK, P.A., PERSHOUSE, M.A., JASSER, S.
A., YUNG, W.K., LIN, H., et al. Identification of a candidate tumour suppressor
gene, MMAC1, at chromosome 10q23.3
that is mutated in multiple advanced
cancers. *Nat. Genet.* **1997**, *15*, 356–362.

60 PODSYPANINA, K., ELLENSON, L.H.,
NEMES, A., GU, J., TAMURA, M., et al. Mutation of Pten/Mmac1 in mice causes
neoplasia in multiple organ systems.
Proc. Natl. Acad. Sci. USA. **1999**, *96*,
1563–1568.

61 STAMBOLIC, V., SUZUKI, A., DE LA POMPA,
J.L., BROTHERS, G.M., MIRTSOS, C., et al.
Negative regulation of PKB/Akt-depen-
dent cell survival by the tumor suppressor PTEN. *Cell.* **1998**, *95*, 29–39.

62 DI CRISTOFANO, A., PESCE, B., CORDON-
CARDO, C., PANDOLFI, P.P. Pten is essential for embryonic development and tumour suppression. *Nat. Genet.* **1998**, *19*,
348–355.

63 MYERS, M.P., STOLAROV, J.P., ENG, C.,
LI, J., WANG, S.I., et al. PTEN, the tumor
suppressor from human chromosome
10q23, is a dual- specificity phosphatase.
Proc. Natl. Acad. Sci. USA. **1997**, *94*,
9052–9057.

64 KANDEL, E.S., HAY, N. The regulation
and activities of the multifunctional ser-
ine/threonine kinase Akt/PKB. *Exp. Cell
Res.* **1999**, *253*, 210–229.

65 GOBERDHAN, D.C., PARICIO, N., GOOD-
MAN, E.C., MLODZIK, M., WILSON, C.
Drosophila tumor suppressor PTEN controls cell size and number by antagoniz-
ing the Chico/PI3-kinase signaling pathway. *Genes Dev.* **1999**, *13*, 3244–3258.

66 HUANG, H., POTTER, C.J., TAO, W., LI,
D.M., BROGIOLO, W., et al. PTEN affects
cell size, cell proliferation and apoptosis
during Drosophila eye development. *Development.* **1999**, *126*, 5365–5372.

67 OGG, S., RUVKUN, G. The C. elegans
PTEN homolog, DAF-18, acts in the insulin receptor-like metabolic signaling
pathway. *Mol. Cell.* **1998**, *2*, 887–893.

68 BHATTACHARJEE, A., RICHARDS, W.G.,
STAUNTON, J., LI, C., MONTI, S., et al.
Classification of human lung carcinomas
by mRNA expression profiling reveals
distinct adenocarcinoma subclasses. *Proc.
Natl. Acad. Sci. USA.* **2001**, *98*, 13790–
13795.

69 HANASH, S.M., BOBEK, M.P., RICKMAN,
D.S., WILLIAMS, T., ROUILLARD, J.M., et
al. Integrating cancer genomics and proteomics in the post-genome era. *Proteomics.* **2002**, *2*, 69–75.

70 HUGHES, T. R., MARTON, M. J., JONES, A. R., ROBERTS, C. J., STOUGHTON, R., et al. Functional discovery via a compendium of expression profiles. *Cell.* **2000**, *102*, 109–126.

71 CARRIER, L., BONNE, G., BAHREND, E., YU, B., RICHARD, P., et al. Organization and sequence of human cardiac myosin binding protein C gene (MYBPC3) and identification of mutations predicted to produce truncated proteins in familial hypertrophic cardiomyopathy. *Circ. Res.* **1997**, *80*, 427–434.

3
Heart Failure: A Genomics Approach
Choong Chin Liew

3.1
Overview of Heart Failure

Heart failure is a syndrome that occurs over time as the heart becomes less and less able to pump blood at a rate that is sufficient to meet the requirements of the body's organs, tissues and cells. For example, cardiomyopathies, which are disorders affecting the heart muscle, are a predominant cause of heart failure [1]. Heart failure may be right sided, left-sided, systolic or diastolic, or both. Clinically, systolic heart failure, an impairment of the right ventricle's ability to eject adequately during systole, is characterized by poor exercise tolerance and easy fatigability – characteristic effects of inadequate cardiac output and impaired tissue perfusion. Diastolic heart failure, the inability of the left ventricle to fill normally, is characterized by dyspnea, orthopnea, hepatomegaly, ascites, and edema, which result from elevated venous pressure. Clinical severity of heart failure varies from mild and asymptomatic to disabling to fatal without heart assist or heart transplant.

Congestive heart failure is rare in patients less than fifty; its prevalence increases to 5% for patients aged 50–70 and may be as high as 15% for patients aged over eighty. Heart failure is twice as common in black than in white populations and black patients tend to develop symptoms at a younger age [2]. Current drug therapies for heart failure include angiotensin-converting enzyme inhibitors, beta adrenergic blockers, diuretics and digoxin. However treatment success remains modest; patients with heart failure are often disabled, and survival is decreased. Having a diagnosis of heart failure increases the risk of death approximately four times, 5-year survival rates are 50% overall with median survival of 1.7–3.1 years in men and women, respectively. The likelihood of survival decreases with age and with more advanced heart failure [3].

A significant and increasing cause of morbidity and mortality, heart failure is becoming a major heath care burden. Over the past two decades the condition has increased by more than 150% and will continue to increase as the population ages and as death rates from acute myocardial infarction decline [2]. Currently, about 4 to 5 million people in the United States suffer from heart failure, resulting in the hospitalization of two million patients each year [4]. Approximately

400,000 new cases of heart failure are diagnosed annually, and the cost of treat-ment is estimated at a low of US \$10 billion annually [5], to as high as US \$21 to \$50 billion per year [2]. Life threatening heart failure unresponsive to drug thera-py may require assistive devices, or heart transplant.

Clearly the burden of heart failure in ill health and mortality, its high cost and increasing prevalence are making it essential to develop new strategies for treat-ing or preventing this common and debilitating disorder. Our understanding of heart failure is undergoing a paradigm shift as molecular, cellular and genetic re-search is developing new insights into the causes and progression of hypertrophy and heart failure. One result of this shift in thinking about heart failure is that drug therapies are beginning to be targeted to interfere in the progression of re-modeling and improving contractility, rather than at symptom relief, as has been the approach in the past. Concurrent with advances at the molecular and cellular level, the genomics revolution is providing exciting new vistas of research to aid in further developing understanding of heart failure at the genomic level of gene-gene interactions and pathways. Although such research is in its infancy and prac-tical genomics-based therapies remain in the far distant future, genomic research is beginning to provide new means to explore the extraordinary complexity of the genetic webs and pathways of this disease.

3.2
Pathophysiology of Heart Failure

A complex, multifactorial condition, heart failure occurs as a result of the interac-tion of environmental, physiological and genetic factors. In the United States, cor-onary artery disease and scarring from myocardial infarction are the most com-mon factors leading to heart failure, followed by cardiomyopathy, and hyperten-sion. Heart failure can also occur in patients with valvular heart disease or sus-tained arrhythmia, or it may be secondary to toxic agents such as alcohol, some chemotherapeutic drugs, or to pulmonary or systemic diseases. Regardless of the initial insult, however, heart failure is the final common pathway of almost any disease or injury that cause cardiac cell and tissue damage (Fig. 3.1).

Over the past twenty years our understanding of the pathophysiology of heart failure has undergone significant changes. The relatively simple hemodynamic model in which heart failure was regarded as a disorder of myocardial contractil-ity has largely been superceded. Research efforts have reconceptualized the failing heart as much a condition of extracardiac neuroendocrine activation, cytokine re-lease, and homeostatic mechanisms as it is a disorder of the heart proper [6]. Re-cently, investigators have been studying inherited channelopathies in hopes of elu-cidating further the pathophysiology of heart failure. The cause of death in heart failure is most likely to be sudden death resulting from cardiac arrhythmia. Long QT syndrome is associated with mutations in HERG and Kq family of genes and myocytes in heart failure have shown action potential prolongation, representing potassium channel downregulation, and repolarization abnormalities [7]. Newer

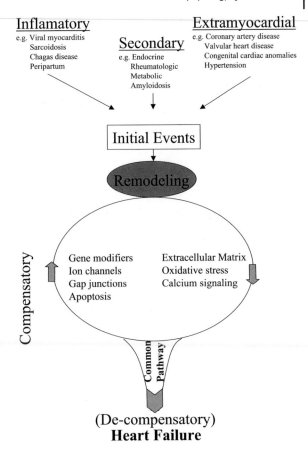

Fig. 3.1   Pathogenesis of heart failure.

concepts of heart failure pathophysiology are providing impetus for the development of more appropriate strategies in the management of heart failure. The use of drugs that act on the cardiac remodeling process (for example, antineuroendocrine agents or beta blockers) is one example of increased rationality in heart failure management. Such approaches may prevent the progression of failure, rather than, as in the past, to simply modify contractility and reduce symptoms [6].

According to the cardiac "remodeling" hypothesis, cardiac hypertrophy and heart failure – at least in earlier stages – represent a classic example of homeostatic defense. Overload, disease or injury to heart tissue, put in motion a set of adaptive mechanisms to preserve the heart's pumping capacity. Thus, factors such as hypertensive pressure, infarction, inflammation, or genetic factors activate mechanisms of the adrenergic nervous system, the renin-angiotensin system and various mediators, including endothelin, cytokines, (tumor necrosis factors and interleukins), nitric oxide and oxidative stress. This activation in turn initiates the progressive alterations called "remodeling". Left ventricular remodeling has been defined by the International Forum of Cardiac Remodeling as: "genome expression, molecular, cellular and interstitial changes that are manifested clinically as changes in the size, shape and function of the heart after cardiac injury" [8]. Remodelling, at first adaptive, over time becomes maladaptive, and sustained cardiac hypertrophy may culminate eventually in heart failure, salt and water retention, congestion, edema, low cardiac output, cardiac dysfunction and eventually death. Exactly how and at what point in the process adaptive hypertrophy becomes maladaptive heart failure is one of the challenges in heart failure research [9].

The course of remodeling from initial event to altered phenotype is highly complex. The cascade of molecular, cellular and biochemical events driving the process are under active investigation. A basic working hypothesis has been proposed [8]. In this model, stress-induced changes in the myocyte, such as stretch, leads to release of norepinephrine, angiotensin, endothelin and other factors. A feedback loop, or heart failure "treadmill" is set up whereby altered protein expression and myocyte hypertrophy lead to a deterioration in cardiac function and then to increased neurohormonal stimulation and further deterioration. Aldosterone and cytokines may stimulate collagen synthesis fibrosis and remodeling of extracellular matrix.

At the level of the cell, left ventricular remodeling involves mainly the cardiomyocyte as the main player. Despite the diversity of stimuli that initate cardiac hypertrophy, the molecular responses of cardiomyocytes to hypertrophic signals are similar. The responses include myocyte hypertrophy and apoptosis, enhanced sarcomeric protein accumulation reorganization of myofibrillar structure, changes in the extracellular matrix composition and functional abnormalities in excitation-contraction coupling. Hypertrophic stimuli are accompanied by induction of espression of immediate early genes, membrane type matrix metalloproteinases (MMP), which is increased in cardiomyopathy. Inhibition of MMP can attenuate left ventricular dilation in heart failure and targeted deletion of MMP-9 also showed limited ventricular enlargement and collagen accumulation after experimental myocardial induction in mice [10]. Kim et al. [11] demonstrated by overexpressing human MMP-1 in cardiac ventricles of mice that destruction of the collagen network in the myocardium caused cardiac hypertrophy and dysfunction. This animal model mimics human heart failure in that initially an adaptive response is seen followed by a progressive loss of function. Moreover this animal model provides direct evidence of the role of the extracellular matrix in the process of cardiac remodeling.

At the gene level hypertrophic stimuli results in induction of immediate-early genes such as c-fos, c-myc, c-jun, and Egr1 and reprogramming of gene expression in the adult myocardium, such that genes encoding fetal protein isoforms like β-myosin heavy chain (MHC) and a-skeletal actin are up-regulated, whereas the corresponding adult isoforms, a-MHC and a-cardiac actin, are down-regulated. The natriuretic peptides, atrial natriuretic peptide and brain natriuretic peptide, which decrease blood pressure by vasodilation and natriuresis, are also rapidly up-regulated in the heart in response to hypertrophic signals.

In addition, myocardial remodeling as well involves the orchestration of a variety of mediators including circulating hormones (endocrine effect), hormones acting on neighboring cells of different types (paracrine effect), and those affecting the cell of origin itself (autocrine effect). These mediators include arginine vasopressin, natriuretic peptides, endothelin, peptide growth factors (e.g., transforming growth factor-β, platelet-derived growth factor), cytokines (e.g., interleukin-1 β, interleukin-6, tumor necrosis factor-α, leukemia inhibitory factor, cardiotrophin-1), and nitric oxide. Each and all of these mediators act on the failing heart thorough a complex web of signaling pathways [12].

Increasing evidence suggests that enhanced production of reactive oxygen species together with accompanying oxidative stress has both functional and structural effects on remodeling. The myocardium is equipped with a variety of endogenous enzymatic and nonenzymatic antioxidant systems that are sufficient to metabolize oxygen free radicals generated during normal cellular activity. In particular dismutation of superoxide anions by cytosolic copper/zinc and mitochondrial manganese-containing superoxide dismutase (CuZnSOD and MnSOD, respectively) and the degradation of H_2O_2 by glutathione peroxidase (GPX) and catalase limit the cytotoxic effect of reactive oxygen metabolites. Dieterich et al. demonstrated that no differences in gene expression of MnSOD, CuZnSOD and GPX exist between failing and nonfailing hearts, whereas catalase gene expression was upregulated at both the mRNA and protein levels in failing hearts, possibly as a compensatory response [13]. Siwik et al. [14] showed that increased intracellular superoxide resulting from inhibition of CuZnSOD has profound effects on the cell growth, hypertrophic phenotype and apoptosis in neonatal rat cardiac myocytes in a graded manner. De Jong et al. [15] reported that xanthine oxidoreductase activity was elevated in failing but not in hypertrophic ventricles, suggesting its potential role in the induction of heart failure. Myocardial energetics has also been shown to be altered in heart failure. The hallmark of the change in myocardial metabolism in cardiac hypertrophy and the failing heart is a switch of the chief myocardial energy source from fatty acid B-oxidation to glycoysis resulting in down regulation of mitochondrial fatty acid oxidation cycle and medium chain acyl-coenzyme A dehydrogenase gene [16].

As the heart remodels, it becomes larger, rounder, and its walls stiffen: gross phenotypic changes that may affect cardiac function. Remodeling is regarded as an adverse sign in the progression to heart failure, and patients with major remodeling show worsening of function. Thus therapeutic efforts have been directed towards slowing or reversing remodeling early in the course of heart failure through the use of such agents as angiotensin converting enzyme inhibitors and beta blockers [8].

3.3

Genomic Approach to Heart Failure

The researcher involved in exploring the molecular pathophysiology of heart failure at the gene level faces a daunting challenge. Understanding heart failure involves working out the hundreds, if not thousands of genes, gene pathways, gene-gene and gene-protein interactions at every stage in the process that contribute to hypertrophy and eventual heart failure. The transition from compensated cardiac hypertrophy to decompensated heart failure is accompanied by marked changes in the expression of an array of genes in the heart. What is the marker for early myocardial decompensation and can we intervene in its progression? This discovery is the target for genomics-based studies (Fig. 3.2) [9].

Traditionally, the geneticist's search for disease causing gene mutations was a lengthy and resource-consuming process. Genetics investigators searched for single mutated nucleotide(s) in the DNA sequence. This methodology, though highly fruitful for locating single gene mutations, is not adequate for elucidating the complexities of multi-gene based diseases such as heart failure in which multiple genes and gene clusters may be involved. As described above, multiple pathways are involved at different clinical stages and in different types of heart failure and several gene mutations or protein defects have been identified in cardiomyopathy.

The Human Genome Project and related projects have paved the way for future studies in genomics. In February 2001, completion was announced of the first map of the whole human genome, comprising more than 90% of the approximately 3 billion bases of the human genome [17, 18]. The Human Genome Project database will serve as an invaluable resource for identifying the complete structure of each gene, interindividual variations of gene structure related to biodiversity, the clustering of genes in each chromosome, and gene polymorphisms related to structure or function. Furthermore, we can exploit gene expression profiles and genome-wide linkage analyses to identify genes contributing to individual variations in disease risk, to develop more efficacious treatments by targeting novel genes or unidentified pathways, and to tailor individual responses to treatment on the basis of genetic constitution.

The technology of the Human Genome Project and the genomics revolution are beginning to transform our approach to such complex disorders as heart failure and cardiomyopathy by enabling investigators to look at the differential expression of hundreds of thousands of genes simultaneously, and to compare gene expression during disease development and over the course of disease progression [19]. The genomics approach by contrast to the single gene approach is a "holistic" technology that attempts to investigate the relationship between clusters of candidate genes expressed or active during the complex processes involved in disease [20, 21].

An important development of the Human Genome Project has been the application of EST methods of gene discovery. ESTs (Expressed Sequence Tags) are sequences derived from complementary DNA libraries, representing particular tissues or organs [22, 23]. Although ESTs are relatively short in sequence (hundreds

a)

A 10,000 element cDNA array containing:
over 3000 known, characterized genes
over 7000 uncharacterized EST clusters

e.g. *Profiled DCM and HCM heart tissue*

b)

DCM (dilated cardiomyopathy) versus
normal (non-failing) heart samples

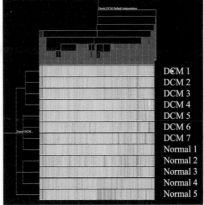

DCM 1
DCM 2
DCM 3
DCM 4
DCM 5
DCM 6
DCM 7
Normal 1
Normal 2
Normal 3
Normal 4
Normal 5

c)

Real Time RT-PCR analysis:

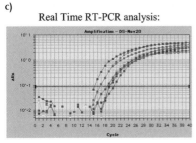

ANF
BNP
Phospholamdan
Calcequestrin
Troponin I
SERCA

SNP analysis:

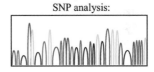

Fig. 3.2 a) Microarray analysis; b) hierarchical clustering; c) verification of individual genes.

of base pairs), they contain enough information to identify the transcript corresponding to the cDNA clone using simple nucleotide and protein database searching algorithms (e.g., BLAST). EST technology is convenient for large-scale transcript profiling through the generation of partial DNA sequences (ESTs) from randomly selected cDNA clones. Using these clones, we can generate an expression profile useful for performing genome-wide comparisons between two or more transcript populations and allow the identification of expression differences at the level of the single gene [24]. Moreover, identifying cDNA clones through EST generation provides a valuable genomic resource that is useful for the identification of protein homologs, chromosome mapping, exon identification in genomic sequences, single nucleotide polymorphism identification and as a substrate for cDNA microarrays [25].

With the publication of the complete sequence of the human genome, the number of genes encoded by the genome is estimated to be 32,000–38,000 [17, 18]. The next step is to determine how many genes are involved in specific cells, tissues, organs. In a recent publication, the first to describe the number of genes expressed in a single organ system, our team took a global approach to address the issue. Using three approaches (a current EST database, the complete nucleotide sequences of chromosomes 21 and 22 and cDNA microarray hybridization) we estimated that the number of genes expressed in the cardiovascular system ranges between 21,000 and 27,000 [26]. The high end of the estimate is a number close to the total number of genes in the human genome, suggesting that the majority of genes in the human genome function in the normal maintenance and function of an organ regardless of its specific function, while only a small proportion are allocated to cell-specific functions.

With access to the human genome map and with the aid of ESTs, we can investigate simultaneously the genes involved in physiologic or pathologic states and begin to develop a modern approach to puzzles of genomics. Our laboratory has applied the EST approach in cardiovascular disease and heart failure. Utilizing EST technology, our laboratory has generated a compendium of genes expressed in the human cardiovascular system, with the ultimate goal of assembling the intricacies of development and of disease, particularly the pathways leading to heart failure [9]. Through a computer-based *in silico* strategy, we have been able to identify on a large scale both known and previously-unsuspected genetic modulators contributing to the growth of the myocardium from fetal through adult, and from normal to a perturbed hypertrophic phenotype. Our laboratory employed a whole-genome approach using ESTs to characterize gene transcription and to identify new genes overexpressed in cardiac hypertrophy [24]. A detailed comparison of individual gene expression identified 64 genes potentially overexpressed in hypertrophy, and analysis of general transcription patterns revealed a proportional increase in transcripts related to cell/organism defense and a decrease in transcripts related to cell structure and motility in the hypertrophic heart. This approach is in striking contrast to the earlier time consuming and cumbersome gene-by-gene approach in elucidating the genes and mechanisms involved in complexities of development and disease.

Another genomic advance is the DNA microarray chip. DNA microarray chip technology makes use of EST data and represents a major advance in genomics research. In microarray analysis, individual cDNA clones are amplified by polymerase chain reaction (PCR). A micro-sample of each cDNA clone is then chemically bonded onto a glass surface or nylon membrane in an array format. The chip can also be made from oligonucleotides synthesized in situ on the surface of the array. Probes labeled with different fluorophores can be used to identify differential gene expression. The strength and specificity of the interaction can be increased or decreased by altering the length of the oligonucleotides or by varying the conditions of the hybridization reactions. Fluorescence signals representing hybridization to each arrayed gene are then analyzed to determine the relative abundance in the two studied samples of mRNA corresponding to each gene. Currently, there are two other techniques available for large-scale monitoring of gene expression, or transcript profiling: differential display and serial analysis of gene expression which are discussed elsewhere [21].

Microarray is increasingly being used to investigate patterns of gene expression. Recently, microarray technology has been employed as a means of large-scale screening of vast numbers of genes – if not whole genomes – that possess differential expression in two distinct conditions. While new and exciting developments have arisen in such fields as cancer [27] and yeast, [28] very few cardiovascular based microarray studies have been published [29]. In animal studies of myocardial infarction, Friddle et al. [30] used microarray technology to identify gene expression patterns altered during induction and regression of cardiac hypertrophy induced by administration of angiotensin II and isoproterenol in a mouse model. A total of 55 genes were identified during induction or regression of cardiac hypertrophy. They confirmed 25 genes or pathways previously shown to be altered by hypertrophy, and further identified a larger set of 30 genes whose expression had not previously been associated with cardiac hypertrophy or regression. Among the 55 genes, 32 genes were altered only during induction, and 8 were altered only during regression. This study, using a genome-wide approach, demonstrated that a set of known and novel genes was involved in cardiac remodeling during regression and that these genes were distinct from those expressed during induction of hypertrophy.

In other studies, to examine the host gene expression involved during different phases of viral myocarditis in a coxsackievirus B-infected mouse model, Taylor et al. [31] showed that 169 known genes of about 7,000 clones initially screened had a level of expression significantly different at one or more postinfection time points (days 3, 9 and 30) as compared with baseline. The genes were sorted according to their functional groups and the portrait thus produced showed gene regulatory processes during viremic, inflammatory, and healing phases of the myocarditic process. The same group also utilized differential mRNA display method to assess gene expression at the transcription level in a mouse enteroviral model, and found 2 up-regulated (*Mus musculus* inducible GTPase, mouse mitochondrial hydrophobic peptide) and 3 down-regulated (mouse β-globin, *Homo sapiens* cAMP-regulated response element binding protein binding protein, *Mus musculus* Nip21 mRNA) candidate genes. Microarray studies in artery and vein have also yielded differential gene expression profiles. In one study using cDNA array analysis to explore the genes in the vasculature system, Adams et al. [32] described a set of 68 genes that were consistently differentially expressed in aortic media as compared with vena cava media.

In the first reported human microarray study in end stage heart failure, Yang et al. [33] examined gene expression in 2 failing human hearts using oligo-based arrays. The investigators used high density oligonucleotide arrays to investigate failing and nonfailing human hearts (end stage ischemic and dilated cardiomyopathy). Similar changes were identified in twelve genes in both types of heart failure, which the authors maintain, indicate that these changes may be intrinsic to heart failure. They found altered expression in cytoskeletal and myofibrillar genes, in genes involved in degradation and disassembly of myocardial proteins, in metabolism, in protein synthesis and genes encoding stress proteins. While the "Affychip" in this study offers a carefully-controlled systematic method of analysis, its current lack of user flexibility in its design hinders novel gene discovery currently available in tissue-specific arrays.

Our laboratory has taken a different approach to microarray technology. Taking advantage of our vast previously-acquired resources, we have constructed a first generation custom-made cardiovascular-based cDNA microarray, which we term the "CardioChip" [34]. Its practicality and flexibility has allowed us to conceptualize the molecular events surrounding end-stage heart failure. The current Cardiochip contains 10,368 redundant and randomly sequenced expressed sequence tags, derived from several human heart and artery cDNA libraries. The Cardiochip has been used to develop a profile of previously-suspected candidates involved in molecular events surrounding the pathology of heart failure; more importantly, this method identifies novel candidates that may, with further verification at the functional level, be responsible for contributing to the demise of myocardial function.

In our recent study of dilated cardiomyopathy (DCM), the Cardiochip verified several expected candidate genes and identified some novel candidates [35]. Atrial naturietic peptide showed an intense level of up-regulation across the DCM patient samples, confirming at the microarray level its pivotal role as a circulating marker of cardiac muscle stress [36]. Indeed, its presence in our analysis lent a degree of credence to the validity of our study. Despite its significant up-regulation versus non-failing samples, the level of atrial naturietic peptide was highly variable among the patients.

In addition, we observed a consistent up-regulation of selected sarcomeric and extracellular matrix proteins (i.e., β-myosin heavy chain, α-actinin, α-cardiac actin, troponin I, tropomyosin, collagen, etc.). Evidence in knockout mice and human studies has offered insight into the putative role of these proteins in maintaining sarcomeric integrity [34–43]. Mutations of proteins associated with α-actinin, namely MLP, cardiac α-actin, desmin and titin, have been shown to be present in certain forms of human DCM [44–48]. Ambiguities exist in the literature regarding the expression of collagen and other members of the extracellular matrix; nonetheless, regulation of the extracellular matrix is important in the formation of fibrosis and impaired contractile function [49, 50].

Calcium signaling has recently become an important area of interest in the investigation of heart failure [51]. A decrease in calcium cycling genes has been shown to result in reduced contractility in mice whose β-adrenergic stimulation is blunted leading to decreased phospholamban phosphorylation [33]. Ca^{2+}ATPase is key in regulating contractility, and its ~2-fold average down-regulation in our DCM samples lends credence to its involvement. This is supported by a recent study in which the transfer of the Ca^{2+}ATPase gene into the rat myocardium prevents certain features of heart failure [52]. The presence of Ca^{2+}/calmodulin-dependent kinase in our analysis, despite showing only about 1.1-fold down-regulation, is particularly intriguing, as it is known to phosphorylate phospholamban [53]. In addition, inositol 1,4,5-triphosphate receptor (a member of the calcium channel family) which may be responsible for calcium release from intracellular stores, [54] was also significantly down-regulated (1.86-fold). Inositol 1,4,5-trisphosphate 3-kinase was recently cloned [55] and may be another key component in this regulation (1.86-fold). Our findings suggest that the role of Ca^{2+} signaling down-regulation may be of crucial significance in the evolution of heart failure and would warrant further investigation.

A number of novel ESTs were also identified from our study to be differentially regulated. Verification with quantitative real-time RT-PCR confirmed this expression. It is an intriguing prospect that these among other transcripts, after full-length sequencing, represent novel cardiac-specific genes encoding proteins that are potentially key to solving the puzzle of the molecular pathophysiology of heart failure. Indeed, our microarray analysis not only serves as a genomic model for a more complete understanding of DCM, but also as a focused target for possible therapeutic interventions specific to the cardiac tissue. Investigations are currently underway to elucidate the function of these candidates.

In a similar study [56] our laboratory developed comparative microarray portraits of DCM and hypertrophic cardiomyopathy (HCM). Overall, our results showed that 192 genes were highly expressed in both DCM and HCM (atrial natriuretic peptide, CD59, decorin, elongation factor 2 and heat shock protein 90) and that 51 genes were downregulated in both conditions (elastin, sarcomeric/reticulum Ca^{2+}-ATPase). Differentially expressed genes as determined quantitatively by RT-PCR (Fig. 3.3) included a B-crystallin, antagonizer of myc transcriptional activity, beta dystrobrevin, calsequestrin, lipocortin and lumican). What this study shows is that although having similar clinical features, the gene defects leading to DCM and HCM differ. DCM, a cytoskeletalopathy, and HCM, a sarcomyopathy, are common forms of cardiomyopathy that result in end stage heart failure through different remodelling and molecular pathways. Our microarray portrait of DCM demonstrated that more genes involved in cell and organism defence were upregulated, especially immune system response genes. By contrast, protein and cell expression genes were downregulated. Hypertrophic processes are evident in the increase in ribosomal genes upregulated in HCM, whereas cell signalling and cell structure genes were downregulated.

These reports describe the most informative cDNA microarray-based analysis of end-stage heart failure derived from DCM and HCM currently available. These in-

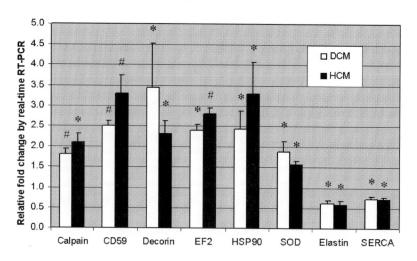

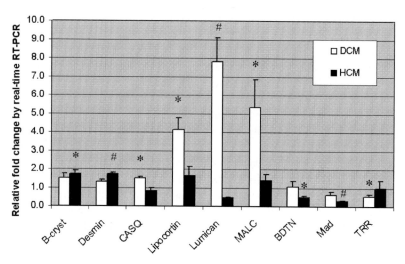

Fig. 3.3 Real-time RT-PCR confirmed commonly up-regulated or down-regulated (A) and differentially expressed (B) genes in DCM and HCM. The fold change is displayed as relative to normalized normal adult heart samples. * denotes $p<0.05$, #: $p<0.01$. The atrial natriuretic peptide was increased more than 20-fold in both DCM and HCM, not shown in the bar graph. Calpain: calcium acti-vated neutral protease, EF2: elongation factor 2, HSP 90: heat shock protein 90, SOD: copper/zinc superoxide dismutase, SERCA: sarcoplasmic/reticulum calcium-ATPase, B-cryst: α B-crystallin, CASQ: calsequestrin, MALC: atrial myosin alkali light chain, BDTN: β-dystrobrevin, Mad: antagonizer of *myc* transcriptional activity, and TRR: thioredoxin reductase.

vestigations are not exhaustive in that they do not attempt to fully characterize the molecular basis of heart failure. Their intention is to provide a preliminary portrait of global gene expression in complex cardiovascular disease using cDNA microarray and QRT-PCR technology, and to highlight the effectiveness of our ever-evolving platform for gene discovery.

As these intriguing findings show, microarray data offer a holistic view of the interrelated gene network during disease processes. Genes which are either over-expressed or under-expressed in a diseased tissue or organ present prima facie evidence that they are involved in disease pathogenesis. Using the genomic approach, previously unrecognized alterations in the expression of specific genes can be identified and novel genes can also be discovered, leading to a clearer understanding of the gene pathways. However, nature functions by integration, and proteins do not work in isolation but are instead involved in interrelated networks. The challenge of genomics lies not only in identifying genes, but also in understanding their function, the latter also referred to as "functional genomics". In the short term, the goal is to assign putative function to each of the genes using systematic, high-throughput approaches to the database. As functional information accumulates, the knowledge gap will be filled out in the form of expression profile studies, protein microcharacterization and their post-translational modifications, protein-protein interactions, computational approaches, and the response to loss of function by mutation [57]. Unraveling these networks and their interactions will be vital to an integrated mapping between genotype and phenotype. Gene chip technology may eventually be used to diagnose, stage or classify clinical conditions by detecting genetic markers associated with disease states in biopsy or blood samples. Indeed, gene expression microarray technology is a powerful tool with enormous potential in the years to come.

3.4
Conclusion

Heart failure is a complex syndrome. It may involve either the right or left ventricle, progress from compensated to decompensated stages, result from ischemic or nonischemic etiology, affect mainly systolic or diastolic function, and pertain to high cardiac output or low cardiac output status. The mechanism by which its genetic machinery controls these responses remains to be fully elucidated. The complexity of this disease raises numerous questions. Do all kinds of heart failure share a common final pathway? The transition from compensated cardiac hypertrophy to decompensated heart failure is accompanied by marked changes in the expression of an array of genes in the heart. What is the marker for early myocardial decompensation and could we intervene in its progression? The genome-wide approach will help us to integrate our current understanding of the pathophysiological pathways associated with heart failure. Large-scale DNA sequencing and the use of microarray technology have provided biomedical researchers with powerful tools to handle the vast database of the Human Genome Project. Structural biology and computational technology will further refine the structure prediction method and help decipher the complexity of sequence-structure-function in biological science. With the expanding database of newly discovered novel genes and functional annotations, therapeutic modalities aimed at specific molecular targets may be more effective and closer than previously imagined.

3.5
Acknowledgements

I would especially like to thank Isolde Prince for her help in preparing this manuscript. I would also like to thank David Barrans, Adam Dempsey, Jim Hwang, and Dimitri Stamatiou for their comments and suggestions.

3.6
References

1 TOWBIN, J. A., BOWLES, N. E. The failing heart. *Nature.* **2002.** *415,* 227–233.

2 MILLER, L. W., MISSOV, E. D. Epidemiology of heart failure. *Cardiology Clinics* **2001.** *19,* 547–555.

3 HO, K. K. L., ANDERSON, K. M., KANNEL, W. B., GROSSMAN, W., LEVY, D., et al. Survival after the onset of congestive heart failure in Framingham Heart Study subjects. *Circulation* **1993.** *88,* 107–115.

4 NOHRIA, A., LEWIS, E., STEVENSON, L. Medical Management of advanced heart failure. *J.A.M.A.* **2002.** *287,* 628–640.

5 KARON, B. L. Diagnosis and outpatient management of congestive heart failure. *Mayo Clinic Proceedings.* **1995.** *70,* 1080–1085.

6 FRANCIS, G. S. Pathophysiology of chronic heart failure. *Am. J. Med.* **2001.** *110(7A):* 37S–46S.

7 MARBAN, E. Cardiac channelopathies. *Nature.* **2002.** *415,* 213–218.

8 COHN, J. N., FERRARI, R., SHARPE, N. Cardiac remodeling concepts and clinical implications: a consensus paper from an international forum on cardiac remodelling. *J. Am. Coll. Cardiol.* **2000.** *35,* 569–582.

9 HWANG, J. J., DZAU, V. J., LIEW, C. C. Genomics and the pathophysiology of heart failure. *Current Cardiol. Rep.* **2001.** *3,* 198–207.

10 DUCHARME, A., FRANTZ, S., AIKAWA, M., RABKIN, E., et al. Targeted deletion of matrix metalloproteinase – 9 attenuates left ventricular enlargement and collagen accumulation after experimental myocardial infarction *J. Clin. Invest.* **2000.** *106,* 55–62.

11 KIM, H. E., DALAL, S. S., YOUNG, E., LEGATO, M. J., et al. Disruption of the myocardial extracellular matrix leads to cardiac dysfunction. *J. Clin. Invest.* **2000.** *106,* 857–866.

12 GIVERTZ, M. M., COLUCCI, W. S. New targets for heart-failure therapy: endothelin, inflammatory cytokines and oxidative stress. *Lancet* **1998.** 352 Suppl. *1,* 34–38.

13 DIETERICH, S., BIELIGK, U., BEULICH, K., HASENFUSS, G., et al. Gene expression of antioxidative enzymes in the human heart: increased expression of catalase in the end-stage failing heart. *Circulation* **2000.** *101,* 33–39.

14 SIWIK, D. A., TZORTZIS, J. D., PIMENTAL, D. R., CHANG, D. L., et al. Inhibition of copper-zinc superoxidase dismutase induces cell growth, hypertrophic phenotype and apoptosis in neonatal rat cardi-

ac myocytes in vitro. *Circ. Res.* **1999.** *85,* 147–153.

15 De Jong, J. W., Schoemaker, R. G., De Jonge, R., Bernocchi, P., et al. Enhanced expression and activity of xanthine oxidoreductase in the failing heart. *J. Mol. Cell. Cardiol.* **2000.** *32,* 2083–2089.

16 Sack, M. N., Kelly, D. P. The energy substrate switch during development of heart failure: gene regulatory mechanisms. *Int. J. Mol. Med.* **1998.** *1,* 17–24.

17 Venter, J. C., Adams, M. D., Myers, E. W., Li, P. W., et al. The sequence of the human genome. *Science* **2001.** *291,* 1304–1351.

18 International Human Genome Sequencing Consortium. Initial sequencing and analysis of the human genome. *Nature* **2001.** *409,* 860–921.

19 Chien, K. R. Genomic circuits and the integrative biology of cardiac diseases. *Nature* **2000.** *407,* 227–232.

20 Pratt, R. E., Dzau, V. J. Genomics and hypertension: concepts, potentials and opportunities. *Hypertension* **1999.** *33 part 2,* 238–247.

21 Dempsey, A. A., Ton, C., Liew, C. C. A cardiovascular EST repertoire: progress and promise for understanding cardiovascular disease. *Mol. Med. Today* **2000.** *6,* 231–237.

22 Liew, C. C. A human heart cDNA library-the development of an efficient and simple method for automated DNA sequencing. *J. Mol. Cell. Cardiol.* **1993.** *25,* 891–894.

23 Ton, C., Hwang, D. M., Dempsey A. A., Tang, H. C., et al. Identification, characterization and mapping of expressed sequence tags from an embryonic zebrafish heart cDNA library. *Genome Research* **2000.** *10,* 1915–1927.

24 Hwang, D. M., Dempsey, A. A., Lee, C. Y., Liew, C. C. Identification of differentially expressed genes in cardiac hypertrophy by analysis of expressed sequence tags. *Genomics* **2000.** *66,* 1–14.

25 Collins, F. S., Patrinos, A., Jordan, E., Chakravarti A., et al. New goals for the US Human Genome Project: 1998–2003. *Science* **1998.** *282,* 682–689.

26 Dempsey, A. A., Dzau, V. J., Liew, C. C. Cardiovascular genomics: estimating the total number of genes expressed in the human cardiovascular system. *J. Mol. Cell. Cardiol.* **2001.** *33,* 1879–1886.

27 Alizadeh, A. A., Eisen, M. B., Davis, R. E., Ma, C., et al. Distinct types of diffuse large B-cell lymphoma identified by gene expression profiling. *Nature* **2000.** *403,* 503–511.

28 Gasch, A. P., Spellman, P. T., Kao, C. M., Carmel-Harel, O., et al. Genomic expression programs in the response of yeast cells to environmental changes. *Mol. Biol. Cell* **2000.** *11,* 4241–57.

29 Stanton, L. W., Garrard, L. J., Damm, D., Garrick, B. L. et al. Altered patterns of gene expression in response to myocardial infarction. *Circ. Res.* **2000.** *86,* 939–945.

30 Friddle, C. L., Koga, T., Rubin, E. M., Bristow, J. Expression profiling reveals distinct sets of genes altered during induction and regression of cardiac hypertrophy. *P.N.A.S.* **2000.** *97,* 6745–6750.

31 Taylor, L. A., Carthy, C. M., Yang, D., et al. Host gene regulation during coxsackievrus B3 infection in mice: assessment by microarrays. *Circ. Res.* **2000** *87,* 328–334.

32 Adams, L. D., Geary, R. L., McManus, B., Schwartz, S. M. A comparison of aorta and vena cava medial message expression by cDNA array analysis identifies a set of 68 consistently differentially expressed genes, all in aortic media. *Circ. Res.* **2000** *87,* 623–631.

33 Yang, J., Moravec, C. S., Sussman, M. A., DiPaola, N. R., et al. Decreased SLIM1 expression and increased gelsolin expression in failing human hearts measured by high-density oligonucleotide arrays. *Circulation* **2000.** *102,* 3046–3052.

34 Barrans, J. D., Stamatiou, D., Liew, C. C. Construction of a human cardiovascular cDNA microarray: portrait of the failing heart. *Biochem. Biophys. Res. Comm.* **2001.** *280,* 964–969.

35 Barrans, J. D., Allen, P. D., Stamatiou, D., Dzau, V. J., Liew, C. C. Global gene expression of end-stage dilated cardiomyopathy using a human cardiovascular-based cDNA microarray *Am. J. Pathol.* **2002.** (in press).

36 Vikstrom, K. L., Bohlmeyer, T., Factor, S. M., Leinwand, L. A. Hypertrophy,

pathology and molecular markers of cardiac pathogenesis. *Circ. Res.* **1998**. *82*, 773–778.

37 Towbin, J.A., Bowles, N.E. Genetic abnormalities responsible for dilated cardiomyopathy. *Curr. Cardiol. Rep.* **2000**. *2*, 475–480.

38 Towbin, J.A. The role of cytoskeletal proteins in cardiomyopathies *Curr. Opin. Cell. Biol.* **1998**. *10*, 131–139.

39 Elliott, K., Watkins, H., Redwood, C.S. Altered regulatory properties of human cardiac troponin I mutants that cause hypertrophic cardomyopathy *J. Biol. Chem.* **2000**. *275*, 22069–22074.

40 Kimura, A., Harada, H., Park, J.E., Nishi, H., et al. Mutations in the cardiac troponin I gene associated with hypertrophic cardiomyopathy *Nat. Genet.* **1997**. *16*, 379–382.

41 Redwood, C., Lohmann, K., Bing, W., Esposito, G.M., et al. Investigation of a truncated cardiac troponin T that causes familial hypertrophic cardiomyopathy: Ca(2+) regulatory properties of reconstituted thin filaments depend on the ratio of mutant to wild-type protein. *Circ. Res.* **2000**. *86*, 1146–1152.

42 Geisterfer-Lowrance, A.A., Kass, S., Tanigawa, G., Vosberg, H.P., et al. A molecular basis for familial hypertrophic cardiomyopthy: a beta cardiac myosin heavy chain gene missense mutation. *Cell* **1990**. *62*, 999–1006.

43 Thierfelder, L., Watkins, H., MacRae, C., Lamas, R., et al. Alpha-tropomyosin and cardiac troponin T mutations cause familial hypertrophic cardiomyopathy a disease of the sarcomere. *Cell.* **1994**. *77*, 701–712.

44 Arber, S., Hunter, J.J., Ross, J. Jr., Hongo, M., et al. MLP-deficient mice exhibit a disruption of cardiac cytoarchitectural organization, dilated cardiomyopathy and heart failure. *Cell.* **1997**. *88*, 393–403.

45 Dalakas, M.C., Park, K.Y., Semino-Mora. C., Lee, H.S., et al. Desmin myopathy a skeletal myopathy with cardiomyopathy caused by mutations in the desmin gene. *N. Engl. J. Med.* **2000**. *342*, 770–780.

46 Satoh, M., Takahashi, M., Sakamoto, T., Hiroe, M., et al. Structural analysis of the titin gene in hypertrophic cardiomyopathy: identification of an novel disease gene. *Biochem. Biophys. Res. Commun.* **1999**. *27*, 411–417.

47 Olson, T.M., Michels, V.V., Thibodeau, S.N., Tai, Y.S., Keating M.T. Actin mutations in dilated cardiomyopathy, a heritable form of heart failure. *Science* **1998**. *280*, 750–752.

48 Schonberger, J., Seidman, C.E. Many roads lead to a broken heart: the genetics of dilated cardiomyopthy. *Am. J. Hum. Genet.* **2001**. *69*, 249–260.

49 Francis, G.S. Changing the remodeling process in heart failure: basic mechanisms and laboratory results. *Curr. Opin. Cardiol.* **1998**. *13*, 156–161.

50 Rao, V.U., Spinale, F.G. Controlling myocardial matrix remodeling: implications for heart failure. *Cardiol. Rev.* **1999**. *7*, 136–143.

51 McKinsey, T.A., Olson, E.N. Cardiac hypertrophy: sorting out the circuitry *Curr. Opin. Genet. Dev.* **1999**. *9*, 267–274.

52 Miyamoto, M.I., del Monte, F., Schmidt, U., DiSalvo, T.S., et al. Adenoviral gene transfer of SERCA2a improves left-ventricular function in aortic-banded rats in transition to heart failure. *Proc. Natl. Acad. Sci.* **2000**. *97*, 793–798.

53 Tada, M., Yabuki, M., Toyofuku, T. Molecular regulation of phospholamban function and gene expression. *Ann. N.Y. Acad. Sci.* **1998**. *853*, 116–129.

54 Marks, A.R. Cardiac intracellular calcium release channels: role in heart failure. *Circ. Res.* **2000**. *87*, 8–11.

55 Dewaste, V., Pouillon, V., Moreau, C., Shears, S., et al. Cloning and expression of a cDNA encoding human inositol 1,4,5-triphosphate 3-kinase C. *Biochem. J.* **2000**. *352 Pt 2*, 343–351.

56 Hwang, J.J., Allen, P.D., Tseng, G.C., Lam, C.W., Lameh, F., et al. Microarray gene expression profiles in dilated and hypertrophic cardiomyopathic end-stage heart failure. *Physiological Genomics.* **2002**. (in press).

57 Attwood, T.K. The babble of bioinformatics. *Science* **2000**. *290*, 471–473.

4

Principles of cDNA Microarrays as Applied in Heart Failure Research

Sara Arab, Mansoor Husain, and Peter Liu

4.1
The Clinical Problem

Symptomatic heart failure (HF) affects 4.7 million patients in the US with approximately 550,000 new cases of heart failure identified annually [1–3] . Proportionally, there are 428,000 patients in Canada, incurring $ 8 billion in annual hospital costs alone. This burden expected to *double* in the next 2 decades [4–6]. Other forms of cardiovascular disease are plateauing, but the incidence of heart failure is increasing. The one year mortality rate *is between 25–40%* [5]. This is partly the consequence of our success in treating myocardial infarction and sudden deaths, and partly due to the aging population.

Meanwhile, cardiac dysfunction often occurs in the patient long before symptoms manifest. Thus the best opportunity to curb the tide of heart failure is likely presented by intervention early in the disease process [7, 8]. What is required therefore is a concerted effort to identify the underlying pathogenesis of the disease, and the timely application of this information for the development of *early prognostic indicators and therapies.*

4.2
The Need for a New Paradigm

Currently, only three classes of agents: angiotensin converting enzyme inhibitors, beta blockers and aldosterone antagonists can modify the disease process and alter the natural history of heart failure [6, 7, 9]. These agents can improve survival, reduce hospitalization and enhance quality of life in HF patients. However, many challenges remain, because the one year mortality is still high in the presence of state of art therapy, and increasing numbers of patients in severe heart failure have a tremendously compromised quality of life. One such challenge is the need to find new targets for the treatment of HF. The traditional approaches have been limited to biochemical analysis of peripheral blood or myocardial biopsies. Many of the targets to date have been selected by reference to other fields of investigations, including nephrology, hypertension and cancer biology. Other targets deriving from previous hypothesis driven research have unfortunately failed to deliver in recent investigations. Agents that have been studied modulate calcium sensitivity or calcium release, cytokine inhibitors, vasopeptidase inhibitors as well as endothelin antagonists.

With the birth of microarray technology, and the near completion of the human genome [10], we now have an unprecedented opportunity to supplement the hypothesis driven approach with so called "knowledge driven" approach to science, such that genome based discovery takes place side by side with the traditional routes of discovery. Microarrays let us to generate new knowledge without requiring a prior known paradigm. For example, recently, Dr. Josef Penninger and Peter Backx from our Centre have discovered the function of a novel angiotensin converting enzyme (ACE2). This is partially derived from *in silico* analysis of genome and microarray databases, illustrating the power of this novel approach to discovery [11].

4.3
The Potential Role of the Microarray

Microarray technology has the singular power to provide a system based description of a global change of state in specific experimental conditions. This capapability is a great advantage over traditional molecular approaches that expire a single pathway in a complex network of interacting biological activities (Fig. 4.1). The older approaches metaphorically shine a single beam of light in a darkroom in order to determine the dynamic contents of the room. Microarray allows a system-wide or modular approach to analysis of change in an experimental condition, akin to turning on of multiple lighting systems to illuminate our dark room.

The first time investigator may find, that the information derived from microarray technology is overwhelming and difficult to comprehend. However, as experience accumulates, and the bioinformatics tools become more sophisticated, the strength of the method to decode changing patterns of gene expression becomes more apparent. When the experiments are done carefully, the data can be extremely consistent from experiment to experiment. This consistently adds a level of robustness to the observations beyond that previously available with technologies to study single molecular target techniques.

As the information database grows signals are more easily differentiated from the noise and the patterns emerge across different model systems. Such capabilities give us insight to the key regulators of a number of these systems. The ability of the bioinformatics system to provide this important insight is continuing to improve, thus making the additional experiments aimed at deciphering the function for novel gene targets more meaningful.

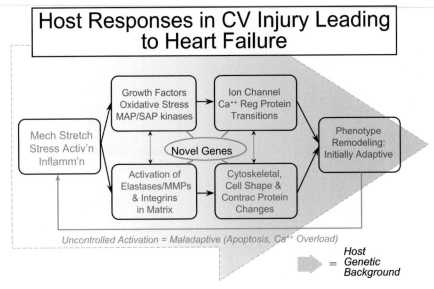

Fig. 4.1 The complex known pathways contributing to the process of heart failure. The traditional approach of investigating a single molecular target is intrinsically limited due to the network nature of the disease process. The microarray will permit the scientist to examine multiple targets simultaneously, to determine potential molecular interactions, and identify novel members of participating pathways.

4.4
Strengths of Microarray Technology

DNA microarray enables researchers for the first time to visualize global patterns of gene expression under different conditions. A long way to express a simple thought. The variation in the expression of a single gene is richer than the allelic variation in its sequence. Finding the link between the expression variation and phenotypic variation may provide clues to the biological roles of gene(s) to identify the molecular basis of phenotypic variation among cells and individuals. To generate a gene expression profile of a condition (s) is to identify the up- or down-regulation of gene (s) expression which can be both the cause and effect in a disease state. To study the global changes involving thousands of genes is not possible without microarray. A large portion of the genome can be interrogated simultaneously to identify the common cluster of genes with similar expression pattern which may reflect a similar function. The exceptions or unknown members may help to identify functionally important novel genes. Moreover, the temporal sequence of gene expression can be followed with microarray technology. Finally, microarray technologies can be used to develop diagnostic tools, either to identify the markers or to use the global geneexpression information for classifying different stages of the disease state, by using a subset of global candidate genes relevant to the disease (Fig. 4.2). The aggregate database is also useful to develop new

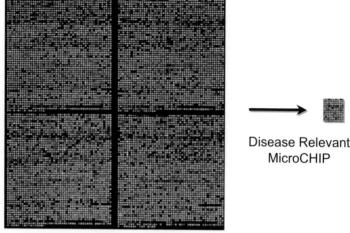

Disease Relevant
MicroCHIP

Potential Global Candidates
in Heart Failure

Fig. 4.2 One of the potential applications of microarrays in the setting of heart failure is by developing disease relevant "chips" in the future, once the relevance of these genes are known. For example, from a global set of can-didate genes, one may ultimately identify 100 or so critical genes that are differentially regu-lated in not only heart failure, but also reflect the specific stage or etiology of heart failure.

drugs based either on novel genes thus identified, or on a knowledge of expression levels of mRNA in different pathological states.

A remarkable growth has occured in the source of the chips and the related components of microarray techniques such as clone-set, arrayer, scanner, and analysis software. Technologies are expected to evolve significantly in the next few years. New microarray methods combined with bioinformatics will continue to provide increasing insight into the molecular basis of biological events.

4.5
Caveats of Using the Microarray Technology

On the other hand, microarray technologies have limitations, some of which reflect its relative early stages of development, and others of which relate to basic principles of good scientific investigation.

1. *Proper experimental design.* Because microarray techniques are costly and complex, much thought must be given to experimental design if results are to be interpreted. Experiments must have advanced planning, rigorous controls, and a hypothesis facilitated process that allows focus and subsequent validation. It is only when traditional hypotheses driven experiments are validated, that one can begin the exploratory component of the dataset.

2. *Ultrahigh quality of RNA* is required for reproducible results. This is the most fundamental requirement for any microarray experiment, otherwise "noise in, noise out". Worse still, the artifacts or inappropriate interpretation of the results may lead to blind-ended follow-up experiments that should never have been.

3. *Inadequate quality control of the chipsets.* Even though chip production is now mostly automated and generally of high quality, there are still many opportunities for errors from improper spotting to the wrong sequence selection. This type of error can become magnified when it involves a large number of the genes from multiple experiments.

4. *Inability to detect low abundance gene transcripts.* The abundance of a gene transcript does not relate directly to its functional importance. Indeed, small critical regulatory messages may be present in low quantity. Chip manufacturers such as Affymetrix and Incyte have increased their sensitivity to approximately one gene/100,000 of genes in TRNA population [12]. This level of sensitivity unfortunately is still poorer than that of classical methods such as Northern blot. Therefore, the microarray still can miss rare gene expression events, which may be important in the overall pathophysiology.

5. *Artifacts* from different sources during the various processing steps, including sample preparation, RNA extraction, labeling, hybridization, cDNA spotting or chip scanning, pose additional problem. However, obsessive care during the processing and multiple replications of the experiments can reduce these problems to a minimum.

6. *A surfeit of data* can only be analyzed by dedicated data mining tools or bioinformatics analysis packages, along with the experts in the field with a biological insight. At the present time, confirmation of microarray results with classical techniques, such as RT-PCR, or real time or other quantitative PCR techniques is still necessary. However, this most likely will no longer be necessary when the techniques mature further and results become more reproducible.

7. *Expectation of immediate functional insight.* Little functional information is typically derived from the first analysis of the expression arrays. Functional insights can only be gained through further evaluation of the known literature, the context of a gene being activated or silenced, and the partners with which the gene putatively interacts.

4.6
Experimental Design

The structure of the microarray data, the appropriate types of analysis, and the quality of the results are influenced by the experimental design. Careful forethought and planning is needed to design a successful experiment. Because microarray systems are so sensitive any small changes in sample-to-sample treatment, RNA extraction, sample handling, probe labeling, and other steps in the processing are likely to affect the results. Every effort must therefore be taken in experimental design, such that variations in the data are due to conditions under investigation and are not due to artifacts.

Whilst it is recognized that microarrays are relatively expensive, it is still important to incorporate a finite number of biological replicates in each experiment to ensure ultimate statistical robustness. This is particularly important in a system that has much inherent biological variation. It is also needed to identify low abundance transcripts and/or small changes in expression levels. The most important points to consider when designing an experiment are:

- what is the question that we are trying to answer using this technique,
- gene expression level of what conditions are compared,
- are there any known expression level for genes in these condition which can be used as a reference marker to confirm the fidelity of our result,
- identify the areas which can be the source of variation and try to eliminate them as much as possible, and finally
- what is the maximum of the replicate that is possible to use to allow us the use of statistical method.

The simplest microarray experimental design is to determine the changes in gene expression patterns across a single factor of interest, e.g. temporal frame shifts, genetic manipulation of a single gene, or effects of drug treatments. However, experiments can now also be designed in a multi-factorial fashion to assess their interactions in one set of microarray experiments. Statistical methods are now available to determine the appropriate number of replicates, or to assist the researcher to design appropriately powered experiments [13].

4.7
Tissue Preparation and Preservation

This is arguably the most important step for RNA stabilization. Reliable results depend the integrity of RNA. Immediate RNA stabilization in the heart (or any biological) sample is necessary, because changes in gene-expression pattern occur rapidly due to specific and non-specific RNA degradation. There are several methods to stabilize the RNA. The simplest and most widely used is snap freezing of the tissue in liquid nitrogen within minutes of removal. The sample then must be preserved at a very low temperature in a nuclease free environment to ensure to sterility and freedom from nuclease contamination, specifically RNAase contamination. Alternatively, samples can be submerged in the RNA stabilizing reagent (e.g. RNA*later*, QIAGEN) immediately after harvesting and stored up to 4 weeks at 8 °C or archived at −20 °C or −80 °C. This RNA stabilizing reagent is specific for animal tissues and can be used for cell-culture and white blood cells, but not for whole blood, plasma, or serum. The advantage of this method over snap freezing is the convenience in cutting and handling of the tissue, that can be transported and stored at close to ambient temperatures.

4.8
RNA Isolation

This is a crucial step for preparing high quality RNA free of RNase contamination. Glassware must be treated with DEPC-H_2O (0.1% DEPC in H_2O) before use. Both the quality and quantity of RNA yield improves if samples are handled with care to avoid RNAase contaminats. Heart tissue is considered a challenging tissue for the isolation of RNA because it is fibrous and thus difficult to homogenize and process for RNA isolation. Several methods are available to purify RNA (either total or mRNA) from the heart tissue. In addition to several commercially available kits, the major method of RNA isolation uses a Guanidine containing reagent, such as TRIZOL (Gibco). Frozen, or RNA*later* stabilized tissue samples are disrupted mechanically in a reagent or under liquid Nitrogen and homogenized. At this point, an additional step is necessary to eliminate the fibrous tissues from the sample before proceeding to phase separation with Chloroform. After phase separation, the TRNA is precipitated using Isopropanol and is washed in 70 ethanol (DEPC H_2O). It is estimated that approximately 0.1–1 pg of TRNA is present in a single cell [14]. TRNA may vary with sample condition, viability of cells, functional status, and phenotype of the cells. The concentration of RNA can be determined by measuring the OD at 260 nm (A260) in 10mM Tris.Cl, pH 7.5. TRNA with A260/A280 ratio of 1.9–2.1 is used for microarray experiments. Integrity and size distribution of TRNA must be checked using denaturing agarose gel electrophoresis. In case of small samples, such as those obtained by laser capture microdissection or biopsies, spectrophotometer reading can be omitted because too low RNA concentration may produce false negative OD values. The best method of RNA characterization in these cases is the RNA Bioanalyzer (Agilent). The system permits rapid screening of RNA samples with each disposable RNA chip to determine the concentration and purity/integrity of 12 RNA samples with a total analysis time of 30 minutes. Purified RNA may be stored at –80 °C in water, with no detectable degradation after one year.

4.9
RNA Amplification

Results of the human genome project [15–17] have laid the foundation for the microarray gene expression profiling [18, 19]. However, broader utilization of microarray methods is limited by the amount of RNA required (typically 10 µg of TRNA or 2 µg of poly (A) RNA) [12]. This is especially a problem with limited samples, such as endomyocardial biopsies and laser capture microdissection (LCM) can be obtained. An important frontier in the development of microarray for expression profiling involves reduction of the required amount of RNA. Methods aiming at intensifying the fluorescence signal have resulted in an improvement [20]. Significant increase in detection level can be achieved by amplifying poly(A) RNA or cDNA [21, 22]. There are two primary approaches which can be employed to overcome RNA limitations. One is PCR-based amplification and reproducible yield, but the relative abundance of the cDNA products is not well correlated with the starting mRNA level. The second approach avoids PCR and utilizes one or more rounds of the robust linear amplification based on cDNA synthesis and a template-directed in vitro transcription reaction [14]. This is a recommended method for amplifying the low abundance mRNA or even gene expression profiling of a single cell by orders of magnitudes from nanograms (1–50 ng) of TRNA or poly(A) RNA in one or two round(s) of amplification(s). This method combines a reverse transcription step with an oligo (dT) primer that contains a T7 RNA polymerase promoter. The first-strand cDNA is then used for synthesis of second-strand DNA by DNA polymerase, DNA ligase, and RNaseH. The resulting double-strand cDNA functions as a template for in vitro transcription step (one or two rounds) which results in a linear amplification of RNA. Fidelity of this mRNA amplification method was assessed using microarray technology [23].

The combination of powerful microarray technology with precise amplification techniques promises to be especially important for small samples of heart biopsies. However, assessment of the yield of labeled mRNA, representation in amplifying various transcripts (fidelity), linearity of amplification, and finally the sensitivity and reproducibility of the method in individual laboratory is essential. Again, the relative efficiency of in-vitro transcription of specific size of mRNA may later correlate with startup levels of mRNA.

4.10
Probe Labeling

Two principal types of arrays, spotted arrays (robotic deposition of nucleic acids) and *in situ* synthesis (using photolithography) are used in gene expression monitoring [24, 25]. Labeled material can prepared by the "one" and "two color" system.

The "one color" system is the method that is used for *in situ* synthesized chips. The RNA can be labeled directly with psoralen-biotin derivative or with a Biotin carrying molecule. The labeled nucleotides are incorporated into cDNA during reverse transcription of poly(A) RNA [24, 26]. Alternatively, cDNA with a T7 promoter at its 5′ end can be generated to serve as template for the subsequent step in which the labeled nucleotides are incorporated into cRNA. Commonly used dyes are fluorescent cyanine based Cy3 and Cy5 and nonfluorescent biotin (Amersham).

The second method is the "two color" system of probe labeling which is often used with cDNA chips. Equal amounts of cDNA from two different conditions are labeled with different fluorescent dyes, usually Cy3 and Cy5, mixed and hybridized to a chip [25]. The information on ratio (relative concentration) of mRNA from two samples is obtained. There are direct and indirect methods of incorporating the dyes into cDNA. In direct method, the labeled nucleotide is incorporated into the cDNA, whereas, in indirect method, an amino-allyl modified nucleotide analogue such as amino-allyl-dCTP is incorporated into the cDNA to which the dyes are subsequently coupled chemically. In addition to systematic variations in direct dual color labeling, Cy3 and Cy5 exhibit different quantum yields. Thus an

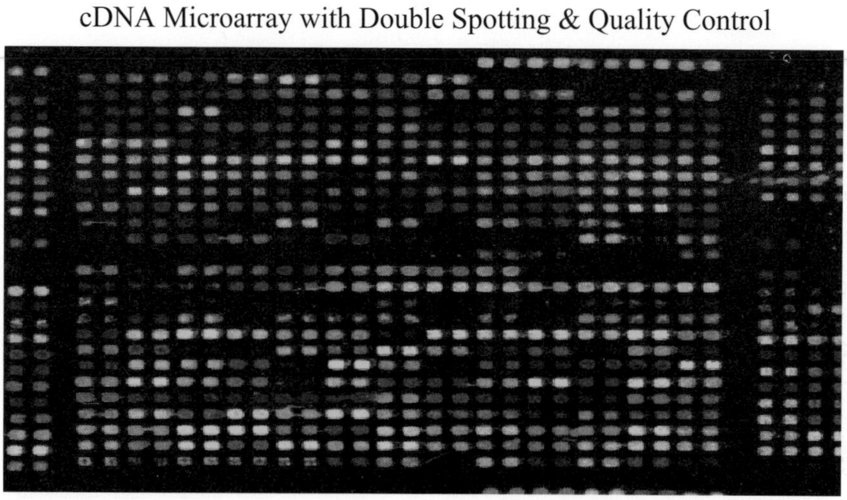

cDNA Microarray with Double Spotting & Quality Control

Fig. 4.3 To ensure high degree of reproducibility, the cDNA microarray here is doubly spotted, and only spots show concordant up or down regulation is included in the final analysis. Fluor-flip is another technique to ensure that the differences observed is due to true expression differences and not due to artifact.

additional chip with exchange dyes (or commonly called Fluor-Flip) is required to obtain a reliable data (Fig. 4.3). After hybridization and washing, the array is scanned at two different wavelengths to determine the relative transcript abundance for each condition and data analysis.

4.11
Data Analysis and Bioinformatics

The basic techniques in microarray experiments from cDNA synthesis to hybridization and washing are conventional methods that have been used in the laboratory for years. Data analysis is the most demanding part in the use of this extraordinary tool because we deal with an unprecedented volume of data. For the most challenging part of this technology, the data analysis, an increasing number of software tools are available [27]. Two basic steps in microarray data analysis and resources are:

- data collection (collecting raw data from images, correction for the background and normalization), target (differentially expressed genes) detection and target intensity extraction.
- Analysis and Bioinformatics with multiple image analysis and data visualization (e.g. clustering methods to identify unique pattern of gene expression).

The care in assuring accurate reproducibility of the data is of paramount importance (Fig. 4.4).

BA

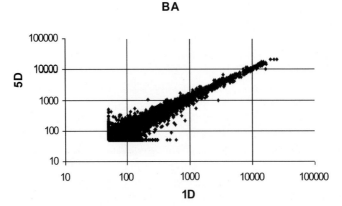

Fig. 4.4 The reliability of microarray experiments is dependent on the reproducibility of the data set not only within the same subject, but also between subjects in an experiments exposed to the same conditions. Here we illustrate normalized gene expression changes between the first and fifth subject of a single experiment of aortic banding in a mouse model, demonstrating high degree of correlation and reproducibility between these two hearts subjected to the same stress.

Data collection: differential gene expression is assessed by scanning the hybridized arrays using either a confocal laser scanner (GSI Scan array) producing 16-bit TIFF images, or a photomultiplier tube (PMT) laser scanner (Axon Scanner) capable of interrogating both the Cy3- and Cy5-labeled probes and producing the ratio image of 24-bit composite RGB (Red-Green-Blue) or capable of detecting additional dye up to 4 wavelengths simultaneously. The ratio image typically represents the level of two cDNAs (Control and Test) that is hybridized to the array in a "two color" system. A great advantage to this approach is its capacity to demonstrate a dynamic pattern of gene expression. These images then must be processed or be converted to numerical representations in order to calculate the relative expression levels of each gene and to identify differentially expressed genes. In image processing, first the spots representing the arrayed genes are identified and distinguished from nonspecific contamination (such as dust), or artifacts. The second step in image analysis is background calculation and subtraction to reduce the effect of nonspecific fluorescence. Different data analysis algorithms utilized are employed by various software tools to quantify the images. For all ratio calculations that require background subtraction, the median background value is usually used (in GenePix Pro, Axon).

Because of multivariate nature of the microarray experiments, it is not easy to compare data from different experiments. To improve the comparison across many microarrays, data normalization is required. Different software packages offer various methods for normalization (Commercial software: GinPix, Axon; GeneSpring, Silicon Genetics; Affymetrix microarray suite and Data mining tool, Affymetrix, Inc; Spot fire, Spotfire Inc and free software: DNA-Chip analyzer, SAM, Stanford; Treeview, Eisen, etc. most of these can be found). Increasing numbers of researchers prefer scaling to normalization. The difference between scaling and normalization relates to the mehtod used to pick the target intensity. For scaling a number that represents the average signal from a large set of arrays is used. For normalization the target intensity is defined as the average signal on the baseline array and then all experimental arrays are adjusted to that value. In addition to per chip normalization or scaling, there must be a per gene normalization in order to bring the data to a relative scale.

Normalized or scaled data are typically analyzed to identify genes that are differentially expressed. Most published studies have used a cutoff of two fold up- or down-regulation to define differential expression; however this can not be true for all genes, because different genes may have different levels of sensitivity.

Multiple statistical methods can be used first to filter the most statistically significant data and then to perform further analysis, data mining, and bioinformatics in order to extract the most reliable information from microarray data. Different software packages offer various statistical methods for data filtering such as: parametric test (assume variance equal) or student's t-test/ANOVA and Welch t-test/Welch ANOVA (do not assume variance equal) or nonparametric test or Wilcoxon-Mann-Whitney test. In addition to filtering by standard deviation, p-values, etc., multiple testing corrections can be added to the above methods to increase the accuracy of filtered data.

Sophisticated bioinformatics tools are required to extract accurate information from the avalanche of data and to draw a logical and reliable conclusion from the massive volume of information that is generated from the microarray experiments. The objective is to reduce complexity and extract or mine as much useful and relevant information as possible. For microarray data analysis both data mining and bioinformatics are required. Data mining has been defined as "the extraction of implicit, previously unknown, and potentially useful information from data", whereas bioinformatics is used for sequence-based extraction of specific patterns or motifs with the ability of specific pattern matching. Currently they exist as separate approaches but eventually, data mining and bioinformatics will be indistinguishable. Most data analysis software is equipped with bioinformatics rather than data mining tools. When the size of the data set is reduced to a manageable volume of statistically significant data, it is possible for the scientist to identify emerging patterns.

There are several popular methods to analyse and visualizae gene expression data:

- *Hierarchical Clustering* is used to visualize a set of samples or genes by organizing them into a phylogenetic tree, often referred to as a dendrogram. One way of analyzing microarray data is to look at the cluster (group) of genes with a similar pattern of expression across many experiments. The co-regulated genes within such groups are often found to have related functions. The distance between two branches of a tree is a measure of the correlation between any two genes in the two branches. This is an exceedingly powerful method and is used most widely. It allows a researcher to find experimental conditions (e.g. various drug treatments, classification of disease states) that have similar effects.

- *K-means Clustering* divides genes into distinct groups based on their expression patterns. Genes are initially divided into a number (k) of user-defined and equally-sized groups. Centroids are calculated for each group corresponding to the average of the expression profiles. Individual genes are then reassigned to the group in which the centroid is the most similar to the gene. Group centroids are then recalculated, and the process is iterated until the group compositions converge. A wide selection of similarity measures (parametric and non-parametric correlations, Euclidean distance, etc.) is available in different software.
- *Self-Organizing Maps (SOMs)* are tools for exploring and mapping the variations in expression patterns within an experiment. This method is similar to k-means clustering, but with an additional feature where the resulting groups of genes can be displayed in a rectangular pattern, with distance representing the level of similarity (adjacent groups being more similar than groups further away).
- *Principal components analysis (PCA)* is standard protection technique that explores the variability in gene expression patterns and finds a small number of themes in expression pattern. These themes can be combined to make all the different gene expression patterns in a data set.
- *Multidimentional Scaling (MDS)* is a method that represents the measure of similarity between pairs of objects. In the clustering section distance matrix is typically used as a similarity matrix between all pairs of samples in one experiment design. Two dimensional scaling plots are used to examine the similarity amongst all samples.

There is no software so far that can extract all the useful data and prevent possible masking of some clusters by transcriptional "noise". Software has become much more powerful in the past few years, but expert data-miners and bioinformaticians are still needed.

4.12
Application: New Classification of Disease

System wide explorations of gene expression patterns provide a unique insight into the internal environment of the cells of a particular organ. This may reflect both genetic predisposition towards the disease, and environmental stresses that lead the disease phenotype. Heart failure is a classic condition in which diverse stimuli may dissimilarly and/or similarly challenge the ability of the heart to adapt [28]. When adaptation becomes limited, disease phenotypes evolve. Indeed, we have traditionally classified heart failure in terms of clinical etiology, for exam-

Fig. 4.5 The ability for microarray expression data to distinguish the gene expression pattern of normal (N), and ischemic cardiomyopathy (ICM) can be seen from this hierarchical clustering and dendrogram display. The genes up and down regulated include both known genes as well as completely novel genes of currently unknown function. This may provide both diagnostic and prognostic information in the future.

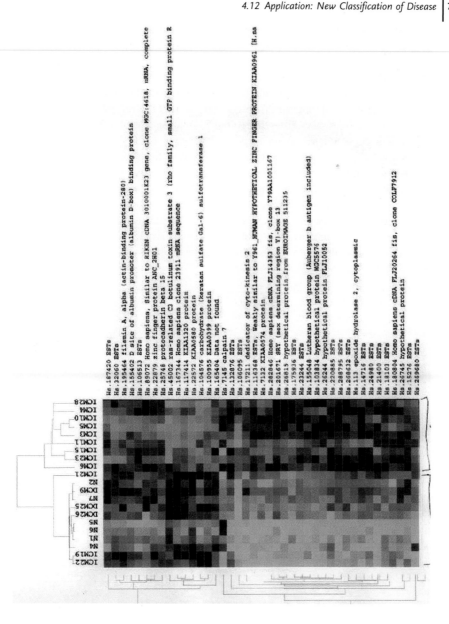

ple, dilated cardiomyopathy, ischemic cardiomyopathy, viral myocarditis, etc. We assume also that the disease progression, the prognosis and response to therapy will be predicated on the disease etiology. However, the ability of microarrays to provide a broad insight into the disease process directly within the tissues provide a unique insight into the intracellular perturbations of the cell organization and function (Fig. 4.5) and an entirely unique new perspective on the heart failure process. Commonalities and differences at the molecular level will identify critical pathways of pathogenesis and/or response to therapy.

This approach have been very successfully applied to the field of cancer biology. In the study of breast cancer, expression microarrays have provided an important insight into the biology of the disease, as well as prognostic markers for favourable vs. poor outcomes. Such methods have also been applied to leukemias, as well as prostate cancer (Fig. 4.6). The advantage in cancer biology is that the tumour is often excised, providing a direct source of tissue to correlate with pathology, as well as opportunity to explore patterns of pathway activation. However,

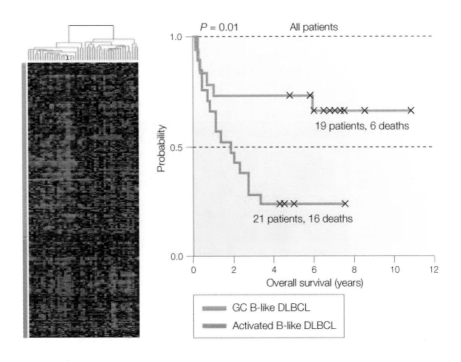

Nature Reviews | Genetics

Fig. 4.6 The pattern of expression microarrays characterizes the biology of the tissues. This is best illustrated in cases of cancer, where the expression patterns between different types of tumours (e.g. lymphoma) not only can differentiate one type of lymphoma from another, but also can be associated with differential prognosis. This will likely become more refined as the database becomes more enriched with samples in time.

with the availability of myocardial biopsies, similar opportunities exist for the study of heart failure.

4.13
Application: Pathogenesis of Disease

Despite its direct relevance, the direct evaluation of human heart failure samples using microarray technology also has significant limitations. The most obvious is the end stage nature of the samples, which may represent a convergence of phenotypes that have no relevance to the pathogenetic mechanisms. Furthermore, many of these patients also have co-morbidities and concomitant medications, which all serve to skew the gene expression patterns. Nevertheless, clinical validation in a patient population is always important to ensure relevance and concordance amongst biological models.

To obviate these concerns, and to capture the potential early triggers of the heart failure phenotype, animal models of heart failure have been used to give interesting insights. Taylor et al. examined the viral model of myocarditis, and identified specific groups of genes relevant to the viral, inflammatory and healing phases of the myocarditic process [29]. Aronow et al. showed that no single gene program is common to all the models of heart failure. However, there does appear to be a group of programs, linked specifically to each etiology, that predisposes to heart failure [30]. In addition, we have performed *serial microarray analysis* of gene families potentially relevant in the setting of heart failure in animal models of heart failure. By using carefully controlled experimental designs, where the animals subjected to the same injury can be synchronously followed and analyzed, and compared to age matched controls, the critical differences in host responses leading to heart failure can be precisely identified (Figs. 4.7, 4.8). In the future, the applications of these concepts to either biopsy or preferentially blood based analysis would be of most important interest in studies of heart failure.

4.14
Application: Early Disease Markers and Prognosis

To identify disease early in its process, critical specific markers will be useful in the diagnoses and in delineating desease etiology. Currently, the best example of an early diagnostic marker is brain natriuretic peptide (BNP), which is elaborated by the ventricular myocardium under stress. The spill-over of this marker into the blood has given a useful marker in early diagnosis of heart failure, and also provides prognostic information. However, it is elevated irrespective of heart failure etiology. Thus we must ask whether there are etiologically specific marker that can be found. We have recently identified 5 potentially useful markers for the early diagnosis of myocarditis leading to dilated cardiomyopathy (unpublished observations). This may indeed represent the beginning for future applications of

Cell Cycle Program

Normal

Heart Failure

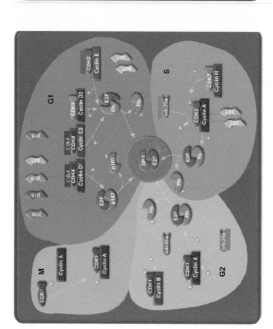

Fig. 4.7 The expression array information can be functionally clustered to provide biological insight into disease processes, with the brightness of the colour of each molecule representing the relative levels of expression. For example, comparing heart tissues from models of heart failure with that of normal condition, we see that Cdc25a and Cdk4 are significantly up regulated, while Cdk2 is significantly down regulated. This pattern suggests that there is increased propensity towards cell cycle progression, but ultimately is blocked.

Cell Stress Response

Normal

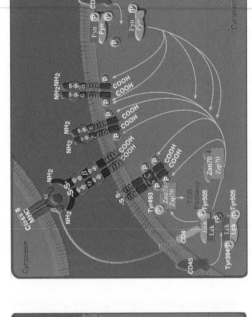

Heart Failure

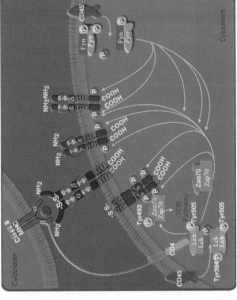

Fig. 4.8 Expression arrays are also useful to determine host stress response signaling pathways in the presence of heart failure. In this experiment, we can see that intracellular cytokine signaling pathways of lck and fyn are both up regulated in the setting of heart failure. This is consistent with the activation of cytokine and immune signaling pathways in the setting of heart failure.

this technology to systematically identify novel ethology specific markers of disease.

This strategy has also been recently employed in prostatic cancer, where early and accurate diagnosis is particularly useful for determining course of action. Using microarray methods, an early diagnostic pattern was indeed recognized, two signaling molecules appear to be unique to early prostatic cancer – hepsin and pim-1, both are serine-threonine kinases [31].

In addition, with the increasing utilization of ventricular assist devices, there is access to tissues from heart failure patients has improved. In addition, gathering data has now become available on the potential changes and reversibility of the abnormalities observed in heart failure, and the opportunity of using this information in the future for prognosis of the patients, and in turn determining which patient may come off the assist device support, and which patient will need heart transplantation.

4.15
Application: Therapeutic Insights

How a drug acomplishes its therapeutic effect in the clinical setting has always been difficult. A drug may be developed for the purpose of targeting a specific pathway. However, in many occasions, it is the unintended effects of a drug that ultimately determine its overall biological profile in the clinical setting. An example is the recent publication of the effect of beta blockers in the setting to understand heart failure. It appears that only in patients when the beta blockade can alter the cardiac contractile protein such as the myosin isoforms that the patient will benefit with respect to reverse remodeling and improvement in clinical outcome. Therefore the availability of microarrays will now offer an unprecedented opportunity to define the diverse targets that are affected by a certain therapeutic intervention, and define the true biological impact of a certain treatment strategy. This will be particularly important in the setting of heart failure, involving a large complex of pathophysiological pathways.

4.16
Acknowledgements

The work is supported in part by grants from the Heart and Stroke Foundation of Ontario, and the Canadian Institutes of Health Research (CIHR).

4.17
References

1 HO KK, PINSKY JL, KANNEL WB, LEVY D. The epidemiology of heart failure: The Framingham Study. *Journal of American College of Cardiology.* **1993**; 2:6A–13A.

2 O'CONNELL J, BRISTOW MR. Economic impact of heart failure in the United States: time for a different approach. *Journal of Heart and Lung Transplantation.* **1994**; *13*:107–112.

3 HALDEMAN GA, RASHIDEE A, HORSWELL R. Changes in mortality from heart failure – United States, 1980–1995. *Mortality Morbidity Weekly Review.* **1998**; *47*:4–7.

4 JOHANSEN H, STRAUSS B, WALSH P, MOE G, LIU PP. Congestive Heart Failure: the Coming Epidemic. *Canadian Medical Association Journal.* **2002** (in press).

5 INSTITUTE FOR CLINICAL EVALUATIVE SCIENCES (ICES). Cardiovascular health & services in Ontario: An ICES atlas. **1999**; 111–122.

6 LIU P, ARNOLD M, BELENKIE I, HOWLETT J, HUCKELL V, IGNAZEWSKI A, LEBLANC MH, MCKELVIE R, NIZNICK J, PARKER JD, RAO V, ROSS H, ROY D, SMITH S, SUSSEX B, TEO KK, TSUYUKI R, WHITE M, BEANLANDS D, BERNSTEIN V, DAVIES R, ISSAC D, JOHNSTONE D, LEE H, MOE G, NEWTON G, PFLUGFELDER P, ROTH S, ROULEAU J, YUSUF S. The 2001 canadian cardiovascular society consensus guideline update for the management and prevention of heart failure. *Canadian Journal of Cardiology.* **2001**; *12*:5E–48E.

7 ADAMS KF, BAUGHMAN K, LIU PP, et al. HFSA guidelines for management of patients with heart failure caused by left ventricular systolic dysfunction – pharmacological approaches. *Journal of Cardiac Failure.* **1999**; *5*:357–382.

8 COHN J. The prevention of heart failure –a new agenda. *New England Journal of Medicine.* **1992**; *327*:725–7.

9 HUNT SA, BAKER DW, CHIN MH, CINQUEGRANI MP, FELDMANMD AM, FRANCIS GS, GANIATS TG, GOLDSTEIN S, GREGORATOS G, JESSUP ML, NOBLE RJ, PACKER M, SILVER MA, STEVENSON LW, GIBBONS RJ, ANTMAN EM, ALPERT JS, FAXON DP, FUSTER V, GREGORATOS G, JACOBS AK, HIRATZKA LF, RUSSELL RO, SMITH SCJ. ACC/AHA Guidelines for the Evaluation and Management of Chronic Heart Failure in the Adult: Executive Summary. *Circulation.* **2001**; *104*:2996–3007.

10 HWANG DM, DEMPSEY AA, WANG RX, REZVANI M, BARRANS JD, DAI KS, WANG HY, MA H, CUKERMAN E, LIU YQ, GU JR, ZHANG JH, TSUI SKW, WAYE MMY, FUNG KP, LEE CY, LIEW CC. A genome-based resource for molecular cardiovascular medicine: Toward a compendium of cardiovascular genes. *Circulation.* **1997**; 96:4146–4203.

11 CRACKOWER M, OUDIT GY, BACKX P, PENNINGER J. Functional analysis of a novel angiotensin converting enzyme – ACE2. *Nature.* **2002**;*250*:(in press).

12 LOCKHART DJ, et al. Expression monitoring of hybridization to high-density oligonucleotide arrays. *Nature Biotechnology.* **1996**; *14*:1675–1680.

13 KERR MK, CHURCHILL GA. Experimental design for gene expression microarrays. *Biostatistics.* **2001**; *2*:183–201.

14 PHILLIPS J, EBERWINE JH. Antisense RNA amplification: a linear amplification method for analyzing the mRNA population from single living cells. *Methods of Enzymology.* **1996**; *10*:283–288.

15 COLLINS FS. New goal for US Human Genome Project. *Science.* **1998**; *282*:682–689.

16 COLLINS FS. Shattuck lecture – Medical and societal consequences of the Human Genome Project. *New England Journal of Medicine.* **1999**; *341*:28–37.

17 VENTER JC, etc. Completion of the human genome proejct. *Science.* **2001**; *291*:1304–1351.

18 LANDER ES. The new genomics: global view of biology. *Science.* **1996**; *274*:536–539.

19 BROWN PO, BOTSTEIN D. Exploring the new world of genome with DNA microarrays. *Nature Genetics.* **1999**; *14*:457–460.

20 STEARS RL, GETTS RC, GULLANS SR. A novel, sensitive detection system for high-density microarrays using dendri-

mer technology. *Physiological Genomics.* **2000**; 3:93–99.

21   EBERWINE JH, et al. Analysis of gene expression in single live neurons. *Proceedings of National Academy of Sciences USA.* **1992**; 89:3010–3014.

22   LUO L, et al. Gene expression profiles of laser-captured adjacent neuronal subtypes. *Nature Medicine.* **1999**; 5:117–122.

23   WANG E, MILLER LD, OHNMACHT GA, LIU ET, MARINCOLA FM. High-fidelity mRNA amplification for gene profiling. *Nature Biotechnology.* **2000**; 18:457–459.

24   LIPSHUTZ RJ, FODOR SP, GINGERAS TR, LOCKHART D. High density synthetic oligonucleotide arrays. *Nature Genetics.* **1999**; 21:20–24.

25   SCHENA M, SHALON D, R.W. D, BROWN PO. Quantitative monitoring of gene expression patterns with a complementary DNA microarray. *Science.* **1995**; 270:467–470.

26   WODICKA L, DONG H, MITTMANN M, HO M, LOCKHART D. Genome-wide expression monitoring in Saccharomyces cerevisiae. *Nature Biotechnology.* **1997**; 15:1359–1367.

27   BOWTELL DD. Options available-from start to finish-for obtaining expression data by microarray. *Nature Genetics.* **1999**; 21:25–32.

28   LIU P. The path to cardiomyopathy: cycles of injury, repair and maladaptation. *Current Opinion in Cardiology.* **1996**; 11:291–292.

29   TAYLOR LA, CARTHY CM, YANG D, SAAD K, WONG D, SCHREINER G, STANTON LW, MCMANUS BM. Host gene regulation during coxsackievirus B3 infection in mice: assessment by microarrays. *Circulation Research.* **2000**; 87:328–334.

30   ARONOW BJ, TOYOKAWA T, CANNING A, HAGHIGHI K, DELLING U, KRANIAS EG, MOLKENTIN JD, DORN III GW. Divergent transcriptional responses to independent genetic causes of cardiac hypertrophy. *Physiological Genomics.* **2001**; 6:19–28.

31   DHANASEKARAN SM, BARRETTE TR, GHOSH D, SHAH R, VARAMBALLY S, KURACHI K, PIENTAK KJ, RUBIN MA, CHINNAIYAN AM. Delineation of prognostic biomarkers in prostate cancer. *Nature.* **2001**; 412:822–826.

5

Gene Profiling in the Heart by Subtractive Hybridization

CHRISTOPHE DEPRE

Although it has long been considered that myocardial gene expression is not subject to major regulation because the cardiomyocytes do not divide, it is now widely accepted that the myocardium can develop a remarkable plasticity at the gene level [1–9]. This adaptation of the nuclear activity is the direct consequence of variations in physiological conditions, and is strongly associated with the adaptation of both metabolic [10–14] and contractile [15] properties of the heart. A large-scale analysis of the regulation of gene expression in response to changes in physiological parameters is the goal of functional genomics.

Because of its fundamental function for the organism, the heart is extraordinarily receptive to variations of extracellular conditions, which are often referred to as a "stress" for the cardiac myocyte: increased contractile performance [16], oxygen deprivation [17, 18], burst of free radicals [19], endothelial dysfunction [20], increased preload [21, 22], cellular stretch [23] are just a few examples of the changes in extracellular conditions that directly affect the cardiac cell. All these stimuli, either physiological or deleterious, are detected by different sensors (receptors, ion channels or transmembrane proteins) [24], relaying the information to transmitters (signaling pathways) [23], which in turn regulate the activity of the effectors (enzymes and transcription factors) [25, 26]. The effectors are directly responsible for the adaptation of the expression of different genes in response to the initial stimulus (Fig. 5.1). In other words, the gene expression profile is the "end-point" of the effects of any stimulus on the heart (Fig. 5.1). Determining this profile helps deciphering the fundamental characteristics of a specific stimulus. Comparing the profiles in response to different stimuli also helps to determine the similarities and differences between these stimuli. The accumulation of gene profiles in response to different conditions leads to the elaboration of a compendium of cardiac gene expression. This compendium can subsequently be used, for instance, to analyze the effects of a drug on cardiac gene expression in different physiological conditions [27].

Genomics and proteomics are complementary, but they also address very different questions. The analysis of gene expression as an "end-point" of the effects of a specific stimulus on the heart is in sharp contrast with the concept of the gene as a "starting point" leading to the expression of specific proteins (Fig. 5.1). The "end-point strategy" consists in looking at the gene not as a protein provider, but

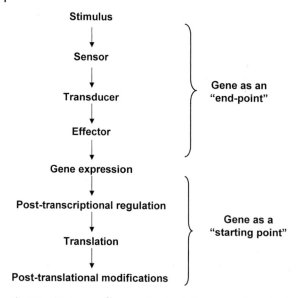

Fig. 5.1 Genome profile as an "end-point" or a "starting point". In the "end-point" approach, the genomic profile is the result of the transmission of extracellular stimuli to the nucleus. In the "starting point" approach, the gene leads to the production of proteins, which needs to be further studied by proteomics.

as an integrator of incoming information. Considering that a cell is viable as long as it has a functional nucleus, gene expression profiling is a way to evaluate the nuclear activity of a specific cell type in a specific milieu. We know that, in turn, the increased expression of a gene does not automatically result in the increased expression of the corresponding protein, and that the expression of a specific protein can be regulated by non-transcriptional mechanisms. We also know that a transcript is not necessarily targeted to the translational machinery, as shown recently with the concept of gene silencing by RNA interference [28, 29].

The "end-point" strategy is a way to better understand the mechanisms of cardiac adaptation. For example, we often refer to the concept of "stress response" to characterize the response of the heart to conditions as diverse as transient ischemia, aortic banding or catecholamines infusion. Although these conditions can all induce the increased expression of "stress" gene markers (such as the atrial natriuretic factor or proto-oncogenes), each of them can activate the expression of specific subsets of genes, such as specific heat-shock proteins, anti-apoptotic proteins, metabolic enzymes, activators of protein translation or others. Therefore, the response of the myocardium to "stress" depends on the context and the stimulus.

5.1
Strategies and Limitations of Genome Profiling

Several parameters must be taken into consideration to perform functional genomics and gene profiling. These include, but are not restricted to, a biological problem, a good model and a reliable technique.

5.1.1
The Biological Problem

The biological problem, or the question asked, is usually obscured by the fact that
functional genomics is perceived as a "fishing expedition". Although it is true that
genomics studies are not always hypothesis-driven, very expensive and time-consuming experiments can rapidly turn into deep frustration if they are conducted
in the absence of an underlying biological question. Depending on the question
or the hypothesis, two broad approaches can be used, the horizontal approach and
the vertical approach (Fig. 5.2). In the horizontal approach, the tools of functional
genomics are used to determine how a change in the function of the organ results in a change of gene expression. This approach is horizontal, because it will
analyze a large number of genes at one level of investigation (usually the mRNA
expression) and lead to the determination of a genomic profile. In the vertical
approach, the question asked to functional genomics is what is the function of a
specific gene (Fig. 5.2). In that approach, the investigator concentrates on one
gene and conducts his experiments by the combination of various techniques

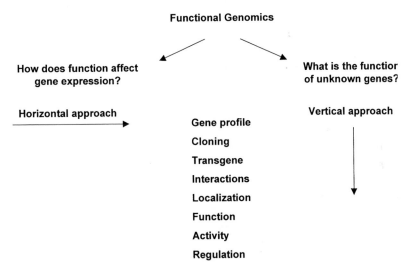

Fig. 5.2 The horizontal and vertical approaches of functional genomics. The horizontal approach shows how gene expression
is affected by function. The vertical approach
shows the function of a specific gene product.

(mRNA expression, protein-protein interaction, cellular localization, transfection, genetically-modified animals. . .). This last approach is used for the characterization of unknown and novel genes, which still represent a substantial part of our genome [30, 31]. Both approaches are complementary. As we will see below, the subtractive hybridization methodology is used to investigate a large array of genes in a specific condition (horizontal approach), but also brings some information about unknown genes, which in turn determines the vertical approach. Whereas the horizontal approach leads to the discovery of new concepts, the vertical approach leads to the discovery of new functions. Combining both approaches provides a link between description and mechanism.

5.1.2
The Model

A good model of investigation is equally important. Most of the experiments of functional genomics are performed on samples from *in vivo* studies. This adds to the complexity of the experiments because, in addition to the potential problem of reproducibility and sample-to-sample variation, we must also consider the relevance of the model used and the possibility of inter-species differences. For obvious epidemiological reasons, most of us are interested in determining how the human heart develops and ages, and how it can be injured and repaired. We therefore need to develop and use the animal models which are relevant to clinical problems in the human heart. So far, these questions are mainly addressed in rodent models, which points out to the issue of species differences. A classical example of genomic differences between species is the "fetal gene program". The pioneering experiments of gene expression in the heart pinpointed that a rodent heart submitted to increased workload switches the expression of contractile proteins from an "adult" isoform back to an isoform preferentially expressed in the fetal heart. For instance, the cardiac-specific isoform of alpha-actin is partially replaced by the "fetal" skeletal isoform, whereas the alpha-isoform of myosin heavy chain is partially replaced by the beta-isoform [32–34]. This protein switching results from a switch in the expression of the corresponding genes. It has been shown since that this concept can be extended to other categories of genes (such as genes involved in the regulation of calcium handling and metabolism) [11, 12, 35, 36], and is also reproduced when the workload is decreased rather than increased [13]. However, such "fetal gene program" is hardly reproduced in larger mammals [37]. For instance, the predominant isoform of myosin heavy chain in the adult is already the beta-isoform [38] and other mechanisms of adaptation of contractile performance, such as the phosphorylation or dephosphorylation of myosin light chains [39, 40], is more developed in large mammals. Other adaptive mechanisms are activated in larger mammals that do not necessarily exist in rodents. An obvious reason for the discrepancy between species is that a mouse heart beating at 500 beats per minute has different regulatory mechanisms than a human heart beating at 75 beats per minute, although the overall goal of cardiac function remains the same. Heart rate, cardiac chamber geometry, loading conditions, mechanisms of calcium handling, coronary anatomy and collaterals are just a few examples of biological and anatomical parameters responsible for species differences. It is therefore crucial to determine the relevance of the model for the problem investigated. Rodents, especially mice, are particularly attractive as "test tubes" to study the effects of one specific gene, either overexpressed or knocked-out, on global gene expression. As shown below, larger mammals can be useful for more complex models of ischemic heart disease or heart failure.

5.1.3
The Technological Approach

One of the first techniques used for gene profiling was the differential display [41–43]. Although elegant and simple, this technique is usually extremely time-consuming because it addresses one gene at a time. Currently, the technology of microarrays is extremely popular. One chip contains most of the expressed transcripts from one species. The sample preparation is relatively easy, provided that some RNA of good quality is extracted from the tissue or the cells. The procedure can be started with as low as 5 µg total RNA. Processing the chips is a quick and automated process. The downside is that the signal to noise ratio can be weak in some hybridizations, which necessitates to repeat the experiments or to increase the number of samples in the series. Each chip represents a compromise between sensitivity and specificity. The choice between cDNA and oligonucleotide chips is an illustration of this compromise. Also, the analysis of the data is not necessarily easy, as thousands of genes have to be analyzed and clustered [44]. It results that the statistical tests needed to determine significant differences are far more complex than a simple Student's t test [45]. Biological variations from sample to sample must be taken into account and several recent studies stressed the importance of repeating the experiments in a representative series of samples [46, 47]. Even more importantly, the investigator must often rely on the use of commercial chips, which limits the possibility to control their quality or content. Cross-contaminations between different clones can not be excluded, and the chips are sometimes printed erroneously [48]. Therefore, even if the technique is powerful and attractive, its use requires a validation of the results by alternative molecular techniques. In addition, these chips are so far available for a restricted number of species only. For these reasons, the subtractive hybridization can be an attractive alternative in specific cases.

5.2
Analyzing Gene Expression by Subtractive Hybridization

The subtractive suppression hybridization represents a large-scale, unbiased method for detecting transcriptionally and post-transcriptionally regulated genes, both known and unknown, independently of the prevalence of these transcripts [49]. A fruitful utilization of this technique requires some RNA of excellent quality, a sequencing facility and alternative methods of validation. It can be coupled to the printing and hybridization of microarrays [50].

Methodology

The procedure is summarized in Fig. 5.3. The subtraction is performed between two biological samples from which total RNA and, subsequently, poly-A messenger RNA are extracted. The subtraction is followed by the subcloning of the different subtracted cDNAs. After purification, the clones can be used for sequencing or microarray printing. Querying the sequences in different databases identifies the subtracted genes. The microarrays offer a semi-quantitation of the variation in gene expression between the samples. The results need to be validated by alternative methods, such as Northern blot, quantitative PCR and in-situ hybridization.

1. Experimental samples. Experiments are usually started from frozen tissue. A comparison can be made between an experimental heart and a sham, or, for larger species, between two areas of the same hearts. The last choice allows a paired comparison of the data. About 200–400 mg of tissue is usually enough to collect the amount of RNA needed for the experiment. Compared to other tissues, the mRNA content per gram of cardiac tissue is relatively low. It is recom-

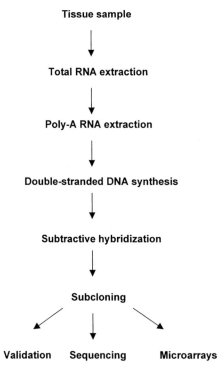

Tissue sample

Total RNA extraction

Poly-A RNA extraction

Double-stranded DNA synthesis

Subtractive hybridization

Subcloning

Validation   **Sequencing**   **Microarrays**

Fig. 5.3   Overview of the different steps of the subtractive hybridization. See text for details.

mended to pool several samples from each group, to ensure that the subtraction is well representative of the condition tested.

2. RNA isolation. This step is probably the most important of the whole procedure, as all the information gathered from the subtraction tightly depends on the quality of the starting material, which is the poly-A RNA. Total RNA extraction is usually performed by the phenol/chloroform extraction [51] and does not require further purification. RNAse-free conditions are mandatory to preserve the structure and the full length of the RNA as well as possible. Purification of the poly-A mRNA requires more attention. Different techniques are available, some insisting more on the exclusion of the contaminating ribosomal RNA, some insisting more on the recovery of rare transcripts. The latter choice is probably the best in this case, to ensure that the library to be subtracted contains as many different transcripts as possible. A contamination by ribosomal RNA will usually not jeopardize the experiment. The quality of the RNA must be verified by the 260/280 absorbance ratio, by Northern blot or by automated methods.

3. Subtractive hybridization. The subtraction between two libraries begins by the reverse transcription of the poly-A mRNA into double-stranded DNA (Fig. 5.4). The reverse transcription is usually performed with an oligo-dT primer, rather than random hexamers, to increase the length of the products. Because of this priming method, the need of good-quality poly-A RNA is self-explanatory. The DNA from both samples is subsequently digested by a 4 base-cutter restriction

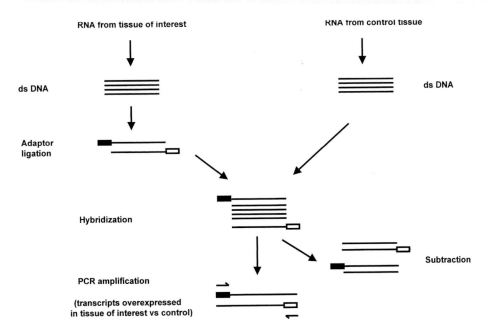

Fig. 5.4 Principle of the subtractive hybridization and amplification of the subtracted products. See text for details.

enzyme (R*sa* I, typically). One of the two libraries is then ligated to specific adapters, before the two preparations are hybridized together. Two successive rounds of hybridization are usually performed. The duration of the hybridization is a crucial aspect of the experiment for the following reason. The hybridization between two complementary strands of DNA is proportional to the concentration (Co) of the strands and the time (t) of hybridization. The product of these two factors (Cot) is specific for each sample of DNA sample. Because of the Cot principle, the more abundant transcripts will hybridize faster. As the subtraction does not affect the concentration (Co) of the cDNAs, adjusting the time of hybridization (t) is a mechanism of normalization, to ensure that all the transcripts (rare or abundant) will hybridize relatively equally. Because of this normalization, the chance to find a transcript in the subtracted library after the experiment is relatively independent of its original abundance. The DNA fragments that do not hybridize correspond to those which are expressed differently in the two preparations (Fig. 5.4). These gene products are further amplified by PCR using a single primer pair that recognizes the adapters, and subcloned in a vector system. The plasmid can be used for sequencing, microarrays, Northern blot or Dot blot (Fig. 5.4). The adapters are ligated to one cDNA library only. For instance, if the adapters are ligated on the cDNA library from an ischemic heart (labeled "tissue of interest" in Fig. 5.4) and if this library is subsequently subtracted from a library from normal myocardium, the experiment will show the genes that are overexpressed in ischemic myocardium. Reciprocally, the adapters can be ligated to the library from normal heart, to show the genes that are downregulated in the ischemic territory. A complete hybridization therefore includes a "forward" and a "reverse" library to show both the up-regulation and downregulation of genes in one condition compared to another.

4. Sequencing and database query. Although the hybridization in itself can be achieved in less than a week, the identification of the subcloned genes is more labor-intensive. Nowadays, the sequencing work is greatly facilitated by the use of automated sequencers using a 96-well or 384-well format. Some investigators prefer to verify first the quality of the library by microarrays. The sequencing also determines the quality of the library by the extent of the different gene products found, and by the redundancy of the same products. Finding new products by sequencing a library follows an exponential curve. The first batches of sequences bring many different gene products, then further sequencing shows the same products coming back again and again. It is reasonable to interrupt the sequencing when the last batch of clones does not bring more than 10% of new information (for instance, if less than 10 new gene products are found in a 96-well sequencing plate). Also, if a specific gene is highly regulated (i.e., if the concentration of the transcript between the two conditions tested shows more than a 10-fold difference), this specific product has a high probability to be sequenced more often.

5. Microarrays. Microarrays represent a reliable technique to verify at a glance which clones from the library are truly regulated. It consists in spotting all the clones isolated on nylon membranes or glass chips and to hybridize them with

labeled RNA from the different experimental groups tested. The labeling can be radioactive (using ^{33}P isotope instead of ^{32}P for a better resolution) or non-radioactive (using fluorescent dyes or chemiluminescence). The isotopic method is limited in its sensitivity. This technique therefore offers some information about the magnitude of changes in expression for each gene product. These chips can also be hybridized with labeled fragments of the mitochondrial genome. By this approach, it is possible to define which clones correspond to mitochondrial DNA, instead of nuclear-encoded genes. This strategy may greatly reduce the number of clones to sequence.

6. Validation. Whatever the results of the hybridization and the identity of the isolated clones, a method of validation is always required. It is necessary to rely on methods that truly quantify the changes in expression of one specific gene of interest. One of the most popular methods is the Northern blot, because the probe is already available in the form of the clone sequenced from the library. The average size of the cDNA fragments obtained during the subtraction (0.6 to 1.2 Kb) corresponds to the recommended size of a usual probe for Northern blot. In addition, the tissue localization of the transcript can be validated further by in-situ hybridization. Again, the subcloned DNA fragment can be used as a probe, although some investigators prefer to use oligo probes. Finally, the quantitative PCR (using specific fluorogenic probes or as a screening technique with the SybrGreen) will be the preferred choice when analyzing a large number of samples, when making a time-course, or when the RNA supply is limited.

5.2.2
Advantages and Disadvantages

The main advantage of the subtractive hybridization consists in its unbiased nature. Potentially, all the genes regulated by a specific condition can be found. This requires a good preparation of RNA and an extensive work of subcloning and sequencing. Ischemia/reperfusion, cardiac hypertrophy and heart failure, in which changes in both workload and ventricular function are found, can affect dramatically the expression of genes in the heart and can trigger the expression of genes that are not normally expressed in myocardium. The subtractive hybridization is therefore an excellent tool for discovery. It can show "unexpected" gene profiles, or lead to the discovery and characterization of novel genes. The technique is therefore excellent for both the horizontal and vertical approaches described above. It starts by a genomic profiling and then, depending on the identity of the clones, can lead to an in-depth analysis of specific products. Another advantage is that the experiment does not depend on the species studied. The different genome projects have substantially increased the number of sequences available in public databases, so that querying sequences from swine, dog or even woodchuck tissues is not a problem in practice.

The inconvenience of the subtractive hybridization is that, due to its random nature, interesting clones can be "missed". The technique is not as comprehensive as a DNA chip (although all the clones collected from a subtraction can be spotted on a heart-specific, custom-made chip). However, the chance to find a specific gene is proportional to the magnitude of its upregulation in one group compared to the other. Another inconvenience is the requirement of an access to a dedicated sequencing facility. Finally, especially in the heart, it is impossible to avoid the "contamination" of the library by non-nuclear genes from the mitochondrial DNA. Due to the large amount of mitochondria in the cardiomyocyte, such contamination can represent 20–50% of the overall library [50, 52]. This contamination can be limited by performing additional rounds of hybridization, at the expense of loosing rarely expressed nuclear transcripts. The mitochondrial genome includes the genes coding for the subunits 1 to 6 of NADH dehydrogenase, the subunits 1 to 3 of cytochrome oxidase, the ribosomal RNAs 12S and 16S, the cytochrome b and the subunits 6 and 8 of ATP synthase. A last inconvenience is that the subtraction requires to start with 2–3 μg of poly-RNA, which may represent about 50 μg of total RNA in the heart. There are some experimental conditions, especially the genomic profiles during development, in which such amount can hardly be obtained.

5.3
Genomics of Myocardial Ischemia

The discovery that gene expression can be affected by changes in workload [21, 32] rapidly led to the investigation of the adaptation of myocardial gene expression in the ischemic heart. Schaper's laboratory provided pioneering studies on alterations in gene expression in large mammalian models of ischemia/reperfusion. These studies were mainly designed to follow the time-course of changes in genes coding for proto-oncogenes and calcium-handling proteins [53–55]. The proto-oncogenes were chosen because they represent a rapid mechanism of transcriptional adaptation to stress, whereas calcium-handling genes were studied to correlate their expression to the prolonged dysfunction that follows ischemia. Another goal was to investigate whether repetitive episodes of ischemia/reperfusion inducing preconditioning are accompanied by a change in gene expression. Several proto-oncogenes (such as c-*fos* or *jun*B) showed an increase either during ischemia or reperfusion. Similarly, the Ca^{2+}-ATPase and calsequestrin were found to increase during post-ischemic dysfunction. These studies were the first to show that even short episodes of ischemia-reperfusion could affect myocardial gene expression. Also, they showed the possibility to investigate myocardial gene expression in large mammalian models of ischemic heart disease. However, the experimental design of this work adopted a "vertical approach", which precluded the genomic profiling of these hearts. Because only limited subsets of genes were investigated, the global impact of these changes could hardly be determined.

More recently, the technology of microarrays has been used in several rodent models to offer a more comprehensive view of the genomic adaptation to ischemia. So far, most studies have been performed in models of myocardial infarction and remodeling [50, 56, 57]. In these studies, the induction of irreversible ischemia mainly affects the expression of genes involved in the synthesis of cytoskeletal proteins and of the extracellular matrix. These studies will lead to a better understanding of the impact of drug therapy (such as angiotensin-converting enzyme inhibitors) on cardiac remodeling [57]. However, it will be important in the future to explore the adaptation to reversible ischemia in these models, to better elucidate the potential protective mechanisms that the heart can develop before ischemia becomes irreversible. Such work will be equally important to develop new therapeutic strategies.

5.4
Subtractive Hybridization of Myocardial Ischemia

5.4.1
Myocardial Stunning

Myocardial stunning refers to the myocardial dysfunction that follows an acute episode of ischemia [58, 59]. A preserved ultrastructure of the myocardium and the absence of cell loss characterize this form of non-lethal, fully reversible ischemia [60]. These characteristics are in striking contrast to other models of ischemia-reperfusion, in which both necrosis and apoptosis are observed [61, 62]. The syndrome of stunning is highly prevalent in different etiologies of clinical ischemic coronary artery disease, including stable or unstable angina pectoris, myocardial infarction, chronic multivessel disease, and post-surgical dysfunction [63, 64]. Due to the major prevalence of ischemic heart disease, stunning is of paramount clinical importance because it corresponds to a condition in which myocardial viability is maintained.

Although it was previously thought that the ventricular dysfunction induced by ischemia could not be brought back to a normal contractile state, it is now clearly demonstrated that the viability of dysfunctional myocardium is preserved [65]. Unraveling the molecular mechanisms of cardioprotection in stunned myocardium can open new avenues to salvage dysfunctional cardiac tissue and prevent cardiac cell loss. Especially, a better understanding of the mechanisms by which the molecular and cellular adaptations maintain cell survival should open new therapeutic opportunities. These mechanisms remain largely unknown in large mammalian models. Yet, the models of ischemic heart disease in large mammals, especially the swine, are most clinically relevant. This is due to the fact that the swine heart and the human heart share the same geometry, heart rate, coronary anatomy and absence of collaterals.

5.4.2
Genomic Profile of Myocardial Stunning

The absence of irreversible cellular damage in stunned myocardium may correspond either to an increased resistance of the heart to ischemia, or to the absence of stimuli triggering the pathways of cell death. The latter possibility is of low probability, because even mild episodes of ischemia/reperfusion can activate different intracellular stress pathways leading to cell death [66–69]. The subtractive hybridization represents a good technique to investigate which cardioprotective mechanisms are activated in the ischemic heart. One hypothesis to be tested by gene profiling is whether myocardial stunning triggers the coordinated expression of different sets of genes acting to protect the myocardium against irreversible damage.

These experiments were performed in a swine model of regional low-flow ischemia, in which the blood flow through the left anterior descending coronary artery is decreased by about 50% during 90 minutes [70]. This ischemic episode is followed by full reperfusion. Despite the normalization of blood flow after reperfusion, the contractile function remains depressed, which reflects myocardial stunning. A full functional recovery is typically observed after 48–72 h, and pathological examination of this myocardium does not show any necrosis or apoptosis. We used this model of reversible ischemia to investigate whether the protection of the myocardium against irreversible damage correlates with a specific gene profile [52]. A subtractive hybridization was therefore performed between the post-ischemic, stunned myocardium and the remote, normal myocardium. Because this is a model of regional ischemia, stunned and normal areas can be compared within the same hearts. In these specific experiments, the contamination of mitochondrial DNA after two rounds of hybridization was about 35%. After database query, 60% of the nuclear-encoded sequences corresponded to known gene products. The remaining 40% could be divided equally between known sequences of unknown function and sequences with no match. Interestingly, we found that more than 30% of the genes that were upregulated in stunned myocardium are involved in different mechanisms of cell survival, including: resistance to apoptosis, cytoprotection ("stress response") and cell growth. Many of them had been implied previously in the survival of other cell types but had not been described before in the heart. In particular, we found an induction in stunned myocardium of several genes not expressed in the normal myocardium and which participate to the development and growth of different forms of tumors. This illustrates the power of the subtractive hybridization to pull out "unexpected" genes and profiles. An example of the validation of these results is shown on Fig. 5.5 for the plasminogen activator inhibitor-1 (PAI-1), a serpin with anti-apoptotic properties [71]. A Northern blot on ischemic and normal samples from five different hearts showed a reproducible and robust induction of this gene during ischemia. It could be confirmed by in-situ hybridization that this induction occurred in cardiomyocytes. Finally, the measurement of PAI-1 by quantitative PCR shows that the expression of this gene starts to increase during ischemia but peaks during post-ischemic stunning and returns back to normal after 12 hours, when the contractile function recovers. The sensitivity of the quantitative PCR also allows to measure separately the sub-endocardium from the sub-epicardium, to show that the gene response is of higher amplitude in the sub-endocardium, where the flow reduction is the most important. Remarkably, this transmural difference was found for all the genes measured [52]. There is therefore a gradient of gene response that matches the gradient of flow reduction, which shows that the nuclear response is not an "all or nothing" phenomenon but is proportional to the intensity of the initial stimulus.

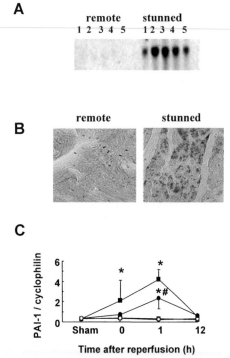

Fig. 5.5 Methods of validation of the subtractive library. This example illustrates the regulation of PAI-1 RNA in stunned myocardium. Panel A shows the reproducibility of the induction of PAI-1 in five different samples from stunned territory. The corresponding control (remote) area in the same hearts does not show any signal. Panel B shows by in-situ hybridization that PAI-1 induction occurs in cardiomyocytes. Panel C shows by quantitative PCR the time-course of PAI-1 induction during stunning, with a maximal increase at one hour reperfusion and a normalization at

5.4.3
Chasing Novel Genes

As mentioned above, about 20% of the gene products found in the subtractive hybridization of stunned myocardium in the pig did not recall any known sequence in public databases. This last group is particularly interesting, because it contains the novel genes. However, because the cDNA synthesis starting the subtraction experiments is primed by oligo-dT, an "unknown" sequence can also correspond to the 3′ UTR of a known gene. This can be due to two reasons. First, many sequencing laboratories are interested in cloning only the coding sequence of the genes ("ORFome"), and the 3'UTR is left undetermined. Second, the 3' UTR is the region of the transcript that varies the most within and between species (alternative splicings, regulatory elements, duplications and deletions). Therefore, characterizing a "novel" gene remains a challenge. Especially, when many sequences from a subtraction library seem to be "novel", it can be difficult to choose which are those with the highest priority for further investigation. Different criteria can be applied to answer that question.

- Sequence length. It is better to start with a long sequence from the subtractive hybridization. Considering two gene products that do not match any known sequence, a fragment of 1.2 Kb is statistically more likely to be novel than a fragment of 250 base pairs.
- Tissue specificity. This is preferable to improve the originality of the work. The tissue selectivity can be assessed easily with a multi-tissue Northern blot. A gene that is expressed only in the heart is probably more interesting than a gene that is expressed ubiquitously. The Northern blot will also determine the size of the transcript and thereby will predict the amount of work needed to achieve a full-length cloning.
- Regulated gene. It is preferable to concentrate on true positives, i.e., genes showing a differential expression between the two experimental groups. Once an interesting target has been identified, the different methodologies of the "vertical approach" (Fig. 5.2) can be applied for a full characterization of the transcript and the corresponding protein. This approach can be applied not only for the novel genes, but also for known genes that had not been described in the heart before.

5.5
Summary

The regulation of myocardial gene expression is highly sensitive to any extracellular or intracellular stimulus that affects contractile function. The recent development of powerful technologies allows to study the broad genomic profile of the heart in various experimental conditions. Although mainly investigated in rodents, gene regulation in the heart needs to be elucidated further in large mammalian models that reproduce the different forms of cardiac disease in humans. The biological questions addressed in these models require the use of appropriate techniques. The use of the subtractive hybridization is particularly attractive because it can be applied to any species. The strength of the subtractive hybridization relies on its unbiased nature and its power to extract even low-abundance transcripts. In addition, the subtraction experiments reveal "unexpected" gene profiles and represent a starting point for the characterization of novel genes.

5.6
Acknowledgements

I am particularly grateful to Stephen F. Vatner for his support and for developing the animal models used in these experiments. I also express my deepest gratitude to James E. Tomlinson who initiated me to the molecular techniques of subtractive hybridization. I also thank Raymond K. Kudej, Song-Jung Kim, Dorothy E. Vatner, Vinciane Gaussin, James N. Topper, Junichi Sadoshima and Maha Abdellatif for their collaboration and fruitful discussions, as well as Erika Thompson, Anna Zajac and Li Wang for their expert technical assistance.

5.7
References

1 NADAL-GINARD, B., MAHDAVI, V. Molecular basis of cardiac performance: plasticity of the myocardium generated through protein isoform switches. *J. Clin. Invest.* **1989**, *84*, 1693–1700.

2 SCHNEIDER, M., ROBERTS, R., PARKER, T. Modulation of cardiac genes by mechanical stress. The oncogene signalling hypothesis. *Mol. Biol. Med.* **1991**, *8*, 167–183.

3 KOMURO, I., YAZAKI, Y. Control of cardiac gene expression by mechanical stress. *Annu. Rev. Physiol.* **1993**, *55*, 55–75.

4 KOMURO, I., KAIDA, T., SHIBAZAKI, Y., KURABAYASHI, M., KATOH, Y., YAZAKI, Y. Stretching cardiac myocyte stimulates proto-oncogene expression. *J. Biol. Chem.* **1990**, *265*, 5391–5398.

5 CHIEN, K.R., ZHU, H., KNOWLTON, K.U., MILLER-HANCE, W., VAN BILSEN, M., O'BRIEN, T.X., EVANS, S.M. Transcriptional regulation during cardiac growth and development. *Annu. Rev. Physiol.* **1993**, *55*, 77–95.

6 CHIEN, K., KNOWLTON, K., ZHU, H., CHIEN, S. Regulation of cardiac gene expression during myocardial growth and hypertrophy: Molecular studies of an adaptive physiologic response. *FASEB J.* **1992**, *5*, 3037–3046.

7 ROBBINS, J. Regulation of cardiac gene expression during development. *Cardiovasc. Res.* **1996**, *31*, E2–E16.

8 SADOSHIMA, S., IZUMO, S. The cellular and molecular response of cardiac myocytes to mechanical stress. *Annu. Rev. Physiol.* **1997**, *59*, 551–571.

9 SCHOENFELD, J., VASSER, M., JHURANI, P., NG, P., HUNTER, J., ROSS, J., CHIEN, K., LOWE, D. Distinct molecular phenotypes in murine muscle development, growth and hypertrophy. *J. Mol. Cell. Cardiol.* **1998**, *30*, 2269–2280.

10 HORNBY, L., HAMILTON, N., MARSHALL, D., SALERNO, T.A., LAUGHLIN, M.H., IANUZZO, C.D. Role of cardiac work in regulating myocardial biochemical characteristics. *Am. J. Physiol.* **1990**, *258*, H1482–H1490.

11 XIA Y., BUJA, L.M., McMILLIN, J.B. Changes in expression of heart carnitine palmitoyltransferase I isoforms with electrical stimulation of cultures rat neonatal cardiac myocytes. *J. Biol. Chem.* **1996**, *271*, 12082–12087.

12 XIA, Y., BUJA, L.M., SCARPULLA, R.C., McMILLIN, J.B. Electrical stimulation of neonatal cardiomyocytes results in the sequential activation of nuclear genes governing mitochondrial proliferation and differentiation. *Proc. Natl. Acad. Sci. USA* **1997**, *94*, 11399–11404.

13 DEPRE, C., SHIPLEY, G., CHEN, W., HAN, Q., DOENST, T., MOORE, M., STEPKOWSKI, S., DAVIES, P., TAEGTMEYER, H. Unloaded heart in vivo replicates fetal gene expression of cardiac hypertrophy. *Nature Medicine* **1998**, *4*, 1269–1275.

14 DEPRE, C., VANOVERSCHELDE, J.L., MELIN, J.A., BORGERS, M., BOL, A., AUSMA, J., DION, R., WIJNS, W. Structural and metabolic correlates of the reversibility of chronic left ventricular ischemic dysfunc-

tion in humans. *Am. J. Physiol.* **1995**, *268*, H1265–H1275.

15 CROZATIER, B. Stretch-induced modifications of myocardial performance: from ventricular function to cellular and molecular mechanisms. *Cardiovasc. Res.* **1996**, *32*, 25–37.

16 SCHWARTZ, K., DE LA BASTIE, D., BOUVERET, P., OLIVIERO, P., ALONSO, S., BUCKINGHAM, M. a-skeletal muscle actin mRNA's accumulate in hypertrophied adult rat hearts. *Circ. Res.* **1986**, *59*, 551–555.

17 SEMENZA, G. Perspectives on oxygen sensing. *Cell* **1999**, *98*, 281–284.

18 SEMENZA, G. HIF-1, O_2 and the 3 PHDs. How animal cells signal hypoxia to the nucleus. *Cell* **2001**, *107*, 1–4.

19 BOLLI, R. Oxygen-derived free radicals and myocardial reperfusion injury: An overview. *Cardiovasc. Drugs Ther.* **1991**, *5*, 249–268.

20 FALLER, D. Endothelial cell responses to hypoxic stress. *Clin. Exp. Pharmacol. Physiol.* **1999**, *26*, 74–84.

21 SCHWARTZ, K., BOHELER, K. R., DE LA BASTIE, D., LOMPRE, A. M., MERCADIER, J. J. Switches in cardiac muscle gene expression as a result of pressure and volume overload. *Am. J. Physiol.* **1992**, *262*, R364–R369.

22 PARKER, T. Molecular biology of myocardial hypertrophy and failure: gene expression and trophic signaling. *New Horizons.* **1995**, *3*, 288–300.

23 SADOSHIMA, J., IZUMO, S. Mechanical stretch rapidly activates multiple signal transduction pathways in cardiac myocytes: potential involvement of an autocrine/paracrine mechanism. *EMBO J.* **1993**, *12*, 1681–1692.

24 SHARP, W., SIMPSON, D., BORG, T., SAMAREL, A., TERRACIO, L. Mechanical forces regulate focal adhesion and costamer assembly in cardiac myocytes. *Am. J. Physiol.* **1997**, *273*, H546–H556.

25 DOEVENDANS, P. A., VAN BILSEN, M. Transcription factors and the cardiac gene programme. *Int. J. Biochem. Cell. Biol.* **1996**, *28*, 387–403.

26 PAWSON, T., SAXTON, T. Signaling networks – do all roads lead to the same genes? *Cell* **1999**, *97*, 675–678.

27 HUGHES, T., MARTON, M., JONES, A., ROBERTS, C., STOUGHTON, R., ARMOUR, C., BENNETT, H., COFFEY, E., DAI, H., HE, H., KIDD, M., KING, A., MEYER, M., SLADE, D., LUM, P., STEPANIANTS, S., SHOEMAKER, D., GACHOTTE, D., CHAKRABURTTY, K., SIMON, J., BARD, M., FRIEND, S. Functional discovery via a compendium of expression profiles. *Cell* **2000**, *102*, 109–126.

28 LIPARDI, C., WIE, Q., PATERSON, B. RNAi as random degradative PCR. *Cell* **2001**, *107*, 297–307.

29 BASS, B. RNA interference. The short answer. *Nature.* **2001**, *411*, 494–498.

30 LANDER, E. Initial sequencing and analysis of the human genome. *Nature.* **2001**, *409*, 860–921.

31 VENTER, J. The sequence of the human genome. *Science.* **2001**, *291*, 1304–1351.

32 MERCADIER, J., LOMPRE, A., WISNEWSKY, C., SAMUEL, J., BERCOVICI, J., SWYNGHEDAUW, B., SCHWARTZ, K. Myosin isoenzymic changes in several models of rat cardiac hypertrophy. *Circ. Res.* **1981**, *49*, 525–532.

33 IZUMO, S., LOMPRE, A.M., MATSUOKA, R., KOREN, G., SCHWARTZ, K., NADAL-GINARD, B., MAHDAVI, V. Myosin heavy chain messenger RNA and protein isoform transitions during cardiac hypertrophy. *J. Clin. Invest.* **1987**, *79*, 970–977.

34 IZUMO, S., NADAL-GINARD, B., MAHDAVI, V. Protooncogene induction and reprogramming of cardiac gene expression produced by pressure overload. *Proc. Natl. Acad. Sci. USA* **1988**, *85*, 339–343.

35 ARAI, M., MATSUI, H., PERIASAMY, M. Sarcoplasmic reticulum gene expression in cardiac hypertrophy and heart failure. *Circ. Res.* **1994**, *74*, 555–564.

36 DEPRE, C., VANOVERSCHELDE, J., TAEGTMEYER, H. Glucose for the heart. *Circulation* **1999**, *99*, 578–588.

37 LIM, D., ROBERTS, R., MARIAN, A. Expression profiling of cardiac genes in human hypertrophic cardiomyopathy. *J. Am. Coll. Cardiol.* **2001**, *38*, 1175–1180.

38 LOWES, B. D., MINOBE, W., ABRAHAM, W.T., RIZEQ, M.N., BOHLMEYER, T.J., QUAIFE, R.A., RODEN, R.L., DUTCHER, D.L., ROBERTSON, A.D., VOELKEL, N.F., BADESCH, D.B., GROVES, B.M., GILBERT,

E. M., Bristow, M. R. Changes in gene expression in the intact human heart. *J. Clin. Invest.* **1997**, *100*, 2315–2324.

39 Morano, M., Zacharzowski, U., Maier, M., Lange, P., Alexi-Meskishvili, V., Haase, H., Morano, I. Regulation of human heart contractility by essential myosin light chain isoforms. *J. Clin. Invest.* **1996**, *98*, 467–473.

40 Schaub, M., Hefti, M., Zuellig, R., Morano, I. Modulation of contractility in human cardiac hypertrophy by myosin essential light chain isoforms. *Cardiovasc. Res.* **1998**, *37*, 381–404.

41 Utans, U., Liang, P., Wyner, L., Karnovsky, M., Russell, M. Chronic cardiac rejection: identification of five upregulated genes in transplanted hearts by differential mRNA display. *Proc. Natl. Acad. Sci. USA* **1994**, *91*, 6463–6467.

42 Knoll, R., Zimmerman, R., Arras, M., Schaper, W. Characterization of differentially expressed genes following brief cardiac ischemia. *Biochem. Biophys. Res. Commun.* **1996**, *221*, 402–407.

43 Singh, K., Sirokman, G., Communal, C., Robinson, K., Conrad, C., Brooks, W., Bing, O., Colucci, W. Myocardial osteopontin expression coincides with the development of heart failure. *Hypertension* **1999**, *33*, 663–670.

44 Abdellatif, M. Leading the way using microarray. *Circ. Res.* **2000**, *86*, 919–920.

45 Rao, J., Bond, M. Microarrays. Managing the data deluge. *Circ. Res.* **2001**, *88*, 1226–1227.

46 Liu, T., Lai, H., Wu, W., Chinn, S., Wang, P. Developing a strategy to define the effects of insulin-like growth factor-1 on gene expression profile in cardiomyocytes. *Circ. Res.* **2001**, *88*, 1231–1238.

47 Pritchard, C., Hsu, L., Delrow, J., Nelson, P. Project normal: defining normal variance in mouse gene expression. *Proc. Natl. Acad. Sci. USA.* **2001**, *98*, 13266–13271.

48 Knight, J. When the chips are down. *Nature* **2001**, *410*, 860–861.

49 Hubank, M., Schatz, D. Identifying differences in mRNA expression by representational difference analysis of cDNA. *Nucleic Acids Res.* **1994**, *22*, 5640–5648.

50 Sehl, P., Tai, J., Hillan, K., Brown, L., Goddard, A., Yang, R., Jin, H., Lowe, D. Application of cDNA microarrays in determining molecular phenotype in cardiac growth, development, and response to injury. *Circulation* **2000**, *101*, 1990–1999.

51 Chomczynski, P., Sacchi, N. Single-step method of RNA isolation by acid guanidium thiocyanate-phenol-chloroform extraction. *Anal. Biochem.* **1987**, *162*, 159–169.

52 Depre, C., Tomlinson, J., Kudej, R. K., Gaussin, V., Thompson, E., Kim, S. J., Vatner, D., Topper, J., Vatner, S. Gene program for cardiac cell survival induced by transient ischemia in conscious pig. *Proc. Natl. Acad. Sci. USA* **2001**, *98*, 9336–9341.

53 Knoll, R., Arras, M., Zimmermann, R., Schaper, J., Schaper, W. Changes in gene expression following short coronary occlusions studied in porcine hearts with run-on assays. *Cardiovasc. Res.* **1994**, *28*, 1062–1069.

54 Frass, O., Sharma, H., Knoll, R., Duncker, D., McFalls, E., Verdouw, P., Schaper, W. Enhanced gene expression of calcium regulatory proteins in stunned porcine myocardium. *Cardiovasc. Res.* **1993**, *27*, 2037–2043.

55 Brand, T., Sharma, H., Fleischmann, K., Duncker, D., McFalls, E., Verdouw, P., Schaper, W. Proto-oncogene expression in porcine myocardium subjected to ischemia and reperfusion. *Circ. Res.* **1992**, *71*, 1351–1360.

56 Stanton, L., Garrard, L., Damm, D., Garrick, B., Lam, A., Kapoun, A., Zheng, Q., Protter, A., Schreiner, G., White, R. Altered patterns of gene expression in response to myocardial infarction. *Circ. Res.* **2000**, *86*, 939–945.

57 Jin, H., Yang, R., Awad, T., Wang, F., Li, W., Williams, S., Ogasawara, A., Shimada, B., Williams, P., de Feo, G., Paoni, N. Effects of Early Angiotensin-Converting Enzyme Inhibition on Cardiac Gene Expression After Acute Myocardial Infarction. *Circulation* **2001**, *103*, 736–742.

58 Heyndrickx, G. R., Millard, R. W., McRitchie, R. J., Maroko, P. R., Vatner,

S. F. Regional myocardial functional and electrophysiological alterations after brief coronary artery occlusion in conscious dogs. *J. Clin. Invest.* **1975**, *56*, 978–985.

59 BRAUNWALD, E., KLONER, R. A. The stunned myocardium: Prolonged, postischemic ventricular dysfunction. *Circulation* **1982**, *66*, 1146–1149.

60 BOLLI, R. Mechanism of myocardial "stunning". *Circulation* **1990**, *82*, 723–738.

61 FREUDE, B., MASTERS, T., ROBICSEK, F., FOKIN, A., KOSTIN, S., ZIMMERMANN, R., ULLMANN, C., LORENZ-MEYER, S., SCHAPER, J. Apoptosis is initiated by myocardial ischemia and executed during reperfusion. *J. Mol. Cell. Cardiol.* **2000**, *32*, 197–208.

62 KAJSTURA, J., CHENG, W., REISS, K., CLARK, W., SONNENBLICK, E., KRAJEWSKI, S., REED, J., OLIVETTI, G., ANVERSA, P. Apoptotic and necrotic cell deaths are independent contributing variables of infarct size in rats. *Lab. Invest.* **1996**, *74*, 86–107.

63 BOLLI, R. Myocardial "stunning" in man. *Circulation* **1992**, *86*, 1671–1691.

64 RAHIMTOOLA, S. H. A perspective on the three large multicenter randomized clinical trials of coronary bypass surgery for chronic stable angina. *Circulation* **1985**, *72*, V123–V135.

65 WIJNS, W., VATNER, S., CAMICI, P. Hibernating myocardium. *New Engl. J. Med.* **1998**, *339*, 173–181.

66 PRZYKLENK, K., KLONER, R. Reperfusioninjury by oxygen-derived free radicals. *Circ. Res.* **1989**, *6*, 86–96.

67 CURRIE, R., TANGAY, R., KINGMA, J. Heatshock response and limitation of tissue necrosis during occlusion/reperfusion in rabbit hearts. *Circulation* **1992**, *87*, 963–971.

68 SUGDEN, P., CLERK, A. "Stress-responsive" mitogen-activated protein kinases (c-Jun N-terminal kinases and p38 mitogen-activated protein kinases) in the myocardium. *Circ. Res.* **1998**, *83*, 345–352.

69 YIN, T., SANDHU, G., WOLFGANG, C., BURRIER, A., WEBB, R., RIGEL, D., HAI, T., WHEELAN, J. Tissue-specific pattern of stress kinase activation in ischemic/reperfused heart and kidney. *J. Biol. Chem.* **1997**, *272*, 19943–19950.

70 KUDEJ, R., KIM, S., SHEN, Y., JACKSON, J., KUDEJ, A., YANG, G., BISHOP, S., SF, V. Nitric oxide, an important regulator of perfusion-contraction matching in conscious pigs. *Am. J. Physiol. Heart Circ. Physiol.* **2000**, *279*, H451–H456.

71 SILVERMAN, G., BIRD, P., CARRELL, R., CHURCH, F., COUGHLIN, P., GETTINS, P., IRVING, J., LOMAS, D., LUKE, C., MOYER, R., PEMBERTON, P., REMOLD-O'DONNELL, E., SALVESEN, G., TRAVIS, J., WHISSTOCK, J. The serpins are an expanding superfamily of structurally similar but functionnally diverse proteins. *J. Biol. Chem.* **2001**, *276*, 33293–33296.

6
DNA Microarray Gene Profiling:
A Tool for the Elucidation of Cardioprotective Genes

Elizabeth Murphy, Cynthia A. Afshari, John G. Petranka, and Charles Steenbergen

6.1
Introduction

This review will address several issues regarding the use of cDNA microarray gene profiling to identify cardioprotective genes. Section 1 will review data in the literature suggesting that there are cardioprotective genes that can be upregulated to protect the myocardium. This section also will discuss how these cardioprotective genes might lead to protection. Section 2 will consider whether different cardioprotective mechanisms act via induction of common genes. For example, are preconditioning and protection in females mediated via similar changes in gene expression? One of the primary rationales for identifying cardioprotective genes is to enhance expression of these genes to elicit cardioprotection. Thus it is important to understand the consequences of long-term activation of these genes. This issue will be discussed briefly in section 3. Section 3 also will consider the role of the level of gene expression. For example, it has been shown that high levels of expression of protein kinase C-epsilon (PKC-ε) lead to hypertrophy, whereas low or moderate levels of expression lead to cardioprotection. Section 4 will discuss possible strategies to identify these cardioprotective genes, and section 5 will discuss practical considerations in the use of microarray gene profiling to identify cardioprotective genes.

6.2
Candidates for Cardioprotective Genes and Possible Mechanism(s) of Protection

6.2.1
Candidate Genes Involved in Cardioprotection

Cardioprotection can be induced by intermittent stress such as preconditioning, or by addition of agonists which stimulate signaling pathways involved in preconditioning [1]. Another related means of studying cardioprotection and/or preconditioning is to modify candidate genes using transgenic approaches. In addition to preconditioning, endogenous protection is reported in females. We will consider candidate cardioprotective genes from all these approaches.

Preconditioning, with brief intermittent periods of ischemia and reperfusion, has been shown to induce protection against injury during a subsequent sustained period of ischemia [1]. The sustained period of ischemia may occur shortly after the preconditioning stimulus (acute preconditioning) or 1–3 days later (delayed preconditioning). It has been reported that acute preconditioning does not require new gene synthesis [2], whereas delayed preconditioning is dependent on new transcription and translation. Delayed preconditioning is reminiscent of other types of stress-induced protection such as the classic heat shock response. In addition to intermittent ischemia, preconditioning can be induced by a variety of stressors such as osmotic stress, redox stress, heat shock, high calcium, and toxins such as monophosphoryllipid A. These data are consistent with the hypothesis that stress-induced signaling pathways enhance transcription of genes that lead to cardioprotection. However, since acute preconditioning occurs within minutes and is reported to occur without new gene transcription, it appears that post-transcriptional changes in signaling pathways can also have a major role in cardioprotection and such changes may not be elucidated by gene profiling.

Regarding changes in gene expression that lead to cardioprotection, Schaper and coworkers have reported that brief ischemia and reperfusion, similar to the preconditioning trigger, induce transcription of many genes [3]. They found increased expression of mRNA for immediate-early genes (fos, junB, rgr-1), heat shock protein 27 (HSP27), genes involved in calcium-handling (Ca-ATPase, calsequestrin, phospholamban), and growth factors. However, in this study, there was no assessment of whether these changes in gene expression were causally involved in protection.

Many cardioprotective gene products have been identified using inhibitors to block preconditioning, or activators to mimic preconditioning. Transgenic animals have also been used to implicate specific genes involved in preconditioning. Bolli and coworkers have shown that mice lacking inducible nitric oxide synthase (iNOS) did not exhibit delayed preconditioning, suggesting an important role for iNOS in delayed preconditioning [4]. Bolli and coworkers further showed that selective inhibitors of cyclooxygenase-2 (COX-2) blocked delayed preconditioning [5]. As nuclear factor-kappaB (NF-κB) enhances transcription of COX-2 and iNOS, Bolli and coworkers also examined the role of the transcription factor NF-κB in preconditioning. An inhibitor of NF-κB blocked the preconditioning induced nuclear translocation of NF-κB and concomitantly blocked cardioprotection [6]. A new mouse model with cardiac specific expression of a mutant I-κB-alpha should provide further information on the role of NF-κB in cardioprotection [7]. In contrast to iNOS, the results of studies to examine the role for endothelial nitric oxide synthase (eNOS) in cardioprotection are mixed. Most studies report that nitric oxide donors are cardioprotective, and hearts from mice null for eNOS have been reported to have increased infarct size compared to wild-type (WT) littermates [8]. However, another study found no difference in infarct size between eNOS null hearts and WT hearts, but they did find that acute preconditioning was blocked in eNOS-knockout (KO) hearts using 2 or 3 cycles of preconditioning, but was not blocked if 4 cycles of preconditioning were employed [9]. Delayed preconditioning was not studied in these eNOS-KO mice. eNOS-KO hearts have been reported to have increased induction of iNOS during ischemia and this may complicate the results [10]. Hearts from mice null for neuronal nitric oxide synthase (nNOS) had infarcts that were similar in size to those in WT littermates [11].

Several investigators, using inhibitors and transgenic mouse models have shown a convincing role for PKC-ε in cardioprotection [12–14]. Interestingly, a dose-dependent effect of PKC-ε overexpression was observed [12]. Low to moderate cardiac specific overexpression of PKC-ε confered reduced susceptibility to ischemic injury [12, 14]. However, high levels of overexpression of PKC-ε resulted in cardiac hypertrophy [15]. Lck, a src family tyrosine kinase, has been shown to co-precipitate with PKC-ε, and Lck-KO hearts did not exhibit preconditioning [16]. Cardiac specific overexpression of the A1-adenosine receptor [17, 18], and the A3-adenosine receptor [19] also were reported to be cardioprotective. Bradykinin B2 receptor knock-out mice were not protected by acute preconditioning [20]. Cardiac specific overexpression of insulin-like growth factor-1 (IGF-1) [21], or fibroblast growth factor-2 (FGF-2) were also cardioprotective [22].

A number of stress-related genes have also been shown to be important in cardioprotection. Overexpression of heat shock protein 70 (HSP 70) was cardioprotective [23], as was overexpression of Cu/Zn superoxide dismutase (SOD) [24, 25] as well as overexpression of MnSOD [26]. Futhermore, Cu/Zn SOD null mice had increased infarcts following ischemia-reperfusion [27]. Two recent studies suggest an important role for heme-oxygenase-1 (HO-1) in cardioprotection. Overexpression of HO-1 reduced infarct size [28], whereas infarct size was increased in a HO-1 null hearts [29]. In addition, HO-1 null hearts did not exhibit acute preconditioning, suggesting a role for HO-1 in preconditioning. Also, hearts from mice null for glutathione peroxidase-1 (GSHPx-1) had increased infarcts and poorer recovery of function [30].

Overexpression of the anti-apoptotic bcl-2 has also been shown to be cardioprotective [31, 32]. Hearts null for leukocyte-type 12-lipoxygenase did not exhibit preconditioning [32], and hearts from mice null for the prostaglandin-I(2) receptor had an increase in infarct size [33]. Hearts from mice null for tumor necrosis factor alpha (TNF-α) showed a decrease in infarct size [34], whereas hearts from mice lacking both TNF-α receptor 1 and 2 had an increase in infarct size [35].

Animal and human studies have suggested that females show decreased susceptibility to myocardial disease [36–39]. Estrogen and other female hormones lead to altered expression of numerous genes. Data in the literature reported that females have increased levels of nuclear phospho-Akt which could be important in cardioprotection [40]. Estrogen was also reported to activate phosphatidylinositol-3-kinase (PI3-kinase) [41], and increase expression of eNOS [41], which could have a role in cardioprotection. Because of the large number of estrogen regulated genes, there are many potential candidates for cardioprotective genes in females.

6.2.2
Potential Mechanisms of Cardioprotection

Many of the genes implicated in cardioprotection are transcription factors which, by inducing other genes, could alter expression of many proteins and thereby lead to cardioprotection. Another major group of genes involved in cardioprotection are typically classified as stress (or heat shock) proteins. These include HSPs, GSHPx-1, SODs and HO. The antioxidant gene products could enhance protection by a reduction in reactive oxygen species, although it is likely that their protective effect is more complex than this. HSPs have been traditionally thought to be involved in regulating protein folding and degradation, although recent studies suggest that they are also involved in nuclear hormone regulation, apoptosis and other functions that could be important in protection [42]. There is also altered expression of several genes involved in growth factor signaling. Growth factors can induce cardioprotection by upregulating survival pathways such as PI3-K/Akt and NF-κB. Growth factors can also activate PKC, which has been shown to be cardioprotective. The precise mechanisms by which activation of these genes leads to cardioprotection are still to be elucidated.

6.3
Is There a Common Set of Cardioprotective Genes?

As discussed above, multiple signals can induce cardioprotection. Preconditioning can be initiated by multiple, diverse signals. The data so far suggest that there is remarkable similarity in the signaling pathways that lead to cardioprotection by very diverse initiators. For example, intermittent ischemia, activators of PKC and adenosine agonists all seem to activate similar signaling pathways involving PKC and a KATP channel. However it is unknown whether there are subtle differences in activation of cardioprotective genes with different initiators. In addition, it is not clear whether estrogen-mediated protection and preconditioning activate similar cardioprotective genes. Among the important questions that can be addressed by high through-put methods such as microarrays are whether there is a common set of cardioprotective genes or whether there are redundant cardioprotective genes. Although it is tempting to assume that there is a common set of cardioprotective genes this may not be the case. A recent study by Dorn and coworkers [43] used microarray to examine whether there were common gene changes in hypertrophy that developed in different transgenic models of hypertrophy (cardiac expression of calsequestrin, calcineurin, $G\alpha_q$, and PKC-ε activating peptide). Atrial natriuretic peptide was the only common gene changes across the different models of hypertrophy.

A related question is whether activation or inactivation of a single gene is cardioprotective? Studies with transgenic mice suggest that alterations in a single gene can lead to cardioprotection, but it is unclear whether these single gene changes result in additional secondary changes in gene expression, particularly in response to a stress such as pressure overload or ischemia.

6.4
Effects of Long-term Activation and Dosage of Cardioprotective Genes

One obvious goal of identification of cardioprotective genes is so that strategies can de devised to activate these genes to induce cardioprotection. However there could be different effects with short versus long-term activation of these genes. For example, we have shown that phosphorylation and inhibition of glycogen synthase kinase-3β (GSK-3β) is cardioprotective [44]. However, studies in neonatal myocytes suggest that phosphorylation and inactivation of GSK-3β can initiate hypertrophy [45, 46]. Also, activation of the PI3K/Akt pathway is suggested to be important in cardioprotection in females [39, 40]; yet long term activation of PI3K is reported to lead to hypertrophy. These data suggest that sustained activation has different effects than acute activation.

Another issue to consider is the level of gene activation or inactivation. It has been shown that low levels of cardiac specific overexpression of PKC-ε are cardioprotective, whereas high levels of PKC-ε overexpression lead to hypertrophy [12]. Similarly, low level expression of adenosine A3 receptor is cardioprotective, whereas high levels result in a dilated cardiomyopathy [19]. Thus, it is important to consider the level of gene expression when assessing models of cardioprotection.

6.5
Approaches to Identify Genes Involved in Cardioprotection

The quest for genes involved in cardioprotection has many facets. From the preceding discussion it is clear that we need to know whether there are common genes or gene patterns that lead to cardioprotection, or whether different models of cardioprotection are mediated by completely different genes or sets of genes. It is well recognized in transgenic animal studies that alteration of a single gene can have very different phenotypes depending on the genetic background [47]. Upregulation of a gene can have different consequences depending on what other genes are expressed. The ability to examine a large number of genes, to discern such differences, is a strength of the microarray technique. Also, there are data suggesting that altered gene expression can have different effects depending on the level of overexpression or duration of expression. Given these issues, how does one identify genes involved in cardioprotection? One obvious approach is to study multiple models of cardioprotection and determine if there are common genes or gene patterns. In section 1, we discussed different models of cardioprotection. We will briefly consider how one might use these models in microarray studies to identify cardioprotective genes.

6.5.1
Preconditioning

Preconditioning is one of the better studied models of cardioprotection. As discussed, preconditioning with brief intermittent period of ischemia and reperfusion has been shown to reduce injury during a subsequent sustained period of ischemia occurring 24 to 72 hours later. This delayed preconditioning requires new gene transcription. One strategy would be to use gene array technology to examine changes in gene expression induced by preconditioning. However, ischemia and reperfusion are likely to lead to many changes in gene expression, and only a small subset of these changes are likely to be causally important in cardioprotection. Perhaps a more direct strategy would be to treat animals with a cardioprotective agent and examine changes in gene expression. Perfused heart models are only viable for a few hours, therefore to study changes in gene expression that occur hours after administration of an agonist would require an intact animal model. Although a number of drugs and agonists can be given to a perfused heart to mimic acute preconditioning, many of these agents are not well-tolerated by the intact animal, and thus it is difficult to examine the change in gene expression occurring over a 24 hour period. Alternatively, one could add agents that mimic preconditioning to a perfused heart or myocytes and examine acute changes in gene expression. A recent study used cDNA microarray profiling to identify changes in gene expression in neonatal rat cardiomyocytes following overnight incubation with IGF-1 [48]. A large number (~60) of gene changes were observed. However, a limitation of this type of study is that it is difficult to distinguish which changes in gene expression are causally important in cardioprotection. A number of inhibitors of delayed preconditioning have been administered to animals. Therefore a promising strategy would be to determine which genes are not activated by preconditioning when inhibitors of protection are administered. This approach will essentially subtract out the effect of ischemia, and potentially may result in a small number of more relevant gene changes.

6.5.2
Transgenic Models of Protection

A similar approach to that of adding agents that mimic or inhibit preconditioning, is to compare levels of gene expression between transgenic models that exhibit cardioprotection and control non-protected wild-type littermates. Programs such as, the NHLBI-Program for Genomic Applications which has as its goal "linking genes to structure, function, dysfunction and structural abnormalities of the cardiovascular system caused by clinically relevant genetic and environmental stimuli", and the NIEHS National Center for Toxicogenomics, that has the goal of identifying mechanisms that underlie a variety of environmentally associated diseases, should help facilitate this type of study. An increasing number of transgenic mouse models exhibit cardioprotection. By comparing differences in gene expression between these transgenic mice and their WT control littermates one can potentially identify gene changes involved in the protection. In addition, there are a number of genetic mouse models that show enhanced susceptibility to injury and an analysis of differences in gene expression between these animals and their WT littermates could also provide insight into gene changes involved in cardioprotection. One consideration in studies of transgenic animals is that although typically a single gene has been changed, there can be other compensatory changes in gene expression. One of the challenges is to determine which gene changes (initial or compensatory) are causally involved in protection. An additional strategy that may be useful to uncover cardioprotective gene changes in transgenic animals is to examine gene changes that occur when the animals are stressed or injured. It is possible that the gene modified in the transgenic animal leads to protection by altering the genes that are induced or repressed during the response to injury. These gene changes would only be uncovered by stressing the animals (e.g. via pressure overload or an ischemic insult).

6.5.3
Cardioprotection in Females

There are several recent studies in animal models, as well as data in humans, suggesting that, particularly under conditions of increased stress, females show decreased susceptibility to ischemic injury and heart failure [36–39, 49]. One could therefore examine differences in gene expression between males and females. This approach has the advantage that it can be applied to humans as well as animal models. However, the disadvantage of this approach is that even in the heart there are likely to be a large number of male/female differences in gene expression that are unrelated to cardioprotection.

6.5.4
Human Studies

As the goal of these studies is to identify genes which are cardioprotective in humans, it would be useful to perform studies on human myocardium. One potential approach is to determine what genes are altered in diseases, such as ischemia or heart failure. The idea is that by identifying the gene changes that cause injury, we can better understand how to prevent the injury.

In summary, microarray technology could potentially elucidate genes or set of genes that could be manipulated to enhance cardioprotection. The challenge with this type of study is to distinguish which changes in gene expression are causally involved in protection. By comparing changes in gene expression across multiple models of cardioprotection it is hoped that epigenetic changes can be eliminated and the important gene changes will emerge. This assumes that the mechanism of protection in the different models is due to similar changes in gene expression, and this may not be the case. If there is not a common pathway for cardioprotection then comparison between these models will not show common gene changes.

6.6
Practical Considerations in Microarray Technology

There are a number of practical issues that are important in microarray studies. Many of these will be covered in other chapters. This section will highlight a few important issues, including, sample selection, the necessity for replicates, cross-hybridization, validation, the relationship of RNA to protein, and bioinformatics. Details of the cDNA microarray method are described elsewhere [50].

6.6.1
Sample Selection

To identify cardioprotective genes by comparing differences in gene expression between cardioprotected and non-protected models, it is important to make comparisons between animals with the same genetic background. Comparisons in the same strain or among littermates are desirable. Also, as females appear to have some inherent cardioprotection, it is important to compare males to males and females to females, unless one is specifically studying gender differences in cardioprotection. As in others studies, it is also important to compare animals of similar age and with similar treatments. It is also important to sample from the same location, for example left ventricle versus right ventricle.

For studies with human samples there are a number of additional challenges. There have been a number of recent studies utilizing explanted human hearts from patients in heart failure undergoing transplantation. The availability of control tissue is an important issue. Control heart tissue is typically obtained from donor hearts that could not be used for technical reasons, such as age of donor, death of the recipient and other logistical problems. The control donor hearts are frequently on inotropic support such as dopamine or norepinephrine prior to tissue harvest. Thus, the "control" hearts are more appropriately termed non-failing hearts. Rapid autopsy samples are another potential source of control tissue, but it is difficult to obtain heart autopsy samples in less than two to three hours after death, and there can be substantial RNA degradation during this time. Thus, a major limitation of studies of human heart tissue is the lack of suitable control tissue.

Explanted failing hearts are often obtained from patients with ischemic heart disease and thus contain fibrotic areas. In sampling failing hearts, it is important to avoid areas of fibrosis. It is also important to perform histology on both failing and non-failing hearts in order to correlate gene expression changes against a phenotypic description. Another issue with human heart samples is that both non-failing and failing patients are commonly treated with numerous medications that could influence gene expression. As discussed, non-failing patients are typically on inotropic support. The heart failure patients are also typically on a variety of medications. Thus changes in gene expression due to these medications will be superimposed on any changes in gene expression related to heart failure.

6.6.2
Variability in Hybridization

There can be variability within an array due to differences in labeling and hybridization. For example in cDNA array experiments using a direct labeling method, one of the fluorophores (Cy5) is more bulky and sometimes does not incorporate into newly transcribed cDNA as well as the other commonly used fluorophore (Cy3). This can lead to some differences in labeling and hybridization. One way to minimize this problem is to do multiple hybridizations, and label the control with the Cy5 fluorophore for half of the hybridization and with Cy3 for the other half. This "dye reversal" cancels out variation due to differential labeling with Cy3 and Cy5.

6.6.3
Cross-hybridization of Sequences

Cross hybridization may occur between sequences that are similar to each other within the coding sequence. By using a large portion of the 3' untranslated sequence for each clone one can reduce potential cross hybridization. To determine which other genes might be hybridizing with the sequence on the chip one can "BLAST" the sequence against GenBank databases. In addition, one can follow up with Northern analysis to determine if the chip probe recognizes more than one transcript.

6.6.4
Importance of Replicate Hybridizations and Validation of Significantly Changed Genes

There have been recent papers stressing the importance of replicates [51]. The output for a gene array experiment is a list of genes that are significantly induced or repressed. There are multiple methods to determine these genes. Some groups use an arbitrary cut-off. We have implemented a confidence interval described by Chen et al. [52]. However, application of a confidence interval when there are thousands of genes on a chip is accompanied by a certain probability of a chance occurrence of a gene appearing as induced or repressed. The probability of detecting a change in gene expression by random chance can be calculated by binomial probability [53]. For example, by random chance one would expect 240 outliers when performing a single hybridization using a chip with 2,000 genes; triplicate hybridizations reduces the expected number of random outliers to 0 where a gene must be significantly changed in all 3 hybridization. To be sure that changes in gene expression observed on the microarray are correct, it is important to validate changes in representative or important genes by northern blot or quantitative real time-PCR. In addition, it is also important to resequence and validate the original cDNA clone for verification of identification if possible.

6.6.5
Changes in Gene Expression versus Protein or Activity

One issue to consider with gene profiling data is how well the changes in gene expression reflect changes in protein. For example, it is possible that increased degradation of a protein would lead to increased RNA expression. In this case interpreting an increase in gene expression as an increase in protein would lead to the wrong conclusion. Even if gene expression and protein levels change in the same direction, there could be differences in the magnitude of the change. Furthermore, even measurements of protein levels may not tell the whole story. Recent studies have shown that protein targeting and post-transcriptional modification are important for understanding the activity of proteins [16, 54].

6.6.6
Bioinformatics or How to Tell the Forest from the Trees

As mentioned above, there are multiple methods to determine significantly changed genes within an experiment. One should implement a strategy to balance sensitivity with conservatism to reduce error. The most robust models require replication of both chips as well as biological samples.

Once the significantly altered genes from a single array or biological sample have been determined, higher order analyses may be used to visualize patterns or trends across multiple samples One popular form of analysis is clustering [55]. Clustering of genes or samples across a set of significantly changed genes or across genes of a similar function, may allow clear visualization of similarity of samples or activation/inactivation of a biological process. For example we clustered data derived from analysis of human idiopathic cardiomyopathic hearts using gene changes that were statistically significant. This visualization allowed us to determine those changes that were in common across patients that might be implicated in the disease process. In addition, we clustered the data across all the apoptotic genes on the chip (~250 genes). This approach indicated an obvious change in specific apoptotic signaling pathways [56]. Although in the microarray analysis many of the gene changes in this pathway were not statistically significant, we were able to observe statistically significant differences in expression of many of these genes using RT-PCR. Thus clustering of all the genes in a pathway can provide insights that might not be available if only a small fraction of the genes are included. However, with this approach it is essential to confirm changes in gene expression. After identification of a pathway by microarray data, one can then determine whether these changes occur in additional hearts. This approach uses microarray as an unbiased screening tool to identify pathways, but then confirms the importance of these pathways using other methods in a larger number of hearts. This type of approach may have application for identifying pathways that are important in cardioprotection.

6.7
Summary

Gene profiling by microarray is a powerful new technology that promises to provide new insights into genes involved in cardioprotection. At present, we have much to learn regarding gene expression and cardioprotection. Do all or most models of cardioprotection alter expression of a few common genes or are cardioprotective genes as diverse as cardioprotective agents? We know that genetic background or context can modify cardioprotection, but we do not understand this at the level of specific genes. We also know that depending on the level of expression a gene, such as PKC-ε can be cardioprotective or can induce hypertrophy. Gene profiling is an ideal technique to address these complex issues.

6.8
References

1 MURRY C, JENNINGS R, REIMER K. Preconditioning with ischemia: a delay of lethal cell injury in ischemic myocardium. *Circulation.* **1986**; *74*:1124–1136.

2 THORTON J, STRIPLIN S, LIU G, SWAFFORD A, STANLEY A, VAN WINKLE D, DOWNEY J. Inhibition of protein synthesis does not block myocardial protection afforded by preconditioning. *Am. J. Physiol.* **1990**; *259*:H1822–H1825.

3 DEINDL E, SCHAPER W. Gene expression after short periods of coronary occlusion. *Mol. Cell. Biochem.* **1998**; *186*:43–51.

4 GUO Y, JONES W, XUAN Y, TANG X, BAO W, WU W, HAN H, LAUBACH V, PING P, YANG Z, QUI Y, BOLLI R. The late phase of ischemic preconditioning is abrogated by target disruption of the inducible NO synthase gene. *Proc. Natl. Acad. Sci.* **1999**; *96*:11507–11512.

5 SHINMURA K, TANG X, WANG Y, XUAN Y, LIU S, TAKANO H, BHATNAGAR A, BOLLI R. Cyclooxygenase-2 mediates the cardioprotective effects of the late phase of ischemic preconditioning in conscious rabbits. *Proc. Natl. Acad. Sci.* **2000**; *97*:10197–10202.

6 XUAN Y, TANG X, BANERJEE S, TAKANO H, LI R, HAN H, QIU Y, LI J, BOLLI R. Nuclear factor-κB plays an essential role in the late phase of ischemic preconditioning in conscious rabbits. *Circ. Res.* **1999**; *84*:1095–1109.

7 DAWN B, XUAN X, MARIAN M, FLAHERTY M, MURPHREE S, SMITH T, BOLLI R, JONES W. Cardiac-specific abrogation of NF-kappaB activation in mice by transdominant expression of a mutant I kappa B alpha. *J. Mol. Cell. Cardiol.* **2001**; *33*:161–173,.

8 JONES S, GIROD W, PALAZZO A, GRANGER D, GRISHAM M, JOURD'HEUIL D, HUANG P, LEFER D. Myocardial ischemia-reperfusion injury is exacerbated in the absence of endothelial cell nitric oxide synthase. *Am. J. Physiol.* **1999**; *276*:H1567–H1573.

9 BELL R, YELLON D. The contribution of endothelial nitric oxide synthase to early ischaemic preconditioning: the lowering of the preconditioning threshold. An investigation in eNOS knockout mice. *Cardiovas. Res.* **2001**; *52*:274–280.

10 KANNO S, LEE P, ZHANG Y, HO C, GRIFFITH B, SHEARS Ln, BILLIAR T. Attenuation of myocardial ischemia/reperfusion by superinduction of inducible nitric oxide synthase. *Circulation.* **2000**; *101*:2742–2748.

11 JONES S, GIROD W, HUANG P, LEFER D. Myocardial reperfusion injury in neuronal nitric oxide synthase deficient mice. *Coron. Artery Dis.* **2000**; *11*:593–597.

12 PASS J, ZHENG Y, WEAD W, ZHANG J, LI R, BOLLI R, PING P. PKC-epsilon activation induces dishotomous cardiac phenotypes and modulates PKC-epsilon-RACK

interactions and RACK expression. *Am. J. Physiol.* **2001**; *280*:H946–H955.

13 DORN GN, SOUROUJON M, LIRON, T, CHEN C, GRAY M, ZHOU H, CSUKAI, M, WU G, LORENZ J, MOCHLY-ROSEN D. Sustained in vivo cardiac protection by a rationally designed peptide that causes epsilon protein kinase C translocation. *Proc. Natl. Acad. Sci.* **1999**; *96*:12798–12803.

14 CROSS H, MURPHY E, BOLLI R, PING P, STEENBERGEN C. Expression of activated PKC epsilon protects the ischemic heart, without attenuating ischemic H+ production. *J. Mol. Cell. Cardiol.* **2002**; *34*:361–367.

15 TAKEISHI Y, PING P, BOLLI R, KILPATRICK D, HOIT B, WALSH R. Transgenic overexpression of constitutively active protein kinase C epsilon causes concentric cardiac hypertrophy. *Circ. Res.* **2000**; *86*:1218–1223.

16 PING P, SONG C, ZHANG J, GUO Y, CAO X, LI R, WU W, VONDRISKI T, PASS J, TANG X, PIERCE W, BOLLI R. Formation of protein kinase C(epsilon)-Lck signaling modules confers cardioprotection. *J. Clin. Invest.* **2002**; *109*:499–507.

17 YANG Z, CERNIWAY R, BYFORD A, BERR S, FRENCH B, MATHERNE G. Cardiac overexpression of A1-adenosine receptor protects intact mice against myocardial infarction. *Am. J. Physiol.* **2002**; *282*:H949–H955.

18 MORRISON R, JONES R, BYFORD A, STELL A, PEART J, HEADRICK J, MATHERNE G. Transgenic overepxression of cardiac A(1) adenosine receptors mimics ischemic preconditioning. *Am. J. Physiol.* **2000**; *279*:H1071–H1078.

19 BLACK R, JR, GUO Y, GE Z-D, MURPHREE S, PRABHU S, JONES W, BOLLI R, AUCHAMPACH J. Gene dosage dependent effects of cardiac-specific overexpression of the A3 adenosine receptor. *Circ. Res.* **2002**:in press.

20 YANG X, LIU Y, SCICLI G, WEBB C, CARRETERO O. Role of kinins in the cardioprotective effect of preconditioning: study of myocardial ischemia/reperfusion injury in B2 kinin receptor knockout mice and kininogen-deficient rats. *Hypertens.* **1997**; *30*:735–740.

21 LI B, SETOGUCHI M, WANG X, ANDREOLI A, LERI A, MALHOTRA A, KAJSTURA J, ANVERSA P. Insulin-like growth factor-1 attenuates the detrimental impact of non-occlusive coronary artery constriction on the heart. *Circ. Res.* **1999**; *84*:1007–1019.

22 SHEIKH F, SONTAG D, FANDRICH R, KARDAMI E, CATTINI P. Overexpression of FGF-2 increases cardiac myocyte viability after injury in isolated mouse hearts. *Am. J. Physiol.* **2001**; *280*:H1039–H1050.

23 TROST S, OMENS J, KARLON W, MEYER M, MESTRIL R, COVELL J, DILLMAN W. Protection against myocardial dysfunction after a brief ischemic period in transgenic mice expressing inducible heat shock protein 70. *J. Clin. Invest.* **1998**; *101*:855–862.

24 WANG P, CHEN H, QIN H, SANKARAPANDI S, BECHER M, WONG P, ZWEIER J. Overexpression of human copper, zinc-superoxide dismutase (SOD1) prevents postischemic injury. *Proc. Natl. Acad. Sci.* **1998**; *95*:4556–4560.

25 KARLINER J, HOMBO N, EPSTEIN C, XIAN M, LAU Y, GRAY M. Neonatal mouse cardiac myocytes exhibit cardioprotection induced by hypoxia and pharmacological preconditioning and by transgenic overexpression of human Cu/Zn superoxide dismutase. *J. Mol. Cell. Cardiol.* **2000**; *32*:1779–1786.

26 CHEN Z, SIU B, HO Y, VINCENT R, CHUA C, HAMDY R, CHUA B. Overexpression of MnSOD protects against myocardial ischemia/reperfusion injury in transgenic mice. *J. Mol. Cell. Cardiol.* **1998**; *30*:2281–2289.

27 YOSHIDA T, MAULIK N, ENGELMAN R, HO Y, DAS D. Targeted disruption of the mouse Sod I gene makes the hearts vulnerable to ischemic reperfusion injury. *Circ. Res.* **2000**; *86*:264–269.

28 YET S, TIAN R, LAYNE M, WANG Z, MAEMURA K, SOLOVYEVA M, ITH B, MELO L, ZHANG L, INGWALL J, DZAU V, LEE M, PERRELLA M. Cardiac-specific expression of heme oxygenase-1 protects against ischemia and reperfusion injury in transgenic mice. *Circ. Res.* **2001**; *89*:168–173.

29 YOSHIDA T, MAULIK N, HO Y, ALAM J, DAS D. H(mox-1) constitutes an adaptive response to effect antioxidant cardiopro-

tection: A study with transgenic mice heterozygous for targeted disruption of the Hene oxygenase-1 gene. *Circulation.* **2001**; *1031*:1695–1701.

30 YOSHIDA T, MAULIK N, ENGELMAN R, HO Y, MAGNENAT J, ROUSOU J, FLACK JE 3RD, DEATON D, DAS D. Glutathione peroxidase knockout mice are susceptible to myocardial ischemia reperfusion injury. *Circulation.* **1997**; 96(9 Suppl II):216–220.

31 CHEN Z, CHAU C, HO Y-S, HAMDY R, CHUA B. Overexpression of Bcl-2 attenuates apoptosis and protects against myocardial I/R injury in transgenic mice. *Am. J. Physiol.* **2001**; *280*:H2313–H2320.

32 BROCHERIOU V, HAGEGE A, OUBENAISSA A, LAMBERT M, MALLET M, DURIEZ M, WASSEF M, KAHN A, MENASCHE P, GILGENKRANTZ H. Cardiac functional improvement by a human Bcl-2 transgenic mouse model of ischemia/reperfusion injury. *J. Gene Med.* **2000**; *2*:326–333.

33 XIAO C, HARA A, YUHKI K, FUJINO T, MA H, OKADA Y, TAKAHATA O, YAMADA T, MURATA T, NARUMIYA S, USHIKUBI F. Roles of prostaglandin I(2) and thromboxane A(2) in cardiac ischemia-reperfusion injury: a study using mice lacking their respective receptors. *Circulation.* **2001**; *104*:2210–2215.

34 MAEKAWA N, WADA H, KANDA T, NIWA T, YAMADA Y, SAITO K, FUJIWARA H, SEKIKAWA K, SEISHIMA M. Improved myocardial ischemia/reperfusion injury in mice lacking tumor necrosis factor-alpha. *J. Am. Coll. Cardiol.* **2002**; *39*:1229–1235.

35 KURRELMEYER K, MICHAEL L, BAUMGARTEN G, TAFFET G, PESCHON J, SIVASUBRAMANIAN N, ENTMAN M, MANN D. Endogenous tumor necrosis factor protects the adult cardiac myocyte against ischemic-induced apoptosis in a murine model of acute myocardial infarction. *Proc. Natl. Acad. Sci.* **2000**; *97*:5456–5461.

36 ADAMS KJ, SUETA C, GHEORGHIADE M, O'CONNOR C, SCHWARTZ T, KOCH G, URETSKY B, SWEDBERG K, MCKENNA W, SOLER-SOLER J, CALIFF R. Gender differences in survival in advanced heart failure: Insights from the FIRST study. *Circulation.* **1999**; *99*:1816–1821.

37 CROSS H, MURPHY E, KOCK W, STEENBERGEN C. Male and female mice overex-

pressing the beta(2)-adrenergic receptor exhibit differences in ischemia/reperfusion injury: role of nitric oxide. *Cardiovas. Res.* **2002**; *53*:662–671.

38 CROSS H, LU L, C S, PHILIPSON K, MURPHY E. Overexpression of the cardiac Na+/Ca2+ exchanger increases susceptibility to ischemia/reperfusion injury in male, but not female, transgenic mice. *Circ. Res.* **1998**; *83*:1215–1223.

39 GUERRA S, LERI A, WANG X, FINATO N, DILORETO C, BELTRAMI C, KAJSTURA J, ANVERSA P. Myocyte death in the failing human heart is gender dependent. *Circ. Res.* **1999**; *85*:856–866.

40 CAMPER-KIRBY D, WELCH S, WALTER A, SHIRAISHI I, SETCHELL K, SCHAEFER E, KAJSTURA J, P A, SUSSMAN M. Myocardial Akt activation and gender: increased nuclear activity in females versus males. *Circ. Res.* **2001**; *88*:1020–1027.

41 SIMONCINI T, HAFEZI-MOGHADAM A, BRAZIL D, LEY K, CHIN W, LIAO J. Interaction of oestrogen receptor with the regulatory subunit of phosphatidylinositol-3-OH kinase. *Nature.* **2000**; *407*:538–541.

42 GARRIDO C, GUXBUXANI S, RAVAGNAN L, KROEMER G. Heat Schok proteins: endogenous modulators of apoptotic cell death. *Niochem. Biophys. Res. Commun.* **2001**; *286*:433–442.

43 ARONOW B, TOYOKAWA T, CANNING A, HAGHIGHI K, DELLING U, KRANIAS E, MOLKENTIN J, DORN G. Divergent transcription responses to independent genetic causes of cardiac hypertrophy. *Physiol Genomics.* **2001**; *6*:19–28.

44 TONG H, IMAHASHI K, STEENBERGEN C, MURPHY E. Phosphorylation of glycogen synthase kinase-3β during preconditioning through a phosphatidylinositol-3-kinase-dependent pathway is cardioprotective. *Circ. Res.* **2002**; *90*:377–379.

45 HAQ S, CHOUKROUN G, KANG Z, RANU H, MATSUI T, ROSENZWEIG A, MOLKENTIN J, ALESSANDRINI A, WOODGETT J, HAJJAR R, MICHAEL A, FORCE T. Glycogen synthase kinase 3β is a negative regulator of cardiomyocyte hypertrophy. *J. Cell Biol.* **2000**; *151*:117–129.

46 MORISCO C, ZEBROWSKI D, CONDORELLI G, TSICHLIS P, VATNER S, SADOSHIMA J. The Akt-glycogen synthase kinase 3β

pathway regulates transcription of atrial natriuretic factor induced by β-adrenergic receptor stimulation in cardiac myocytes. *J. Biol. Chem.* **2000**; *275*:14466–14475.

47 HUANG T, CARLSON E, KOZY H, MANTHA S, GOODMAN S, URSELL P, EPSTEIN C. Genetic modification of prenatal lethality and dilated cardiomyopathy in Mn superoxide dismutase mutant mice. *Free Radic. Biol. Med.* **2001**; *31*:1101–1110.

48 LIU T, LAI H, WU W, CHINN S, WANG P. Developing a strategy to define the effects of insulin-like growth factor-1 on gene expression profile in cardiomyocytes. *Circ. Res.* **2001**; *88*:1231–1238.

49 HAGHIGHI K, SCHMIDT A, HOIT B, BRITTSAN A, YATANI A, LESTER J, ZHAI J, KIMURA Y, DORN G, MACLENNAN D, KRANIAS E. Superinduction of sarcoplasmic reticulum function by phospholamban induces cardiac contractile failure. *J. Biol. Chem.* **2001**; *276*:24145–24152.

50 NUWAYSIR E, BITTNER M, TRENT J, BARRETT J, AFSHARI C. Microarrays and toxicology: the advent of toxicogenomices. *Mol. Carcinogen.* **1999**; *24*:153–159.

51 LEE M-LT, KUO F, WHITMORE G, SKLAR J. Importance of replication in microarray gene expression studies: Statistical methods and evidence from repetitve cDNA

hybridization. *Proc. Natl. Acad. Sci.* **2000**; *97*:9834–9839.

52 CHEN Y, DOUGHERTY ER, BITTNER ML. Ratio-based decisions and the quantitative analysis of cDNA microarray images. *J. Biomed. Optics.* **1997**; *2*:364–374.

53 BUSHEL P, HAMADEH H, BENNETT L, SIEBER S, MARTIN K, NUWAYSIR EF, JOHNSON K, REYNOLDS K, PAULES R, AFSHARI CA. A Microarray Project System for gene expression experiment information and data validation. *Bioinformatics.* **2001**; in press.

54 MARX S, REIKEN S, HISAMATSU Y, JAYARAMAN T, BURKHOFF D, ROSEMBLIT N, MARKS A. PKA phosphorylation dissociates FKBP12.6 from the calcium release channel (ryanodine receptor): defective regulation in failing hearts. *Cell.* **2000**; *101*:365–376.

55 EISEN M, SPELLMAN P, BROWN P, BOTSTEIN D. Cluster analysis and display of genome-wide expression patterns. *Proc Natl Acad Sci USA.* **1998**; *95*:14863–14868.

56 STEENBERGEN C, AFSHARI CA, PETRANKA JG, COLLINS J, MARTIN K, BENNETT L, HAUGEN A, BUSHEL P, MURPHY E. Alterations in Apoptotic Signaling in Human Idiopathic Cardiomyopathic Hearts in Failure. *Am J Physiol.* **2002**; in press

7
Pitfalls Associated with cDNA Microarrays – A Cautionary Tale

Alfred C. Aplin and Charles E. Murry

7.1
Introduction

The promise of having an entire complement of expressed cDNAs (the transcriptome) of an organism present on a single array leads to many new avenues in genetics, development, and disease pathogenesis research [1–3]. However, in microarray examination of gene expression, it is critical to have a high degree of confidence in the array itself, especially when using the array technology for "gene discovery" experiments where the only identifying marker for a given sequence is an expressed sequence tag (EST) accession number. Similarly, when array technology is used to define an open-ended molecular phenotype, one must be absolutely positive that the cDNAs present on the array are what they claim to be, and this gets to the heart of quality control issues at the array manufacturer. One major problem with using microarrays to analyze the relative expression level of thousands of cDNAs is that it is impractical to cross check all of or even most of the data. If fact, one of the main attractions of using microarrays is the ability to examine expression levels of more cDNAs than possible by any other means.

The following account serves to illustrate some important points when considering whether or not to utilize microarray technology and should help to point out possible problems with the manufacture of arrays. Although our experience is specific to the ResGen gf300 rat cDNA array marketed by Invitrogen, Inc. the issues and problems we encountered are applicable to any microarray technology. We believe that the wider scientific community may have an interest in knowing about this particular array manufacturing problem, because these filters are in wide use in studies of gene expression [4–6]. At the very least, our experience should serve as a warning to anyone using microarrays that proper experiments need to be done to confirm a subset of the array results.

7.2
Phase 1: The Original Experimental Design

Our laboratory recently undertook a series of array-based experiments designed to identify cDNAs that were expressed in different primary cell isolates from the cardiovascular system. Our aim was to profile the complement of cDNAs expressed in cultured rat cardiomyocytes, cardiac fibroblasts, and aortic smooth muscle cells, all isolated from neonatal rats. We hoped to generate a list of marker genes that would make up a molecular phenotype for each of the cardiovascular cell types. Next, we planned to use this information to assay for any "transdifferentiation" events that might take place after overexpression of cardiac-specific transcription factors GATA4, MEF2C, and Nkx2.5 [7–9] in fibroblasts or smooth muscle cells. Furthermore, this approach was expected to identify novel endogenous targets of these transcription factors.

In order to have the most reliable data possible, we controlled for biological variation by independently preparing RNA from three separate primary cell isolations of each cell type. For each array hybridization we used fresh arrays purchased from ResGen. In this way we could control for differences in culture conditions from month to month, and from various lots of printed arrays. We used fresh arrays for every measurement because we had previously determined that ResGen arrays can not be reliably stripped and re-probed.

Our first set of array experiments were designed to establish a "baseline" of gene expression in each to the three cell types. Nine independent RNA isolations were performed, three from each of three cell types, neonatal cardiomyocytes, cardiac fibroblasts, and aortic smooth muscle cells.

We chose to employ microarrays containing 5147 cDNAs spotted onto nylon filters marketed by Invitrogen, Inc. and manufactured by ResGen, Inc. (Fig. 7.1). In the fall of 1999 we purchased a set of filters and the associated probe labeling reagents and hybridization buffers to expedite our studies. These filters were chosen for several reasons, including the low initial startup cost, the fact that the reagents and methods for probe preparation and array visualization were already present in our lab, and that at the time, ResGen had by far the largest commercially available collection of rat cDNAs present on an array. Over the course of a year, we performed a series of 16 hybridizations to the gf300 arrays and analyzed the output data with custom spreadsheets developed in our lab.

Total RNA was extracted from these three primary isolates of neonatal rat cardiovascular cells and used to generate probes for the arrays. Rat gf300 arrays were hybridized to ^{33}P-labeled reverse transcribed probes and all arrays were scanned on a phosphorimager. The images were imported to Pathways 2.0 software package (ResGen, Inc.). This software was used to identify each of the cDNAs spotted onto the array and generate a report on the background and signal intensity of each spot. The raw data reports were exported into Microsoft Excel and processed through a series of calculations to subtract background, and normalize each signal to the average signal of the array. We next took the normalized array measurements from three independent hybridizations, averaged the signal intensities for each

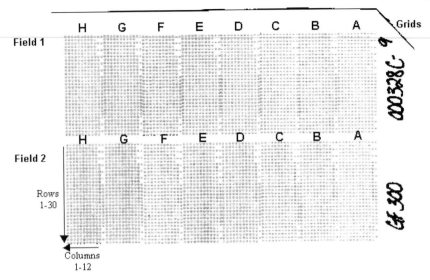

Fig. 7.1 The gf300 array from Research Genetics. 5,123 individual cDNAs are spotted onto the 5 cm×7 cm nylon membrane. Each membrane is cut in the upper right corner for orientation. The cDNA spots are arranged in two fields, 1 and 2, each field containing 8 individual grids laid out right to left, A through H. Each grid is made up of individually spotted cDNA clones arranged in 12 columns of 30 rows. Along with the 1,683 known genes and 3,440 ESTs present on the gf300 array, a series of controls are present including 228 total genomic DNA spots and 228 beta actin spots used for normalization of the signal and for orientation and alignment of the membrane when using the software from Research Genetics. The spacing between each spot is ~750 microns from center to center.

cDNA and calculated the standard error of the mean. Differential expression between the RNA isolated from two cell types was determined by performing an unpaired Student's T-test for each spot, and $p < 0.05$ was taken as statistically significant. Each of the three cell types under study was analyzed in a pair-wise manner with the others.

We identified several cell-type specific cDNAs from our array analysis and published the results as an abstract at the 2000 meeting of the American Heart Association [10]. Many of the genes we identified as being differentially expressed in a cell-type specific manner agreed with previously published and well known cardiovascular cell markers. In RNA isolated from cardiomyocytes, we detected specific expression of transcripts that encode atrial natriuretic factor, desmin, and cardiac isoforms of troponin and myosin [11–14]. In smooth muscle cells, transcripts for the SM22 protein and osteopontin were identified, again agreeing with well described expression markers for this cell type [15, 16]. These results gave us confidence that the microarray methodology we were using was working as predicted, and that the additional novel transcripts we identified as cell-type specific were also correct. Interestingly, we had identified multiple ESTs specifically expressed in cardiomyocytes and aortic smooth muscle, and these transcripts were to be the focus of further study in our lab.

7.3
Phase 2: Validation and Troubleshooting

While preparing a manuscript for publication, and nearing the end of our array experiments, we undertook a series of experiments aimed at confirming the array data with more "traditional" means of northern blotting and RT-PCR [17]. Confirming array cDNA expression data by other independent means is an important control whenever arrays are being used [18]. We selected six cDNAs that appeared to be highly specific for one or more of the three cell types under study and ordered the corresponding bacterial glycerol stocks from ResGen. Clones were identified by accession numbers and were sequence-verified upon arrival in our lab. Oligonucleotide primers specific to each cDNA were designed for use in PCR so

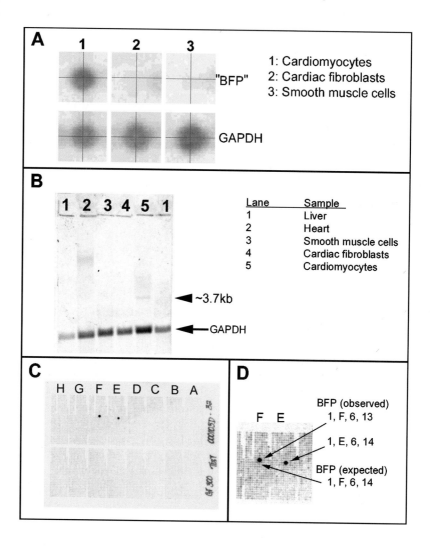

that the resulting products would only contain the insert sequences from each plasmid. These PCR products were gel purified and used to probe northern blots, and later, for the dot blots described below.

In three out of six cases, Northern blotting failed to confirm our array data. One of the cDNAs did not hybridize to RNA on a northern blot with the same cell type distribution as indicated on the array. Another two did not hybridize to anything at all on a Northern blot, even though the genes appeared to be highly expressed on our arrays (Fig. 7.2). RT-PCR, using the same RNA used for array analysis, was also used in an attempt to confirm the expression results seen with the array experiments. Again, in the same three of the six cases, the expression profile we expected was not observed, calling into question the validity of the array results.

Our initial inquiries with the technical support staff at ResGen gave us no indication of how our data analysis could be flawed. The arrays themselves were said to be highly reliable because of the quality control measures employed by ResGen. Additionally, there had been no reports from any of the other customers using these particular arrays that would have suggested any problems with the identity of the spots on the array. After several attempts to validate the clones in question with northern blotting and RT-PCR, we decided to undertake a series of "dot blot" hybridizations to check whether or not the cDNAs that we had selected were indeed present on the gf300 arrays in the grid locations indicated by ResGen.

Our technique was to pick a single cDNA in question, generate a clonal probe by PCR-amplifying the insert, gel purifying the product, labeling it with ^{33}P-dCTP, and hybridizing this product to a fresh gf300 array. Our expectation was

◀────────────────────────────────

Fig. 7.2 A Individual spots representative of hybridization signals seen on the gf300 array corresponding to the location of the Brain Finger Protein (BFP, Accession number AA997188), and glyceraldehyde-3-phosphate-dehydrogenase (GAPDH, accession number AA924111). These images represent the signals seen on three individual arrays hybridized to cardiomyocyte, cardiac fibroblast, and smooth muscle cell probes. A strong BFP hybridization signal is evident with the cardiomyocyte but not with the fibroblast or smooth muscle probes. **B** Northern blot with ~20 µg of total RNA from the indicated samples loaded in each lane hybridized to the BFP probe. GAPDH was included in the hybridization as a control. The ~1.3 kb GAPDH transcript is indicated with an arrow, and the expected location of the BFP transcript at ~3.7 kb is indicated with an arrowhead. Despite the intense signal present at the BFP location on the array when probed with cardiomyocyte RNA, no significant BFP expression was seen by northern blotting. **C** Dot blot using a gf300 test microarray hybridized to the same BFP probe that was used on the northern blot shown in panel A. Two high intensity signals are observed on the array after hybridization instead of a single spot at location field 1, grid F, column 6, row 14 expected for the BFP clone. The observed positions of hybridization signal are 1, E, 6, 14 and 1, F, 6, 13. The same result was obtained from similar arrays from three independent printing lots. Upon sequence verification, the clone at position 1, F, 6, 14 was found to be an unidentified EST unrelated to brain finger protein. **D** Close up of the E–F section of the array showing the location of the two hybridization signals.

that a single clone would produce a single, high intensity hybridization spot on the gf300 array, and furthermore, that the spot would be present in the predicted location on the array grid. Arrays hybridized in this manner were processed according to the standard hybridization and washing protocol for use on a complex probe. This approach, which is similar to a cDNA dot blot, was undertaken on six individual cDNAs. (As an aside, each gf300 array cost ~$1000.00 and we had no expectation that the filters would be re-usable after such a test. The decision to pursue this line of inquiry thus was not taken lightly.)

One example of our array dot blot results is shown in Fig. 7.2 C. We selected one of the cDNAs whose expression by northern did not match the array result (GenBank accession number AA997188, Brain Finger Protein [19]) and hybridized it to a fresh gf300 array. To our surprise, the result gave two discreet hybridization signals on the array grid (Fig. 7.2), and both were in the wrong grid location. The two other cDNAs that were problematic by northern and RT-PCR were similarly problematic on array dot blots. One of the clones hybridized to a different location than expected, and the other clone did not hybridize to any grid location on the array. The three other cDNA clones whose northern analysis agreed with the original array data appeared in the predicted location as a single intense hybridization signals by array dot blot.

7.4
Phase 3: "Postmortem" Analysis at ResGen

As a result of our dot blot experiments, we feared that a large subset of the cDNA spots on the gf300 arrays were incorrectly labeled, and worse, that there would be no way to decide which spots those were. In order to figure out how widespread the misidentification of cDNAs on the gf300 arrays may be, we suggested that ResGen sequence verify all 5147 cDNAs present on the filters and correct the database as needed. At the time, our suggestion was met with skepticism by Res-Gen, because they perform routine quality control to ensure that individual cDNAs are spotted at the correct location on the gf300 array. The quality control methods used by ResGen were designed to identify large scale misalignment of cDNA spots on each lot of printed arrays. A sample from every printed batch of arrays was hybridized to a complex probe that give a predictable pattern of hybridization, and that the filters in question had all passed this quality control test. Additional quality control tests done at ResGen would have revealed if individual plates used by the robotic printer to spot the arrays had been inserted into the robotic printer inappropriately.

We sent the data files and images of our dot blot tests to ResGen and asked that they verify the identity of the clones in question. Over the course of the next two months, ResGen was persuaded to sequence the three clones we had identified as incorrectly spotted on their arrays. They determined that in fact, the clones we questioned were misidentified on the gf300 array. Their interpretation was that some kind of rare spotting error had become incorporated into these arrays, but that the error was restricted to the three clones that we had, unluckily, chosen for detailed study. At that point, ResGen replaced all of our arrays with fresh arrays from a different lot that were presumed to be correct based on their internal quality control tests. Instead of using these fresh arrays for our cell type comparison experiments, we repeated the dot blot experiments with the three cDNAs in question and again found they were not at their predicted locations. This indicated that the gf300 arrays that had been manufactured over a two year period incorporated the same error in printing, and that this error had not been detected by the standard quality control protocols followed at ResGen.

Although still unwilling to sequence all 5147 cDNAs on the array, ResGen was eventually persuaded to sequence a larger subset of cDNAs. From this larger sample they would determine if there was indeed a larger problem that warranted sequencing the entire clone set used to print the gf300 arrays. The limited set of cDNAs were sequence-verified from the plates used to print the arrays, and compared to the sequence of the original stock plates and to the expected GenBank sequences. We submitted a list of 181 cDNA clones from the set of 5147 that were spotted onto the gf300 array. The clones we selected were comprised of cDNAs that appeared interesting to us based on our original data analysis and some others cDNAs picked from across the array at random. The set of cDNA locations were submitted to ResGen in July of 2001.

The first 90 out of 181 (50%) cDNAs returned to us appeared correct, i.e., the identity of the cDNA spotted onto the array matched the predicted cDNA sequence as reported in GenBank for that clone. Furthermore, these clones were spotted at the correct grid location on the gf300 arrays. However, over the course of the next several weeks, a widespread and disturbing pattern of errors was present in all remaining clones. Of the 181 clones we questioned 91 (50%) had some type of error. 53 (29%) were found to be mis-identified on the array, meaning that the cDNA actually spotted onto the array did not match the expected cDNA in the database for this array. At 19 (10%) of the locations we queried, two different cDNAs had been spotted, only one of which was correct. At another 7 (4%) spot locations, there were two or three different clones, none of which were expected at that location. Finally, at 12 (6%) spot locations, the cDNAs were unable to be sequenced at ResGen due to non-viable bacterial stocks.

Some of the cDNAs spotted onto the gf300 array were listed in the set of clones used to produce the gf301 array, a different microarray product containing a different set of rat cDNAs. Although we did not check any clones on the gf301 array, the results of our inquiry also calls into question the identity of clones on the gf301 as well. Several of the cDNAs used to generate our arrays were apparently mis-identified by the cDNA source clone bank prior to being purchased by ResGen for use on their arrays. Unfortunately, the sequences of the entire clone set was not confirmed independently by ResGen prior to printing the gf300 arrays, so that errors in the original clone database carried over to the set of clones used for array printing.

Extrapolating to the entire complement of cDNAs spotted onto the gf300 filter, it is possible that 2,265 of the 5,147 cDNAs present on the array may be misidentified and 308 additional clones may be non-verifiable. To be sure, having an array of even 2,573 cDNAs can be very useful for genomics research. The rub is that one needs to know which loci are reliably identified and which are not. This information is currently not available.

7.5
Phase 4: Future Directions

ResGen has now agreed on the scope of the problem and has begun a large-scale effort to correct the identity and spot locations of the cDNAs present on the gf300 array. Recently, they have discontinued the sale of individual rat cDNA clones and added some information as to the troubles of "misalignment" of their gf300 arrays on the company web site. However, a general recall of the gf300 rat arrays not been undertaken. In fact, ResGen continued to market, produce, and sell the gf300 filter as a "sequence verified" cDNA microarray as of March of 2002, a full nine months after the problem was identified by us and six months after the problems were verified by ResGen. Currently, ResGen is attempting to perform a complete sequence verification of all 5,147 cDNA clones used to manufacture the gf300 arrays.

The final results of our work with the gf300 arrays will not be known until every spot is sequence-verified from the stock plates that were used to print them. In cases where two or more cDNAs are present on a single spot, it will be impossible to use the array data for relative expression levels of either of the cDNAs. We are hopeful that the number of array spots that fall into this category is a small percentage of the total. However, from our limited sample, 14% of the spot locations contain two or three different cDNAs, which extrapolates to over 700 spots on the final array.

The experience we have had with microarray analysis of gene expression has led to several changes in how we are carrying out our experiments. Because large-scale rat arrays are not available from other vendors, we have opted to switch to the mouse and repeat our entire set of experiments. This will permit the use of our institution's newly developed array center. Other issues and ideas are discussed below. We feel that the wider scientific community may have an interest in knowing about this particular array spotting problem as these filters have been used in published studies of gene expression [4–6].

7.6
Caveat Emptor: Suggestions for New Array Users

The following suggestions may help a research scientist who is new to the field of microarray analysis of gene expression to avoid similar problems that can lead to a substantial loss of research time. Our experience is with cDNA arrays that are made by spotting individual cDNAs, made from PCR fragments onto a nylon membrane. The source of our troubles stemmed from a mix-up of the cDNA clones prior to printing the arrays. Oligonucleotide based arrays will not have this type of problem, because the oligos are synthesized directly on the surface of the chip. However, the curation of DNA sequences used to design oligo based arrays is just as important as the curation of individual cDNA clones that are spotted onto an array. A recent recall of arrays manufactured by Affymetrix, Inc. was due to errors in the publicly archived genetic information that went into producing the chips, eventually leading to mistakes in the oligo design used for these chips [20, 21].

As with any scientific experiment, when planning experiments using microarrays it is important to control for experimental, biological, and technical errors. With many arrays having over 10,000 individual cDNA spots, it is impossible to independently verify each result without a gargantuan expenditure of resources and time. The point of having an array of cDNA spots in a small area is to bring the possibility of examining whole transcriptomes from a cell or tissue at once. For this reason, it is vitally important that the array manufacturer takes the necessary steps to ensure the identity of every sequence and the location of each on their arrays. It is equally important that the user understands the specific quality control and testing procedures that are utilized in the array production process. In this respect, purchasing microarrays, even from an in-house distributor based at your university or within your company, is different than purchasing a particular chemical reagent where it would be possible to verify the quality of the reagent with a simple test or two, such as checking the pH. With microarrays, it is not practical for you to do any extensive quality control yourself so you must be confident that the manufacturer is being exceedingly careful of all aspects of microarray fabrication. In our own experience, this sadly was not the case.

Based on our experience, we believe it is of the utmost importance to have at least the following information prior to investing the substantial time and money required to start using array technology. While it is often difficult to obtain the specifics from an array manufacturer, one should demand as much information as possible. Avoid doing business with a vendor that does not give you concise, accurate, and believable answers to the following. These questions are best answered by the director of quality control.

1. How were the cDNAs prepared?
   It is standard practice to use multi-well plates to maintain individual cDNAs in the form of plasmids or PCR products that are used to print out spots on nylon filters or glass slides, although the specific steps may vary from manufacturer to manufacturer. Manipulations of liquids are carried out by a robotic instrument that physically interacts with each sample, repetitively transferring a portion of the original cDNA or glycerol stock from each well to a replicate plate that serves as a template for a PCR reaction. The replica plate is cycled through standard PCR conditions and the resulting PCR products are picked up by a robot and spotted onto the surface of the array. There are several steps during the array printing process that can lead to spotting errors. The individual PCR products can become cross-contaminated with one another during

the PCR process. This could be due to misalignment of the individual template plates, bent pins on the robotic transfer head, or incomplete washing of the pins between cycles. Good quality control practices will reveal the existence of either of these problems, and it is standard practice to check the integrity of the pins. What is important is that at some point, the PCR products are checked for cross-contamination.

2. How are the cDNA clones curated, maintained, sequenced, and checked?
Another problem can arise when the original cDNA clones are misidentified or mislabeled at the source. The only way to asses the true quality of a cDNA array is to have verified the identity of each clone that is spotted on the filter. This needs to be done both with the actual plates used in array printing and on the master set of glycerol stocks. Ultimately, every cDNA used to spot an array must be sequence verified. As our experience showed, it is not sufficient to show consistent results with a complex probe, because this failed to detect the arrays being wrong from the beginning.

3. How often are clones cross-checked with the database and to each other?
Some percentage of the clones from an array need to be selected randomly and sequence-verified for fidelity to the original plate, and to the database. Ideally, this would be a random selection of spots selected from each printed lot of the arrays and should vary from time to time.

4. How does the robot manipulate each array slide, or filter?
Is there a micropipette or capillary transfer, how is the tip cleaned between spots, how many pins are actually employed for each array grid, how is the pin alignment checked and how often?

5. What are the controls for batches of glass slides, nylon membranes etc?
This is perhaps the most critical aspect of array manufacture, and should be completely understood prior to purchasing a given array. Are random arrays pulled from each lot for testing or are the first and last arrays from a printing run always used? How many individual tests does each array undergo? What are the exact conditions of quality control testing and do they closely match your own hybridization conditions? How much error can be tolerated for an array to "pass"? When a test array is visualized, what criteria are used to compare one lot to a different lot?

6. What probes are used as the controls for a given lot of array filters?
Ideally, a series of arrays would be pulled from each lot and hybridized to a set of complex probes. The critical issue here is what the probes consist of. Is the test probe ever varied? How are the test probes themselves generated and tested? If the same set of quality control spots are being checked each time, then by definition the other spots on the array are never checked. Additionally, the arrays tested should be randomly selected from throughout the printing run and not just always be the first or last (or both) arrays from each lot.

7. What is the mechanism for reporting errors and additions to you?
Is there a web site that is updated with information about the arrays? Will you receive notice of problems or updates via email, fax, phone, sales representative, or will you have to search for updates and corrections to the array data-

bases yourself? A responsible manufacturer should notify the user any time a problem is detected.

8. What is the mechanism for reporting errors to the company?

Do you have a direct contact other than the sales representative? Is there a specific person who you can call or send an email to? Ideally there will be a company contact that is a supervisor who has the complete knowledge of your array product and not a group of technical support staff each with varying knowledge about the arrays. It would be a good idea to check out the quality of the tech support by calling the toll free number and asking some of the above questions prior to purchasing an array.

9. What is explicitly stated in the guarantee?

Will arrays be replaced and under what conditions, will allowance be made for personnel costs and the costs of all associated reagents or just for the arrays themselves? Who is responsible for making the decision to reimburse you for a defective array and how will a determination be made?

7.7
Final Thoughts

The main problem with using arrays to analyze the relative expression level of thousands of cDNAs is that it is impractical to cross check all of or even most of the data. If fact, one of the main attractions of using microarrays is the ability to examine expression levels of more cDNAs than possible by any other means. It is critical that the cDNAs on the array are what they claim to be, and this gets to the heart of quality control issues at the array manufacturer. Perhaps in no other molecular endeavor does a user have to place so much trust in an outside manufacturer to obtain meaningful results. As we found through our own experiments, this is not a perfect, "off the shelf" technology. There is a tendency to place more confidence in arrays than they may merit.

Although our experience is specific to the ResGen gf300 rat array marketed by Invitrogen, the issues and problems we encountered are applicable to any cDNA microarray technology. We would like to point out that the underlying technology behind robotic spotting of cDNA clones to nylon filters and subsequent hybridization methodologies is generally sound. We have only examined a subset of cDNAs on the gf300 array. Although statistically unlikely, it is formally possible that full sequence verification may identify a significantly smaller or larger set of misidentified cDNAs on the gf300 arrays than we predict based on simple extrapolation. Additionally, ResGen produces other arrays representing cDNAs from human, mouse, and yeast and we have no evidence for errors in any of these products.

Our experience left us with the difficult position of having detailed data on over 5,100 loci but not being able to unequivocally identify any of these clones. Among other problems, this made grouping of functional categories of gene products impossible. Our experience with the gf300 rat cDNA Arrays from ResGen has ultimately led to our changing our experimental system from rat to mouse, necessitating many months of set up work and the purchase of arrays from a different source. Furthermore, personnel costs in excess of $100,000 were incurred, and approximately 18 months of research had to be abandoned. It is our hope that array printing and misidentification errors will be identified and corrected prior to the widespread use of the arrays in scientific experimentation, and as the technology of array manufacturing progresses, methods to validate the results of array experiments will continue to evolve.

7.8
Acknowledgements

We would like to thank Isa Werny and Ronald Hanson for their expert advice and technical assistance in carrying out these experiments. This work was supported in part by NIH grants, R01 HL61553, PO1 HL-03174, and NSF grant ERC-9529161.

Note added in proof: As of October, 2002 ResGen has discontinued sale of the gf300 rate array.

7.9
References

1 SCHENA, M., SHALON, D., DAVIS, R. W., BROWN, P. O., Quantitative monitoring of gene expression patterns with a complementary DNA microarray. *Science* **1995**, *270*, 467–470.

2 PAONI, N. F., LOWE, D. G., Expression profiling techniques for cardiac molecular phenotyping. *Trends Cardiovasc. Med.* **2001**, *11*, 218–221.

3 HWANG, J. J., DZAU, V. J., LIEW, C. C., Genomics and the pathophysiology of heart failure. *Curr. Cardiol. Rep.* **2001**, *3*, 198–207.

4 BALASUBRAMANIAM, J., DEL BIGIO, M. R., Analysis of age-dependant alteration in the brain gene expression profile following induction of hydrocephalus in rats. *Exp. Neurol.* **2002**, *173*, 105–113.

5 MZIAUT, H., KORZA, G., ELKAHLOUN, A. G., OZOLS, J., Induction of stearoyl CoA desaturase is associated with high-level induction of emerin RNA. *Biochem. Biophys. Res. Commun.* **2001**, *282*, 910–915.

6 IRWIN, L. N., Gene expression in the hippocampus of behaviorally stimulated rats: analysis by DNA microarray. *Brain Res. Mol. Brain Res.* **2001**, *30*, 163–169.

7 KUO, C. T., MORRISEY, E. E., ANANDAPPA, R., SIGRIST, K., et al., GATA4 transcription factor is required for ventral morphogenesis and heart tube formation. *Genes Dev.* **1997**, *11*, 1048–1060.

8 LIN, Q., SCHWARZ, J., BUCANA, C., OLSON, E. N., Control of mouse cardiac morphogenesis and myogenesis by transcription factor MEF2C. *Science* **1997**, *276*, 1404–1407.

9 LYONS, I., PARSONS, L. M., HARTLEY, L., LI, R., et al., Myogenic and morphogenetic defects in the heart tubes of murine embryos lacking the homeo box gene Nkx2-5. *Genes Dev.* **1995**, *9*, 1654–1666.

10 APLIN, A. C., ADAMS, L., SCHWARTZ, S. M., MURRY, C. E., Defining Molecular Phenotypes of Cardiovascular Cells by Microarray Analysis. *Circulation* **2000**, *102, Suppl.,* Abs. #139.

11 FLYNN, T. G., The elucidation of the structure of atrial natriuretic factor, a new peptide hormone. *Can. J. Physiol. Pharmacol.* **1987**, *65*, 2013–2020.

12 van Groningen, J.J., Bloemers, H.P., Swart, G.W., Rat desmin gene structure and expression. *Biochim. Biophys. Acta.* **1994**, *1217*, 107–109.

13 Adamcova, M., Pelouch, V., Isoforms of troponin in normal and diseased myocardium. *Physiol. Res.* **1999**, 48, 235–247.

14 Ruppert, C., Kroschewski, R., Bahler, M., Identification, characterization and cloning of myr 1, a mammalian myosin-I. *J. Cell Biol.* **1993**, *120*, 1393–1403

15 Solway, J., Seltzer, J., Samaha, F.F., Kim, S., et al., Structure and expression of a smooth muscle cell-specific gene, SM22 alpha. *J. Biol. Chem.* **1995**, *270*, 13460–13469.

16 Giachelli, C., Bae, N., Lombardi, D., Majesky, M., Schwartz, S., Molecular cloning and characterization of 2B7, a rat mRNA which distinguishes smooth muscle cell phenotypes in vitro and is identical to osteopontin (secreted phosphopro-tein I, 2aR). *Biochem. Biophys. Res. Commun.* **1991**, *177*, 867–873.

17 Sambrook, J., Fritsch, E.F., Maniatis, T., Molecular Cloning, a laboratory manual. Cold Spring Harbor Laboratory Press, **1989**.

18 Jiang, L., Tsubakihara, M., Heinke, M.Y., Yao, M., et al., Heart failure and apoptosis: electrophoretic methods support data from micro- and macro-arrays. A critical review of genomics and proteomics. *Proteomics* **2001**, *1*, 1481–1488.

19 Inoue, S., Orimo, A., Saito, T., Ikeda, K., et al., A novel RING finger protein, BFP, predominantly expressed in the brain. *Biochem. Biophys. Res. Commun.* **1997**, *240*, 8–14.

20 Marshall, E., Affymetrix Settles Suit, Fixes Mouse Chips. *Science* **2001**, *291*, 2535.

21 Knight, J., When the chips are down. *Nature* **2001**, *410*, 860–861.

8
Application of Biologic Skepticism to Analysis of Expression Phenotypes

LAWRENCE ADAMS, ROGER E. BUMGARNER, and STEPHEN MARK SCHWARTZ

It is all too obvious that expression arrays have been better at producing a 1950's version of the Sears catalog than at constructing an interactive 2002 website. This review will suggest criteria for more useful analysis and publication.

8.1
What Is an Array?

As Fig. 8.1 shows, an expression array is simply a list of values representing the amounts of each mRNA species found in a sample. This concept greatly simplifies the usual red-green display and points out that "arrays" are a simple list of RNA quantities resulting from any measurement technique. Tab. 8.1 shows six different RNA detection methods used to quantify large sets of RNA expression values.

Finally, although not usually thought of as an "array", methods derived from RNA-based representational difference analysis (RNA-RDA) subtraction suppression hybridization (SSH) [1], use selective forms of PCR to eliminate those sequences found in common between two cDNA collections [2].

In the mathematical sense, SSH does produce an array of differentially expressed sequences. This sort of "all or none" expression data provides a compliment to array hybridization methods since the normalization step means that SSH may even detect low abundance, but differentially expressed, sequences. These genes can then be quantified by real time PCR or used to construct expression arrays [3]. Like SAGE or bend sequencing, SSH is unbiased; however, SSH requires extensive sequencing to identify the subtracted clones.

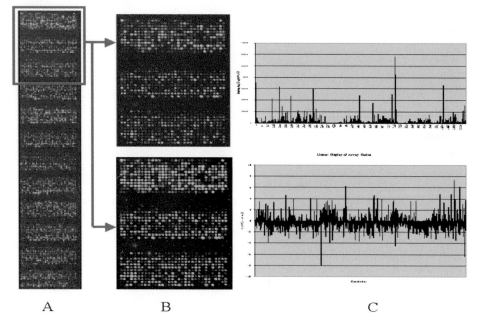

A   B   C

Fig. 8.1 Alternate views of 2-color array data. **A)** A traditional display of DNA array data – data presented as a false color image overlay of the Cy3 and Cy5 data channels. **B)** Both the top and bottom blocks show an expanded view of the same region of A. The difference between the two figures is the result of manipulation of the color mapping functions – e.g. determining what level of green or red is displayed for a given signal intensity in the Cy3 and Cy5 channels. Note that it is possible to present the reader with dramatically different views of the data simply by manipulation of the color mapping functions. **C)** A linear display of array data. The top graph displays the intensity (proportional to expression level) while the bottom graph displays the Cy3/Cy5 ratio for each spot shown in B. The ordered pairs of (intensity, ratio) data contain similar information content to that in either view in B. However, the representation of the data as ordered (intensity, ratio) pairs provides a less subjective view than a false color image.

8.2
Applications of Array Analysis

8.2.1
Catalogs by Function

With a long enough list of differentially expressed genes, it is almost impossible not to find 5–10 on which an author can comment. Moreover, these subjective evaluations are often biased by predefined functions, such as "inflammation" or "growth". Imputing function from changes in RNA level is dangerous. Even if one accepts the assumptions that regulation expression correlates with function, small changes in expression of a group of functionally-related genes may have little to do with how the cell is using the proteins transcribed by these genes.

Tab. 8.1 How are array data obtained?

SAGE [22, 23]	Uses concatemers of short sequences to identify genes represented in a collection of mRNA.	
	Does not require predefined sets of genes; can identify previously unknown sequences; can identify genes by multiple 3′ exons; is totally unbiased by predefined ideas.	Uses PCR-based amplification; only minimal sequence is provided; may detect but not measure frequency of low abundance sequences; existing database of sequence tags exceeds the likely number of total gene transcripts (200,000 vs. 35,000).
Bead-based sequencing [24]	Uses a bead-based sequence to simultaneously provide large numbers of sequences {http://www.lynxgen.de/}.	
	Does not require predefined sets of genes; can identify previously unknown sequences; is totally unbiased by predefined ideas.	Methodology is new and untested; even at 10^6 sequences per run, number of "hits" may not measure frequency of low abundance sequences.
High throughput sequencing of libraries	Although sequencing of whole libraries is impractical, screening methods can eliminate redundancy. For example, PCR Select® {www.clontech.com/products/literature/pdf/brochures/PCR-SelectBR.pdf} is often used to decrease numbers of clones to sequence. Detected sequences can then be array spotted and analyzed for abundance.	
	Identifies new sequences; accurate sequence identification; detection of splice forms; unbiased; able to measure abundance even of rare sequences if enough sequencing is done.	Expensive; depends on accuracy of library construction and amplification.
Sequence analysis by mass	This method, also called gene calling, [25–27] combines the use of a large number of restriction fragments with a database of patterns generated from known transcripts {www.curagen.com/technology/frametech.htm}. A computer analysis is used to generate the set of array data much as mass spectrometry fragment lengths are used to identify proteins [28].	
	Unbiased; able to measure abundance even of rare sequences.	Expensive; requires sophisticated information technology; abundance of rare species very dependent on numbers of fragments and therefore quantity of RNA studied; cannot identify unknown sequences.
cDNA Arrays	Most common method now used.	
	Rapid; able to analyze large numbers of sequences at one time; becoming inexpensive; applicable to organisms for which little or no genomic data exist.	Need to store and validate large numbers of cDNA clones; cross hybridization problems; hybridization signal may not prove target sequence is actually present; rapidly changing technologies; can only analyze what is on the physical array.

Tab. 8.1 (continued)

Oligonucleotide DNA Arrays	Several different kinds of oligonucleotide arrays are available including: Affymetrix© arrays on which oligonucleotides are synthesized directly on a silicon chip via a photo-lithographic process {www.affymetrix.com}; Agilent© inkjet oligo arrays on which oligos are synthesized by spraying reagents onto a glass surface with an inkjet printing head {www.chem.agilent.com/Scripts/Icol.asp?iInd=2}; Arrays constructed by depositing pre-synthesized oligonucleotides onto glass or membrane surfaces probably will be the major technology as costs drop {www.operon.com}.

Oligomers can be custom designed and theoretically will be valid to provide highly accurate data; differences in oligomeric hybridizations may be used to detect mutant or spliced forms; because oligomers can be synthesized as needed, there is no need to store and validate large numbers of cDNA clones; choice of sets of good oligomers or heuristic comparisons done with different sets of oligomers representing each gene should minimize cross hybridization.	Current costs for the most common type (Affymetrix© chip) are high. New methods, e.g. the use of the inkjet printing, should reduce costs; are not applicable to an organism with little or no available sequence data. As in cDNA arrays, oligonucleotide arrays can only analyze what is on the physical array. Moreover, because they are so sequence specific, oligomeric sequences are not as useful as longer clone sequences when crossing species barriers and may be more prone to error if the gene has an unsuspected splice form or mutation.

In part, the problem with making functional inferences from gene expression patterns is a statistical artifact sometimes called the "Kevin Bacon" game [4, 5]. The goal of the game is to relate any actor to Kevin Bacon through a series of co-starring links in six or less steps. Ultimately every actor has some relationship to Mr. Bacon, albeit at various levels of removal. Similarly, any gene can be related to almost any biological state or other gene through a small number of publications in which the genes "co-star". Biologic analysis can be further complicated by looking through the literature for papers that might indicate a "link" between observed differentially expressed genes and the biology of interest.

8.2.2
Functional Genomics

Perhaps the most aggressive effort at using arrays to infer function has been made in yeast which have only ~6,200 genes [6]. Ideker et al. used a combination of DNA microarrays, quantitative proteomics, and databases of known physical interactions to study the yeast galactose-utilization pathway. Access to all these resources allowed the investigators to formulate new hypotheses about the regulation of galactose utilization and molecular interactions between this and a variety of other metabolic pathways. Although such comprehensive data are not available in mammalian systems, the availability of genomic mapping data, in combination with expression data, is already of great help in focusing attention on the sets of relevant genes in human disease. It is worth noting that expression studies alone cannot prove function. For example, a large and functionally miscellaneous set of genes is rapidly upregulated when serum starved cells are fed. Some of these "early response genes" have clear functions in the cell cycle; the role of others – e.g. protinases, heat shock protein, and even tissue factor – may only reflect promoter promiscuity.

8.2.3
Phenotypic Definitions

The potential of arrays for producing phenotypes should be compared with the more traditional approach based on morphology. A moderately high quality digital camera may have three million pixels. Each pixel is capable of showing three channels of color at one byte or 24 bits per channel, or about $16 \times 10^{18,000,000}$ possible combinations in one digital microscope image. In contrast, if we split the range of expression of one gene into 8 different bins, we can calculate the maximum number of potential observable expression states. Thus, the complexity of expression genomics is 8 (different expression levels for each gene) raised to the power of 35×10^3 (the rough number of genes) [7]. However, it appears that cells rarely express more than 2×10^4 genes, and probably less than 10^3 genes actually are involved in any specific cell response. Thus, for any comparison between two cells, it is not likely that one has to think about more than 80,000 different possible phenotypes. This number, 80×10^3, while still impressive, is easier to imagine than $16 \times 10^{18,000,000}$.

Tumor biologists have already begun to use arrays to classify specific neoplasms [8], or to define two functional states of a cell, e.g. comparing a metastatic cell with its non metastatic parent [8]. A particularly impressive paper was published by Golub et al., comparing acute myelogenous leukemia with acute lymphatic leukemia [9]. Where the pathologist might make a decision based on subjective criteria only, the objective diagnoses by Golub et al. were made by looking at these 7,000 transcription units, quantifying the amount of RNA made by all of these, doing this in a large number of cases, and showing that a pattern emerged. Fifty genes were sufficient to define the two types of leukemia.

8.2.4
Methodological Biases in Array Interpretation

One common problem in array analysis is the use of biased arrays that focus on specific sets of genes identified in such cellular processes as inflammation, development, replication, etc. If, for example, one looks at an array limited to "inflammation", we will inevitably find multiple members driven by NFκB. For example, McCaffrey et al. [10] used expression arrays focused on inflammation to show that atherosclerotic plaques are enriched in transcripts driven by the EGR1 and EGR2 transcription factors downstream of TGF-β and PDGF, two growth factors believed to play a critical role in the formation of the plaque [10]. The use of a focused array with only 568 genes created the implicit assumption that the genes on the array are more important to the biology of interest than the other approximately 35,000 genes in the genome.

8.2.5
Usefulness of Clustered Genes

The use of some form of unbiased pattern recognition has become the most common approach to analysis of array data. While the mathematics are beyond the scope of this review, it is useful to offer a somewhat general description of clustering and related analytical methods. Tab. 8.2 shows a hypothetical data set of the sort typically used in cluster analyses. With this kind of data set, one may cluster experiments – e.g. identify experiments (or samples) in which genes behave similarly or identify genes that behave alike across experiments. Experiments 1 and "N" show identical values for all genes and hence these two experiments are more closely related to each other than they are to experiment 2. An alternative approach is to cluster genes; this is equivalent to looking for rows in Tab. 8.2 that show similar patterns. For example, genes 1 and 4 behaved similarly in these experiments and, we presume, have similar promoters.

Clustering may be either biased or unbiased. Biased analyses ask about predetermined lists of genes, e.g. genes belonging to specific functional classes ("show me genes that behave like these"). Unbiased analyses simply look for common expression patterns irrespective of previous knowledge. The former question might also be called "hypothesis-based research" and the latter, "systematic exploration". Biased clustering analysis can be useful to help identify genes which behave similarly to specific genes of interest. This approach must be used with caution as has already been discussed above because of the "Kevin Bacon" effect. We also need to be aware that a gene of interest may not be representative of its function or the function of genes under similar control.

Tab. 8.2 Clustering by rows versus columns. An example of a hypothetical data set that may be used in a clustering analysis. Rows represent genes, columns represent experiments or samples, and values in the table represent absolute or relative expression measurements

		Cluster by experiment		
		Expt. 1	*Expt. 2*	*Expt. 3*
Clustery gy Gene	Gene 1	1	6	1
	Gene 2	2	1	2
	Gene 3	1	6	1
	Gene 4	−1.4	1.0	−1.4
	Gene 5	1.2	1.0	1.2

8.2.6
False Associations

A major concern with all approaches to cluster analysis is the lack of statistical criteria to decide which clusters are valid. The work of Golub et al. cited above is instructive. Their ability to use the set of genes identified by a self-organizing map to differentiate two kinds of leukemia was determined prospectively by demonstrating that this pattern could be used with known cases. It is important to realize, however, that no statistical method existed to give the investigators proof that this cluster was diagnostic. We do not know whether other patterns might have been even more successful. "SAM" (daisy.Stanford.edu/MicroArray/SMD/research.html), is an attempt to deal with this problem by scoring each gene on the basis of expression relative to standard deviation of repeated measurements (in array data) [11]. Unfortunately, SAM and similar methods provide only an arbitrary way of setting a threshold. In effect, one can increase or decrease the number of acceptable genes but we still lack any rigorous test that would allow us to state that the possibility of a specific set of genes being differentially expressed is at a specific p value [12].

8.2.7
Hypotheses Based on Expression Phenotypes

The task of "hypothesis-based" research then becomes identification of the mechanisms controlling an expression pattern. For example, Clark et al. used expression phenotypes to define melanoma cells selected for their metastatic capacity by serial harvesting of cells that metastasize after implantation of the tumor in mice [8]. Besides defining the metastatic phenotype, they were able to identify a specific member of the Rho small GTPase family as being differentially expressed. They were also able to show that RhoC alone was necessary and sufficient to support metastasis and that the closely related RhoA or B were not differentially expressed and were not needed to support metastasis. It would have been very interesting if they had gone on to tell us whether or not the rest of the metastatic expression phenotype was RhoC-dependent.

Another example of the hypothesis-based approach to analysis of array was taken by Svaren et al. in their study of the role of EGR1 in prostate cancer [13]. These authors began by using adenovirus to transfect prostatic cells with EGR1, then analyzed the effects with an array containing 5,600 randomly chosen genes. Genes identified as EGR1-dependent were subsequently confirmed by quantitative PCR in cancer tissue. Intriguingly, many of these mapped to the 11p15 chromosomal locus, an imprinted region implicated in other cancers. One such gene, IGF-II has been implicated in prostate cancer [13].

8.2.8

Multiple Venn Diagrams: Decreasing the Size of the Catalog

A review of the array literature in preparation for this paper suggests that the most successful way to reduce the size of the data set produced by an array study to a more manageable and often more meaningful list is to use multiple comparisons between strategically chosen data sets, including data independent of the array itself. As more comparisons are done, fewer genes will achieve significance. For example, a recent paper in the diabetes literature compared the expression patterns of adipose tissue in several strains of obese rats with the expression patterns of the same rats made hyperglycemic by streptozoticin (Fig. 8.2) [14]. The use of multiple strains of obese animals, even though each strain was studied only once, led to identification of only 92 genes that marked the diabetic-obese animals. It is likely that these numbers would be further reduced by a study that takes more account of animal-to-animal variability, or even more powerful by using additional Venn diagrams to focus attention on those genes that are regulated by diabetes alone and then comparing that group with the two published studies.

As another example, positional cloning within the areas identified by genetic studies [15] is difficult unless there are obvious candidate genes. Fig. 8.3 suggests one possible resolution to this problem. Any over- or under-expressed gene found in a locus associated with the disease phenotype would become an obvious candidate, as would any transcription factor located in such a locus and known to control genes showing a clustered expression pattern in the array. Mirnics and collaborators compared expression of mRNA from brains of hospitalized patients with and without schizophrenia with that of patients with an unrelated psychiatric disorder [16]. Monkeys treated with antipsychotic drugs provided a further control. Only one gene, RGS4, showed consistent change in expression. RGS4 was the only gene in their study to map to the major schizophrenia susceptibility locus 1q21–22.

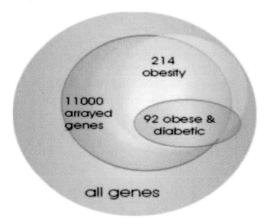

214
obesity

11000
arrayed
genes

92 obese &
diabetic

all genes

Fig. 8.2 Venn Diagram. The use of multiple types of data is a powerful tool to filter expression arrays.

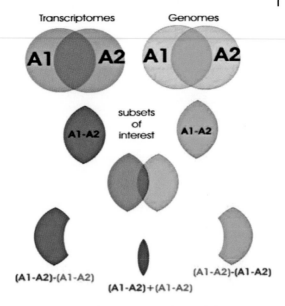

Fig. 8.3 Venn diagram of transcriptome versus genome. Use of transcription patterns to map multiple genomic loci. These Venn diagrams show comparisons of genes (on the right) with transcripts (on the left). The first set of intersects include those genes that are different in sequence (right) versus expression (left), producing the "subsets of interest". The third level set shows the intersection of the genomic subset with the expression subset. Differences in transcription that map to differences in the genome are suspect as sites where mutation has affected a promoter, mRNA stability, or message function. At a more sophisticated level, differences in transcription may soon be mapped to genetic difference marking functional mutations in transcription factors, receptors, or cytokines.

8.3
Analysis of Measurement Errors

No review of microarray technology would be complete without a discussion of appropriate criteria for quality control. Microarray technology is essentially a highly parallel form of a dot blot. Correlations with assays by other methods, e.g. Northern blot, are surprisingly good [17, 18]. Unfortunately, even when most genes are accurately measured, data for individual genes may be discrepant. In one study, four of forty-six differences substantiated by Northern analysis were not demonstrable in the array [19]. This error may reflect the inability of a dot on an array to distinguish between related genes, splice forms, or alternative polyadenylation sites. Another problem is the error due to "non specific" cross hybridization. In one study in our lab we found differential hybridization at a spot that turned out to carry an ALU-rich sequence. The gene spotted at this site apparently was not even present in the RNAs being used to make probes.

Even when arrays accurately detect a species of mRNA, spots that have very little hybridization will of necessity have a high variance due to fluctuation in the

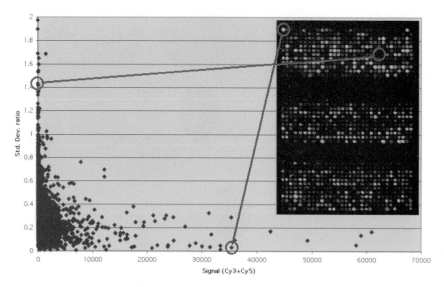

Fig. 8.4 Error in ratio as a function of Expression level for typical array data. Error is expressed as the standard deviation of the ratio in 4 quadrants of the spot; expression level is expressed as the sum of the Cy3 and Cy5 signals. Since most ratios are near 1, the standard deviation * 100 is approximately equal to the percent error in the ratio. The image inset shows a portion of the array data from which this graph was derived. Two spots corresponding to a highly and lowly expressed gene are circled in red in both the image and the graph.

background and other sources of noise. Expression ratios based on such spots may be the result of two values of which neither has, or only one offers, statistical confidence. The resulting number may be meaningless (Fig. 8.4).

A complimentary issue is the assumption that the hybridization curves for any two samples are parallel at all spots. This, of course, is unlikely. Obviously most curves will be at best S-shaped and some data may not even be monotonic. The most extreme example of this problem is recent evidence for differences in hybridization with fluorescent probes based on Cy3 and Cy5 [20]. There are two types of observed discrepancies. First, it appears that the Cy3/Cy5 ratio is not linear with intensity, e.g., if one takes aliquots of the same RNA, labels one in Cy3 and one in Cy5, and hybridizes them to an array, the observed ratio is often a non-linear function of intensity. Second, there are some spots on any given array which appear to be consistently differentially expressed in one channel regardless of the samples that are applied. The exact cause of this effect is not yet proven. Correction for the Cy3/Cy5 error requires either that the array be reciprocally hybridized (e.g. a "flipped color experiment") or that a universal standard be used for one "channel" of arrays having two colors.

8.4
Standards and Statistical Validation

Even if the array measurements are perfect, all data must reflect underlying biological variability. Unfortunately, high costs have meant that few studies are done in replicate. Array papers, moreover, can be very misleading by displaying the error implicit in the measurement method and then comparing data sets based on this error instead of using biological error.

Biologists typically accept a minimum of six measurements to determine a mean because of the t-statistic distribution. It seems reasonable to use a similar number of arrays when comparing any two sets of specimens. The usual biological experiment, however, compares small numbers of variables. Statistical tests, only now being defined, are needed to determine the minimum sample size when large numbers of comparisons are being made, as in our own study of aorta versus vena cava [21]. In the absence of methods to validate all spots on an array, it is possible to validate data only for a few specific genes by individual quantitative RNA measurement, e.g., by Realtime PCR, competitive PCR, solution hybridization, *in situ* hybridization, or Northern blot. Such measurements are expensive; less expensive options include:

1. Standardized probes could be used at different probe concentrations against an array to develop a set of curves, and
2. Reduction of all possible sources of error in the study design can be followed by determining the variance of our data for each gene in the actual data. This approach leaves aside the question of the quantitative validation of each spot on the array, but does provide for valid comparisons between groups of specimens.

Finally, measuring thousands of genes does not remove the obligation to reproduce the measurements in at least a few independent samples.

8.5
Summary

Array papers should provide valid answers to the following questions:

1. Does the paper satisfy normal statistical issues as to population sampling?
2. Does the paper serve to discover genes or to identify a tissue-specific pattern?
   a. If discovery of newer genes is the goal, were valid methods used to corroborate the expression data?
   b. If pattern identification is the goal, were valid statistical methods used and is the sample size large enough to accept the pattern?
3. If mRNA is derived from a tissue made up of multiple cell types, are there criteria for deconvolving the tissue complexity from the mRNA array analysis and was the isolation of each cell type done in a manner likely to preserve *in vivo* expression patterns?

4. If the data are in the form of a large list of values, how valuable is the list in and of itself?

5. Are independent criteria, e.g. genetic [16] or functional [6] data, used to provide a guide to interpreting the expression data?

8.6
References

1 HUBANK, M., SCHATZ, D.G. Identifying differences in mRNA expression by representational difference analysis of cDNA. *Nucleic Acids Res.* **1994**, *22*, 5640–5648.

2 JAIN, R.K., SAFABAKHSH, N., SCKELL, A., CHEN, Y. et al. Endothelial cell death, angiogenesis, and microvascular function after castration in an androgen-dependent tumor: role of vascular endothelial growth factor. *Proc Natl Acad Sci USA.* **1998**, *95*, 10820–10825.

3 YANG, G.P., ROSS, D.T., KUANG, W.W., BROWN, P.O. et al. Combining SSH and cDNA microarrays for rapid identification of differentially expressed genes. *Nucleic Acids Res.* **1999**, *27*, 1517–1523.

4 COLLINS, J.J., CHOW, C.C. It's a small world. *Nature.* **1998**, *393*, 409–410.

5 WATTS, D.J., STROGATZ, S.H. Collective dynamics of "small-world" networks. *Nature.* **1998**, *393*, 440–442.

6 IDEKER, T., THORSSON, V., RANISH, J.A., CHRISTMAS, R. et al. Integrated genomic and proteomic analyses of a systematically perturbed metabolic network. *Science.* **2001** , *292*, 929–934.

7 EWING, B., GREEN, P. Analysis of expressed sequence tags indicates 35,000 human genes. *Nat.Genet.* **2000**, *25*, 232–234.

8 CLARK, E.A., GOLUB, T.R., LANDER, E.S., HYNES, R.O. Genomic analysis of metastasis reveals an essential role for RhoC. *Nature.* **2000**, *406*, 532–535.

9 GOLUB, T.R., SLONIM, D.K., TAMAYO, P., HUARD, C. et al. Molecular classification of cancer: class discovery and class prediction by gene expression monitoring. *Science.* **1999**, *286*, 531–537.

10 McCAFFREY, T.A., FU, C., DU, B., EKSINAR, S. et al. High-level expression of Egr-1 and Egr-1-inducible genes in mouse and human atherosclerosis. *J.Clin.Invest.* **2000**, *105*, 653–662.

11 TUSHER, V.G., TIBSHIRANI, R., CHU, G. Significance analysis of microarrays applied to the ionizing radiation response. *Proc Natl Acad Sci USA.* **2001**, *98*, 5116–5121.

12 BRUTSCHE, M.H., BRUTSCHE, I.C., WOOD, P., BRASS, A. et al. Apoptosis signals in atopy and asthma measured with cDNA arrays. *Clin Exp Immunol.* **2001**, *123*, 181–187.

13 SVAREN, J., EHRIG, T., ABDULKADIR, S.A., EHRENGRUBER, M.U. et al. EGR1 target genes in prostate carcinoma cells identified by microarray analysis. *J Biol Chem.* **2000**, *275*, 38524–38531.

14 NADLER, S.T., STOEHR, J.P., SCHUELER, K.L., TANIMOTO, G. et al. The expression of adipogenic genes is decreased in obesity and diabetes mellitus. *Proc Natl Acad Sci USA.* **2000**, *97*, 11371–11376.

15 LANDER, E.S. Genes and genomes. *Harvey Lect.* **1997**, *93*, 35–48.

16 MIRNICS, K., MIDDLETON, F.A., MARQUEZ, A., LEWIS, D.A. et al. Molecular characterization of schizophrenia viewed by microarray analysis of gene expression in prefrontal cortex. *Neuron.* **2000**, *28*, 53–67.

17 TANIGUCHI, M., MIURA, K., IWAO, H., YAMANAKA, S. Quantitative assessment of DNA microarrays–comparison with Northern blot analyses. *Genomics.* **2001**, *71*, 34–39.

18 McCORMICK, S.M., ESKIN, S.G., McINTIRE, L.V., TENG, C.L. et al. DNA microarray reveals changes in gene expression of shear stressed human umbilical vein endothelial cells. *Proc. Natl. Acad. Sci. USA.* **2001**, *98*, 8955–8960.

19 LEE, M.L., KUO, F.C., WHITMORE, G.A., SKLAR, J. Importance of replication in microarray gene expression studies: statistical methods and evidence from repetitive

cDNA hybridizations. *Proc Natl Acad Sci USA*. **2000**, *97*, 9834–9839.

20 LUO, L., SALUNGA, R.C., GUO, H., BITTNER, A. et al. Gene expression profiles of laser-captured adjacent neuronal subtypes. *Nature Medicine*. **1999**, *5*, 117–122.

21 ADAMS, L.D., GEARY, R.L., MCMANUS, B., SCHWARTZ, S.M. A comparison of aorta and vena cava medial message expression by cDNA array analysis identifies a set of 68 consistently differentially expressed genes, all in aortic media. *Circulation Research*. **2000**, *87*, 623–631.

22 ISHII, M., HASHIMOTO, S., TSUTSUMI, S., WADA, Y. et al. Direct comparison of GeneChip and SAGE on the quantitative accuracy in transcript profiling analysis. *Genomics*. **2000**, *68*, 136–143.

23 ST CROIX, B., RAGO, C., VELCULESCU, V., TRAVERSO, G. et al. Genes expressed in human tumor endothelium. *Science*. **2000**, *289*, 1197–1202.

24 BRENNER, S., JOHNSON, M., BRIDGHAM, J., GOLDA, G. et al. Gene expression analysis by massively parallel signature sequencing (MPSS) on microbead arrays. *Nat.Biotechnol*. **2000**, *18*, 630–634.

25 DE FOUGEROLLES, A.R., CHI-ROSSO, G., BAJARDI, A., GOTWALS, P. et al. Global expression analysis of extracellular matrix-integrin interactions in monocytes. *Immunity*. **2000**, *13*, 749–758.

26 KAHN, J., MEHRABAN, F., INGLE, G., XIN, X. et al. Gene expression profiling in an in vitro model of angiogenesis. *Am J Pathol*. **2000**, *156*, 1887–1900.

27 SHIMKETS, R.A., LOWE, D.G., TAI, J.T., SEHL, P. et al. Gene expression analysis by transcript profiling coupled to a gene database query. *Nat Biotechnol*. **1999**, *17*, 798–803.

28 FIGEYS, D., DUCRET, A., YATES, J.R. III, AEBERSOLD, R. Protein identification by solid phase microextraction-capillary zone electrophoresis-microelectrospray-tandem mass spectrometry. *Nat.Biotechnol*. **1996**, *14*, 1579–1583.

9

Genes Involved in Atherosclerosis and Plaque Formation

Tiina T. Tuomisto and Seppo Ylä-Herttuala

9.1
Introduction

Atherosclerosis is the principal underlying cause of myocardial infarction, stroke and gangrene of extremities and the most common cause of death in Western countries. It affects large and medium-sized arteries. Atherosclerosis is induced by several risk factors including hypertension, smoking, diabetes, elevated serum low density lipoprotein (LDL) and very low density lipoprotein (VLDL) levels. The earliest form of atherosclerotic lesion is "fatty streak", an aggregration of lipid-rich macrophages, foam cells and T-cells within intima. Fatty streaks precede intermediate lesions, which are composed of macrophages and smooth muscle cells (SMC) and develop to more advanced, complex occlusive plaques that contain macrophages, SMC, T-cells, atheromatous core and calcium. Complicated lesions occlude the artery and may be ruptured resulting in thrombus formation [1]. Surgical bypass and angioplasty are the interventional therapies but they are limited by the problems of restenosis and graft occlusions. Smooth muscle cell proliferation and migration are the key factors in the development of restenosis [2]. Even though many important features of atherogenesis have already been characterized, more basic information is still needed to fully understand pathophysiological mechanisms involved in atherosclerosis and develop novel therapeutic strategies [3].

The events in atherogenesis have been importantly clarified by studies in animal models, including rabbits, pigs, non-human primates and rodents [4]. The availability of spontaneously hypercholesterolemic models, Watanabe heritable hyperlipidemic rabbit (WHHL) and mice deficient in apolipoprotein E (apoE) or the low-density lipoprotein receptor develop advanced lesions and are the models most often used in genetic and pathophysiological studies. Also, several features of atherosclerosis can be mimicked in cell culture models, for example by treating cultured macrophages with oxidized low density lipoprotein to generate foam cells [5]. Although animal and cell culture models are very useful in studying atherogenesis, they give a partial figure of the complexity of human disease. On the other hand, the use of human samples is limited by the availability of samples, since fresh arterial samples can only be collected from amputation operations, biopsies, organ donors and fast autopsies.

9.2
Genes Influencing Atherogenesis

Atherosclerotic lesions are heterogenous and they consist of many cell types including SMC, endothelial cells, fibroblasts, macrophages and T cells. These cell types produce many proteins that are involved in atherogenesis. These proteins include lipoprotein receptors, growth factors, cytokines, matrix metalloproteinases and cell adhesion molecules.

Macrophage scavenger receptors are membrane glycoproteins that are involved in internalisation of modified LDL. They mediate accumulation of modified lipoproteins in macrophages and participate foam cell formation [6]. They also anchor macrophages to atherosclerotic lesions and bind glycosylation end products, which is an important problem in diabetes [7].

Cell-cell or cell-matrix interactions are mediated by adhesion molecules that can also function in cell migration, signalling and other vascular responses. Endothelium expresses adhesion molecules like integrins and selectins that increase the adhesion of monocytes and T-lymphocytes to endothelium [8]. Members of immunoglobulin superfamily e.g vascular cell adhesion molecule (VCAM-1) [9], intracellular adhesion molecules (ICAMs) [10] and platelet – endothelial cell molecule (PE-CAM) [11] serve as ligands for integrins. Adhesion molecule expression can be regulated by different cytokines. Leukocyte adhesion is mediated by E-, L- and P-selectins which interact with ligands on leukocytes [12].

Growth factors can stimulate cell proliferation, inhibit proliferation and act as chemoattractans. Peptide growth factors platelet derived growth factor (PDGF), fibroblast growth factor (FGF), vascular endothelial growth factor (VEGF) and insulin like growth factor (IGF-1) are involved in many important cellular processes in atherogenesis. These can induce SMC proliferation and are generally expressed in the normal artery, whereas they are upregulated in atherosclerotic lesions [1, 13]. Leukocytes move into the artery wall and SMC from the media to intima in a process called chemotaxis. Leukocyte chemotaxis is induced e.g. by colony stimulating growth factors (CSFs) and SMC chemotaxis by PDGF and IGF-1 [1, 14]. Cytokines e.g. interleukin 1 (IL-1) and interferon γ (IFNγ) modulate inflammatory processes [1].

Peroxisome proliferator-activated receptors (PPARs) are nuclear receptor-type transcription factors that modulate inflammation and influence lipid metabolism [15]. NF-κB is a transcription factor associated with oxidative stress and inflammation. NF-κB regulates the expression of many important atherogenic genes including VCAM-1 and ICAM-1 [16]. Macrophages, SMC and T-cells in atherosclerotic lesions undergo apoptosis. Apoptosis is controlled by number of different genes or gene families e.g. Bcl-2, caspases and nitric oxide (NO) [17]. Matrix metalloproteinases degrade extracellular matrix components which is essential for matrix remodelling, infiltration of inflammatory cells, plaque rupture and angiogenesis [18].

Number of genes encoded by the human genome has been estimated to be $\sim$ 32,000–38,000 [19]. Not all genes are expressed in cardiovascular system. The total number of expressed sequence tags (ESTs) expressed in cardiovascular system was evaluated by EST sequencing and microarray analysis and estimated as 20,930 –27,160 genes [20]. Genes expressed in human heart have also been studied by EST sequencing, and over 40 000 different ESTs were produced [21].

The analysis of changes in gene expression during atherogenesis enables the studies of the molecular mechanisms behind the pathological process. Several methods are available for studying the gene expression. Every expressed gene in a tissue can be sequenced from a cDNA library of the tissue in a process called expressed sequence tag (EST) sequencing. The SAGE technique (Serial Analysis of Gene Expression) leverages DNA sequencing as a measure of transcript abundance. Real-time quantitative PCR can be used to quantify mRNAs. By differential display it is possible to identify differentially expressed genes between two or more samples. DNA arrays have revolutionized expression studies since they have the ability to profile the expression of thousands of genes in a single experiment [22].

9.3
DNA Arrays: The Principle

DNA arrays have been applied to gene expression studies in bacteria, plants, yeast and mammals. Disease-related changes in gene expression have been identified in cancer, inflammatory diseases and heart diseases. Array analyses have also been used for identification of new drug targets and mechanisms of drug action.

Two types of DNA arrays have been utilized for the profiling of gene expression: cDNA arrays and oligonucleotide arrays. In addition, arrays differ with respect to methods of arraying, chemistry, linkers, hybridization and detection. In a cDNA array, cDNA fragments usually produced by PCR are spotted to microscope slides (microarrays) or nylon membranes (macroarrays) [23, 24]. Oligonucleotide arrays developed by Affymetrix are composed of thousands of oligonucleotides, 25 nucleotides in length [25]. mRNAs from samples of interest are labelled with fluorescent dyes (Cy3 and Cy5) or radioactive nucleotides (33-P) and hybridized with immobilized targets. Two-colour fluorescent detection can be used with glass microarrays. Arrays with radioactive detection are more sensitive than fluorescent arrays, requiring only 0.5 µg of mRNA in contrast to 2–5 µg of mRNA needed for arrays with fluorescent detection. After hybridisation, signal intensities are measured with confocal fluorescent microscopy or phosphoimager, and special software is used for the rapid identification of differentially expressed genes. Clustering methods can be used to find patterns in differentially expressed genes [26].

Comparison of expression data from multiple arrays requires normalization. Use of signal intensities of a subset of genes e.g. housekeeping genes or commercial hybridization controls can be used for the normalization for divergent samples. If intensities of all genes are used, samples have to be closely related [3]. To describe the difference in signal intensities and thus in expression, intensity ratios have to be calculated. We have developed a formula where signals are normalized and intensity scores (fold increase or decrease) are calculated

$$\text{score} = \frac{\text{int } GDA1 + n}{m \times \text{int } GDA2 + n} \, . \tag{1}$$

Where the int GDA 1 and GDA 2 are intensities of filters 1 and 2, m is the average of all intensities on filter 1 divided by the average of intensities on filter 2 and n is $0.2 \times m$. The rationale for using the formula was to avoid false results caused by very low signal intensities which with the current formulation only produce values ~ 1. The formula also takes into account possible differences in the general background of the filters. Genes showing ≥ 1.5–2 – fold increase or decrease should be processed further. Statistical significances of the differences are calculated according to Claverie [27].

Gene expression studies produce similar needs to use bioinformatics irrespective of the array method used. The enormous amount of data has to be analyzed and presented in a meaningful way, which has been identified as the greatest challenge in the array research. Computational biology and mathematical modelling need to be integrated with DNA array related work. Ideally, data from gene expression experiments require a uniform format so that the results can also be used for meta-analyses. The enormous amount of data may be most sophistically published through www interface. One potential alternative would be the establishment of a centralized public data bank having a similar organization as e.g.

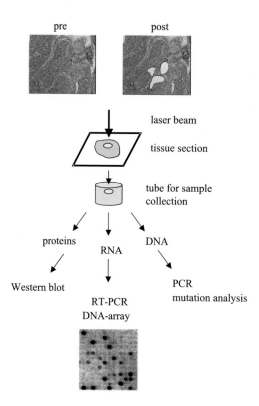

Fig. 9.1 The principle and applications of laser microdissection technology. In laser microdissection laser beam is used to dissect small cell populations from paraffin or frozen sections. Cells can then be used to ordinary DNA, RNA and protein analysis.

Tab. 9.1 Minimal criteria for an array analysis in cardiovascular field

– Findings have to be confirmed by other, independent methods such as RT-PCR, Northern blot, immunohistochemistry or *in situ* hybridisation
– Findings must be reproducible i.e. confirmed with repeated hybridisations (at least 3 times) using multiple sources of RNA
– statistical methods must be used to interpret the data

GenBank. Comparative databases in www should include the possibility to 1) search for the data by name of the gene, 2) find the expression value corrected to a uniform format, 3) have an access to brief annotations of specific genes and links to known biochemical pathways, including interactions at the transcriptional level, 4) download data for further analyses and 5) generate reports [3].

Atherosclerotic plaques can usually be divided to macrophage-rich shoulder areas, fibrous cap and atheromatous core. By manual dissection it is not possible to adequately separate these parts, and one has to be satisfied with heterogenous samples containing various parts of the lesion, media and adventitia. Recently, a novel method named "laser microdissection" has been developed to dissect very small cell populations or even single cells. Thin (<5 μm) frozen or paraffin sections are cut to special plastic slides, stained and dissected by laser beam under microscopic control [28]. RNA, DNA and proteins can be extracted from microdissected cells and used for PCR and DNA-array analyses (Fig. 9.1). Because the amounts of RNA that can be extracted are very low, T7-RNA-polymerase-based amplification is needed before any array analyses can be performed [29].

Tab. 9.1 states the minimal criteria for an array analysis in cardiovascular field. Analyses have to be repeated at least three times and statistical methods need to be used to interpret the data. Findings must be confirmed by complementary methods such as RT-PCR or immunohistochemistry since DNA arrays may still produce false signals due to sequence homologies and repetitive elements present in many genes.

9.4
DNA Arrays in The Research of Atherosclerosis

We have analyzed gene expression in normal arteries and in different types of immunohistologically characterized human atherosclerotic lesions using a nylon-filter-based DNA-array method. Radioactively labelled probes were generated from 1.0 μg of mRNA from normal arteries, fatty streaks and advanced lesions and hybridized to filters of 18,376 cDNA clones (Incyte Genomics). Intensities were analyzed as duplicates with pairs of filters comparing pooled normal samples vs. fatty streaks and advanced lesions. Scores were calculated as described in formula 1. The sensitivity of the filter arrays currently is at the level of one molecule in 100,000, which together with limitations in phosphoimaging allows detection of differential expression in excess of 1.5–2 folds. Only those genes/ESTs detected in three repeated pairs of arrays for each lesion type and showing ≥ 1.5 fold increase or decrease were processed further [30]. We first validated the array by detecting a group of genes (n = 17) that were already known to be connected to atherogenesis.

Tab. 9.2 Some examples of differentially expressed genes in advanced atherosclerotic lesions

Gene	GB access	Function	Score	p-value
a) Genes upregulated in advanced lesions				
Melanoma adhesion molecule MCAM	R79246	Signaling	56.5	p < 0.05
neuronal PAS domain protein	R78870	Expression	49.8	p < 0.05
HS solute carrier family 31 (copper transport) member 2	R68089	Transport	44.8	p < 0.05
EST	R68091	Unknown	40.8	p < 0.05
Proteasome (prosome, macropain) subunit, alpha type 2	H12633	Expression	37.6	p < 0.05
Oligopherin 1	R81942	Unclassified	34.4	p < 0.05
Platelet/ endothelial cell adhesion molecule-1 (PECAM-1) CD31 antigen	R33252	Signaling	29.5	p < 0.05
EST	R36114	Unknown	29.0	p < 0.05
Ubiquitin-conjugating enzyme E2D 1	H12682	Metabolism	21.3	p < 0.05
Lectin, mannose-binding, 1	R62532	Metabolism	21.3	p < 0.05
NADH-ubiquinone oxidoreductase, 51 kDa subunit	R67754	Metabolism	11.6	p < 0.05
Human non-muscle myosin alkali light chain	R70035	Unclass ified	10.1	p < 0.05
Zinc finger protein 7	AA005168	Expression	6.0	p < 0.05
Proteasome, chain 7	R80719	Expression	4.5	p < 0.05
b) Genes downregulated in advanced lesions				
transcript asssociated with monocyte to macrophage differentiation	R34270	Signaling	66.5	p < 0.05
Human cleavage and polyadenylation specifity factor mRNA	R82814	Expression	63.6	p < 0.05
LIM binding domain 2	R36692	Unclassified	61.8	p < 0.05
EST	H02191	Unknown	57.3	p < 0.05
Deoxiribonuclease 1-like 1	R32385	Metabolism	56.7	p < 0.05
growth differentiation factor 11, bone morphogenetic protein 11	T81804	Unclassified	52.4	p < 0.05
EST	R80390	Unknown	51.3	p < 0.05
EST, FLJ1321	R69260	Unknown	40.5	p < 0.05
Diphosphoinositol polyphosphate phosphohydrolase	H13795	Metabolism	37.0	p < 0.05
EST, Weakly similar to reverse transcriptase	H12636	Unclassified	36.7	p < 0.05
RecQ protein like-5	R31058	Unknown	34.4	p < 0.05
EST	R81144	Unknown	31.1	p < 0.05
protein phosphatase 2, regulatory subunit B, alpha isoform PPP2R2A	R65749	Metabolism	30.7	p < 0.05
EST, similar to mus musculus AT3 gene for antithrombin	R77725	Unknown	26.9	p < 0.05

These genes included *e.g.* apoE, CD68, and tissue inhibitor of metalloproteinase (TIMP). Next we detected 150 differentially expressed genes that were previously not connected to atherogenesis. Among these genes we found upregulation of 82 genes, 63 of which were known genes and downregulation of 68 genes, 33 of which were known genes [30]. Tab. 9.2 presents some examples of the genes up or downregulated in advanced lesions. In Fig. 9.2 the expression intensities of all arrayed genes in advanced lesion are blotted against the intensities in normal artery, showing that the expression level of the majority of the genes has not changed.

It is evident that cellular composition of the analyzed lesions has a major impact on the pattern of gene expression. For that reason, arterial samples were immunostained for the presence of SMC, macrophages and T-cells. Samples analyzed using the DNA array were confirmed to represent typical atherosclerotic lesions with only a moderate infiltration of inflammatory cells. Thus, it is likely that if extensive infiltration of macrophages or T-cells had been present the results from

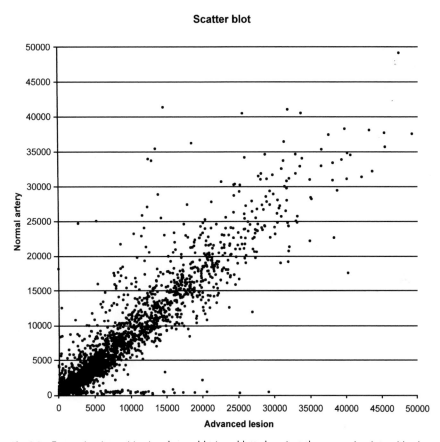

Scatter blot

Fig. 9.2 Expression intensities in advanced lesions blotted against the expression intensities in normal artery.

gene expression profiling could have identified activation of different genes. Since advanced lesions frequently involve microdomains of complex pathology it is likely that laser microdissection of lesions will improve the accuracy of the analysis. Results obtained from atherosclerotic lesions using DNA arrays should always be confirmed by *in situ* hybridization and/or RT-PCR analyses of the same lesions since current DNA arrays may still produce false results. It is also important to obtain information about the localization of mRNA species in different cell types.

Macrophage-rich shoulder areas of human atherosclerotic lesion were laser microdissected and gene expression profiles were compared to normal intima. Upregulation of several macrophage-specific genes e.g. IL-4 and additionally, upregulation of many known e.g. pancreatic lipase and unknown genes were detected (Tuomisto et al 2002, unpublished results).

Studies published so far regarding gene expression profiles in atherogenesis have mainly focused on cultured cells or animal models. Foam cell formation has been mimicked by treating macrophages with oxidized low density lipoprotein and changes in gene expression have been studied using a microarray of 9,808 genes in timepoints ranging from 30 min to 4 days. 268 of the genes showed 2-fold differential expression for at least 1 time points e.g. adipophilin, heparin-binding growth factor like growth factor and thrombomodulin [5]. Cluster analysis was used to further interpret the data. Vascular inflammation was mimicked by treating human vascular SMC with tumor necrosis factor a (TNFa), and subsequent array analysis revealed the importance of eotaxin and its receptor in inflammatory cell recruitment [31]. Endothelial cell senescence was demonstrated using cultured endothelial cells of early and late passages, and it was found that the expression of thymosin β-10 was decreased, which might contribute to the senescent phenotype by reducing the endothelial cell plasticity [32]. Wuttge et al. analyzed the gene expression during atherogenesis in apoE deficient mouse and found that atherogenesis is not a linear process with a maximal expression at advanced lesions stage, instead the gene expression has its peaks at the intermediate lesion stage [33]. The gene expression profiles between aorta and vena cava were studied in macaques, and 68 differentially expressed genes out of 4,048 genes had elevated expression in aorta [34]. The expression patterns in fibrous cap versus adjacent media in human atherosclerotic lesions was studied by array of 588 clones, and the induction of Egr-1 expression was detected also in a mouse model [35]. The expression differences between human stable and ruptured plaque was studied using suppression subtractive hybridization and macroarrays, and upregulated expression of periphilin, a phosphoprotein involved in process of lipolysis, in ruptured plaques was detected [36]. Human activated SMC expression was studied by differential display, and 10 known and 30 novel "smooth muscle activation-specific genes" were identified [37].

Analysis of atherogenesis with genomics techniques is still in a very early stage. However, there is no doubt that DNA array techniques will become a valuable tool for studying gene expression patterns in atherosclerosis and identifying novel candidate genes. Laser microdissection will further improve the accuracy of the technique in the analysis of local microdomains in atherosclerotic lesions.

9.5
Acknowledgements

This study was supported by grants from Finnish Academy and Sigrid Juselius Foundation.

9.6
References

1 Ross R. The pathogenesis of atherosclerosis: a perspective for the 1990s. *Nature* **1993**; *362*:801–809.

2 RUTANEN J, RISSANEN TT, KIVELA A, VAJANTO I, YLA-HERTTUALA S. Clinical Applications of Vascular Gene Therapy. *Curr. Cardiol. Rep.* **2001**; *3*:29–36.

3 HILTUNEN MO, NIEMI M, YLA-HERTTUALA S. Functional genomics and DNA array techniques in atherosclerosis research. *Curr. Opin. Lipidol.* **1999**; *10*:515–519.

4 LUSIS AJ. Atherosclerosis. *Nature* **2000**; *407*:233–241.

5 SHIFFMAN D, MIKITA T, TAI JT, WADE DP, PORTER JG, SEILHAMER JJ, SOMOGYI R, LIANG S, LAWN RM. Large scale gene expression analysis of cholesterol-loaded macrophages. *J. Biol. Chem.* **2000**; *275*:37324–37332.

6 KODAMA T, FREEMAN M, ROHRER L, ZABRECKY J, MATSUDAIRA P, KRIEGER M. Type I macrophage scavenger receptor contains alpha-helical and collagen-like coiled coils. *Nature* **1990**; *343*:531–535.

7 VLASSARA H, BROWNLEE M, CERAMI A. High-affinity-receptor-mediated uptake and degradation of glucose-modified proteins: a potential mechanism for the removal of senescent macromolecules. *Proc. Natl. Acad. Sci. USA.* **1985**; *82*:5588–5592.

8 PRICE DT, LOSCALZO J. Cellular adhesion molecules and atherogenesis. *Am. J. Med.* **1999**; *107*:85–97.

9 DUSTIN ML, ROTHLEIN R, BHAN AK, DINARELLO CA, SPRINGER TA. Induction by IL 1 and interferon-gamma: tissue distribution, biochemistry, and function of a natural adherence molecule (ICAM-1). *J. Immunol.* **1986**; *137*:245–254.

10 POBER JS, GIMBRONE MAJ, LAPIERRE LA, MENDRICK DL, FIERS W, ROTHLEIN R, SPRINGER TA. Overlapping patterns of activation of human endothelial cells by interleukin 1, tumor necrosis factor, and immune interferon. *J. Immunol.* **1986**; *137*:1893–1896.

11 DELISSER HM, CHILKOTOWSKY J, YAN HC, DAISE ML, BUCK CA, ALBELDA SM. Deletions in the cytoplasmic domain of platelet-endothelial cell adhesion molecule-1 (PECAM-1, CD31) result in changes in ligand binding properties. *J. Cell. Biol.* **1994**; *124*:195–203.

12 BEVILACQUA MP, POBER JS, WHEELER ME, COTRAN RS, GIMBRONE MAJ. Interleukin 1 acts on cultured human vascular endothelium to increase the adhesion of polymorphonuclear leukocytes, monocytes, and related leukocyte cell lines. *J. Clin. Invest.* **1985**; *76*:2003–2011.

13 WALTENBERGER J. Modulation of growth factor action: implications for the treatment of cardiovascular diseases. *Circulation* **1997**; *96*:4083–4094.

14 GERSZTEN RE, MACH F, SAUTY A, ROSENZWEIG A, LUSTER AD. Chemokines, leukocytes, and atherosclerosis. *J. Lab. Clin. Med.* **2000**; *136*:87–92.

15 SCHOONJANS K, STAELS B, AUWERX J. The peroxisome proliferator activated receptors (PPARS) and their effects on lipid metabolism and adipocyte differentiation. *Biochim. Biophys Acta* **1996**; *1302*:93–109.

16 COLLINS T, CYBULSKY MI. NF-kappaB: pivotal mediator or innocent bystander in atherogenesis? *J Clin. Invest.* **2001**; *107*:255–264.

17 ROSSIG L, DIMMELER S, ZEIHER AM. Apoptosis in the vascular wall and atherosclerosis. *Basic. Res. Cardiol.* **2001**; *96*:11–22.

18 George SJ. Tissue inhibitors of metallo-proteinases and metalloproteinases in atherosclerosis. *Curr Opin Lipidol* 1998; 9:413–423.

19 Venter JC, Adams MD, Myers EW, Li PW, Mural RJ, Sutton GG, Smith HO, Yandell M, Evans CA, Holt RA, Gocayne JD, Amanatides P, Ballew RM, Huson DH, Wortman JR, Zhang Q, Kodira CD, Zheng XH, Chen L, Skupski M, Subramanian G, Thomas PD, Zhang J, G.L., Nelson C, Broder S, Clark AG, Nadeau J, McKusick VA, Zinder N, Levine AJ, Roberts RJ, Simon M, Slayman C, Hunkapiller M, Bolanos R, Delcher A, Dew I, Fasulo D, Flanigan M, Florea L, Halpern A, Hannenhalli S, Kravitz S, Levy S, Mobarry C, Reinert K, Remington K, Abu-Threideh J, Beasley E, Biddick K, Bonazzi V, Brandon R, Cargill M, Chandramouliswaran I, Charlab R, Chaturvedi K, Deng Z, Di F, V, Dunn P, Eilbeck K, Evangelista C, Gabrielian AE, Gan W, Ge W, Gong F, Gu Z, Guan P, Heiman TJ, Higgins ME, Ji RR, Ke Z, Ketchum KA, Lai Z, Lei Y, Li Z, Li J, Liang Y, Lin X, Lu F, Merkulov GV, Milshina N, Moore HM, Naik AK, Narayan VA, Neelam B, Nusskern D, Rusch DB, Salzberg S, Shao W, Shue B, Sun J, Wang Z, Wang A, Wang X, Wang J, Wei M, Wides R, Xiao C, Yan C, Yao A, Ye J, Zhan M, Zhang W, Zhang H, Zhao Q, Zheng L, Zhong F, Zhong W, Zhu S, Zhao S, Gilbert D, Baumhueter S, Spier G, Carter C, Cravchik A, Woodage T, Ali F, An H, Awe A, Baldwin D, Baden H, Barnstead M, Barrow I, Beeson K, Busam D, Carver A, Center A, Cheng ML, Curry L, Danaher S, Davenport L, Desilets R, Dietz S, Dodson K, Doup L, Ferriera S, Garg N, Glueckmann A, Hart B, Haynes J, Haynes C, Heiner C, Hladun S, Hostin D, Houck J, Howland T, Ibegwam C, Johnson J, Kalush F, Kline L, Koduru S, Love A, Mann F, May D, McCawley S, McIntosh T, McMullen I, Moy M, Moy L, Murphy B, Nelson K, Pfannkoch C, Pratts E, Puri V, Qureshi H, Reardon M, Rodriguez R, Rogers YH, Romblad D, Ruhfel B, Scott R, Sitter C, Smallwood M, Stewart E, Strong R, Suh E, Thomas R, Tint NN, Tse S, Vech C, Wang G, Wetter J, Williams S, Williams M, Windsor S, Winn-Deen E, Wolfe K, Zaveri J, Zaveri K, Abril JF, Guigo R, Campbell MJ, Sjolander KV, Karlak B, Kejariwal A, Mi H, Lazareva B, Hatton T, Narechania A, Diemer K, Muruganujan A, Guo N, Sato S, Bafna V, Istrail S, Lippert R, Schwartz R, Walenz B, Yooseph S, Allen D, Basu A, Baxendale J, Blick L, Caminha M, Carnes-Stine J, Caulk P, Chiang YH, Coyne M, Dahlke C, Mays A, Dombroski M, Donnelly M, Ely D, Esparham S, Fosler C, Gire H, Glanowski S, Glasser K, Glodek A, Gorokhov M, Graham K, Gropman B, Harris M, Heil J, Henderson S, Hoover J, Jennings D, Jordan C, Jordan J, Kasha J, Kagan L, Kraft C, Levitsky A, Lewis M, Liu X, Lopez J, Ma D, Majoros W, McDaniel J, Murphy S, Newman M, Nguyen T, Nguyen N, Nodell M. The sequence of the human genome. *Science* 2001; 291:1304–1351.

20 Dempsey AA, Dzau VJ, Liew CC. Cardiovascular genomics: estimating the total number of genes expressed in the human cardiovascular system. *J. Mol. Cell Cardiol.* 2001; 33:1879–1886.

21 Hwang DM, Dempsey AA, Wang RX, Rezvani M, Barrans JD, Dai KS, Wang HY, Ma H, Cukerman E, Liu YQ, Gu JR, Zhang JH, Tsui SK, Waye MM, Fung KP, Lee CY, Liew CC. A genome-based resource for molecular cardiovascular medicine: toward a compendium of cardiovascular genes. *Circulation* 1997; 96:4146–4203.

22 Stanton LW. Methods to profile gene expression. *Trends. Cardiovasc. Med.* 2001; 11:49–54.

23 Schena M, Shalon D, Heller R, Chai A, Brown PO, Davis RW. Parallel human genome analysis: microarray-based expression monitoring of 1000 genes. *Proc. Natl. Acad. Sci. USA.* 1996; 93:10614–10619.

24 Nguyen C, Rocha D, Granjeaud S, Baldit M, Bernard K, Naquet P, Jordan BR. Differential gene expression in the murine thymus assayed by quantita-

tive hybridization of arrayed cDNA clones. *Genomics* **1995**; *29*:207–216.

25 LOCKHART DJ, DONG H, BYRNE MC, FOL-LETTIE MT, GALLO MV, CHEE MS, MITT-MANN M, WANG C, KOBAYASHI M, HOR-TON H, BROWN EL. Expression monitoring by hybridization to high-density oligonucleotide arrays. *Nat. Biotechnol.* **1996**; *14*:1675–1680.

26 TORONEN P, KOLEHMAINEN M, WONG G, CASTREN E. Analysis of gene expression data using self-organizing maps. *FEBS Lett.* **1999**; *451*:142–146.

27 CLAVERIE JM. Computational methods for the identification of differential and coordinated gene expression. *Hum. Mol. Genet.* **1999**; *8*:1821–1832.

28 KOLBLE K. The LEICA microdissection system: design and applications. *J. Mol. Med.* **2000**;*78*:B24–B25.

29 OHYAMA H, ZHANG X, KOHNO Y, ALEVI-ZOS I, POSNER M, WONG DT, TODD R. Laser capture microdissection-generated target sample for high-density oligonucleotide array hybridization. *Biotechniques* **2000**; *29*:530–536.

30 HILTUNEN MO, TUOMISTO, TT, NIEMI M, BRÄSEN JH, RISSANEN TT, TÖRÖNEN P, VAJANTO I, YLÄ-HERTTUALA S. Changes in gene expression in atherosclerotic plaques analyzed using DNA array. *Athero-sclerosis* **2002**; *165*:23–32

31 HALEY KJ, LILLY CM, YANG JH, FENG Y, KENNEDY SP, TURI TG, THOMPSON JF, SUKHOVA GH, LIBBY P, LEE RT. Overexpression of eotaxin and the CCR3 receptor in human atherosclerosis: using genomic technology to identify a potential novel pathway of vascular inflammation. *Circulation* **2000**; *102*:2185–2189.

32 VASILE E, TOMITA Y, BROWN LF, KOCHER O, DVORAK HF. Differential expression of thymosin beta-10 by early passage and senescent vascular endothelium is modulated by VPF/VEGF: evidence for senescent endothelial cells in vivo at sites of atherosclerosis. *FASEB J.* **2001**; *15*:458–466.

33 WUTTGE DM, SIRSJO A, ERIKSSON P, STEMME S. Gene expression in atherosclerotic lesion of ApoE deficient mice. *Mol. Med.* **2001**; *7*:383–392.

34 ADAMS LD, GEARY RL, MCMANUS B, SCHWARTZ SM. A comparison of aorta and vena cava medial message expression by cDNA array analysis identifies a set of 68 consistently differentially expressed genes, all in aortic media. *Circ. Res.* **2000**; *87*:623–631.

35 MCCAFFREY TA, FU C, DU B, EKSINAR S, KENT KC, BUSH HJ, KREIGER K, ROSEN-GART T, CYBULSKY MI, SILVERMAN ES, COLLINS T. High-level expression of Egr-1 and Egr-1-inducible genes in mouse and human atherosclerosis. *J. Clin. Invest.* **2000**; *105*:653–662.

36 FABER BC, CLEUTJENS KB, NIESSEN RL, AARTS PL, BOON W, GREENBERG AS, KIT-SLAAR PJ, TORDOIR JH, DAEMEN MJ. Identification of genes potentially involved in rupture of human atherosclerotic plaques. *Circ. Res.* **2001**; *89*:547–554.

37 DE VRIES CJ, VAN ACHTERBERG TA, HOR-REVOETS AJ, TEN CATE JW, PANNEKOEK H. Differential display identification of 40 genes with altered expression in activated human smooth muscle cells. Local expression in atherosclerotic lesions of smags, smooth muscle activation-specific genes. *J. Biol. Chem.* **2000**; *275*:23939–23947.

Section 2
Proteomics

10

Gene Expression Profiling in Pulmonary and Systemic Vascular Cells Exposed to Biomechanical Stimuli

Shwu-Fan Ma, Shui Qing Ye, and Joe G. N. Garcia

The pulmonary and systemic circulations are continuously exposed to blood-borne bioactive agonists, hormones and cellular constituents as well as biophysical forces such as shear stress and cyclic strain. Alterations in gene expression represent important adaptive responses which allow vascular cells to either remodel, form new vessels or progress to programmed cell death. Differential gene expression in vascular cells and in blood cells, due to gene-gene and gene-environment interactions (including local hemodynamics), can be considered the molecular basis of relevant vascular pathogenic processes. Fortunately, the ability to elucidate the role of specific genes in pathophysiologic pathways relevant to human disease has been accelerated by the emergence of new technologies such as cDNA/oligonucleotide microarrays and serial analysis of gene expression (SAGE), which allow high throughput gene expression profiling. In this chapter, we provide information on microarray and SAGE techniques in the context of human vascular diseases and the systematic analysis of gene expression of vascular cells, information with important implications for our understanding of the mechanisms of vascular homeostasis and dysfunction.

10.1
Introduction

The Human Genome Project (*www.ncbi.nlm.nih.gov*), well on its way to providing a complete catalogue of all human genes, is arguably the most important biological discovery ever. Annotating and integrating these rivers of genomic data into biologically-relevant processes in order to provide useful translational information is a daunting challenge which will persist for the foreseeable future. This is particularly true for a variety of systemic and pulmonary vascular processes, which are evoked either by rapid-acting chemical and cellular agents or by chronic environmental stimuli. Both normal adaptive and pathobiologic vascular responses to such stimuli involve well-orchestrated alterations in gene and protein expression in vascular cell targets. It is in this context that the field of functional genomics has begun to focus on dissecting the molecular basis of cellular function at the level of the organ and the tissue. Characterization of genes abnormally expressed in diseased vascular tissues offer the potential for the discovery of genes that serve as diagnostic markers, prognostic indicators or targets for therapeutic intervention [1]. Historically, methods such as cDNA subtraction [2], and mRNA differential display [3] have been limited by the capacity to analyze only a limited number of genes without quantitative information. More recently, two technologies, serial analysis of gene expression (SAGE) [4] and gene-specific oligonucleotide [5] or cDNA microarrays [6] now allow researchers to determine the expression pattern of thousands of genes simultaneously and have yielded novel insight into the pathophysiology of human diseases. While the full impact of these technologies on vascular diseases has not yet been realized, the potential for their use in the post-genomic era as diagnostic molecular markers or for identifying novel therapeutic interventions are currently being explored.

In this chapter, we will address relevant literature to the genomic-based studies that have focused on vascular responses with a particular focus on vascular gene expression induced by biophysical perturbations. Although both pulmonary and systemic circulations are exposed to varying level of shear stress, lung endothelial cells also experience variable flow patterns dictated by the unique pulmonary circulation where the distribution of blood flow is non-uniform decreasing from the base to the apex. Furthermore, lung capillaries routinely collapse with cessation of flow, a physiologic event in the lung for example, during hypoxic vasoconstriction. Finally, the lung circulation is uniquely exposed to cyclic strain produced in the course of normal tidal breathing, an exposure exaggerated when patients with respiratory failure are placed on mechanical ventilation. Thus, it might be predicted that systemic and pulmonary vascular cells would share similar gene expression profiles but also retain phenotypic responses under specific environmental or biophysical conditions unique to their respective vascular beds.

10.2
Principles of Array, SAGE and Mini SAGE Technologies

The basis for gene microarrays (addressed in detail elsewhere, see i.e. Chapter 6) embodies the core principle of molecular biology with the specific hybridization between labeled-nucleic acids and the complementary immobilized nucleic acids on a matrix [7]. Gene arrays utilize this basic principle to place a collection of gene-specific nucleic acids on solid supports at defined locations. With the adaptable nature of the fabrication and hybridization methods, microarrays have become one of the most powerful and versatile tools for genomic analysis. Probes such as PCR products, cDNAs or oligonucleotide probes (25- to 70-mers), are robotically deposited (e.g. cDNA microarray) or synthesized by photolithographic techniques (oligonucleotide microarray). Microarrays are relatively easy to use and suitable for high-throughput applications where expression profiling of hundreds of disease samples can be efficiently performed.

The SAGE technology involves a short sequence tag (10 bp) from a unique position within each transcript containing sufficient information to uniquely identify a transcript. The concatenation of tags in a serial fashion allows for increased efficiency of a sequence-based analysis (Fig. 10.1, details of the SAGE technique found in two comprehensive reviews [1,8]). Sequences of the concatemers enable cataloging and quantification of the tags, which delineates the identification and the relative abundance of the transcripts for a given cell line or tissue [4]. The portability of the SAGE data allows direct comparisons to be made, a difficult task to accomplish with microarrays given the differences in formats and normalization methodologies. SAGE accurately determines the absolute abundance of mRNAs and with the potential identification of new genes and alternatively processed transcripts that are unique to a specific cell type. A significant drawback of SAGE is the requirement for large amount of input mRNA (2.5–5.0 µg) equivalent to 250 to 500 µg total RNA. We recently utilized a modified method, mini SAGE, to profile gene expressions of human fibroblasts from 1 µg total RNA without additional PCR amplification [9].

10.3
Analysis of Gene Expression Profiling in Vascular Cells

As early as 1856, Virchow *et al.* [10] recognized that vascular endothelium might participate as a central component in the atherosclerotic disease process. Ku *et al.* [11] reported that atherosclerotic lesions often originate near branches, bifurcations and curvatures of arteries, where the laminar flow pattern is disturbed. In contrast, the unbranched, tubular portions of the arteries that carry uniform laminar flow are relatively protected from the atherogenesis. Ultrasound and magnetic resonance imaging techniques demonstrated a close correlation between flow disturbances and arterial susceptibility to the development of atherosclerosis. Experiments using an *in vitro* flow chamber to simulate regions susceptible to atherosclerosis (flow disturbance) and resistant to atherosclerosis (undisturbed flow), also underscore the spatial relationships between hemodynamic shear stress forces and endothelial monolayer [12].

Endothelial cells exposure to biological stimuli (cytokines, growth factors, hormones, metabolic products) produced by or in conjunction with pulsatile flow, generates a complex interplay of three distinct types of fluid mechanical forces: wall shear stresses, cyclic strain and hydrostatic pressures [13]. The impact of these hemodynamic factors on the structure and function of the cells that comprise the wall of the distributive vascular network is incompletely understood. Exposure of vascular endothelium to various fluid mechanical forces generated by pulsatile blood flow results in alterations in cell morphology [14], metabolic and synthetic activities [12, 15], and gene regulation [16]. Single gene analyses have revealed that a broad spectrum of pathophysiologically relevant genes (PDGF-A and PDGF-B; TGF-*β*), and tissue plasminogen activators (tPA); ICAM, VCAM-1, etc.), are modulated by shear stress in cultured endothelial cells [16, 17]. This regulation is likely accomplished via the binding of transcription factors such as NF-*κ*B and the immediate-early response gene, Egr-1, to shear-stress response elements

1. cDNA synthesis using biotin (◖)-oligo dT

2. Anchoring enzyme (e.g., *Nla*III) digestion
   Streptoavidin beads(▱) binding

3. Add linker 1(▭) and linker 2(▨)

4. Tagging enzyme(e.g., *Bsm*FI) digestion

5. Blunting ends

6. Ligation and PCR amplification

7. Anchoring enzyme digestion
   Concatenation by ligase

8. Cloning and DNA sequencing

9. Data analysis

that are present in the promoters of biomechanically inducible genes [17, 18]. Transcription profiles of the gap junctional connexin 43 gene in individual endothelial cells isolated from both disturbed and undisturbed flow regions exhibited chronically elevated expression in disturbed than in undisturbed flow [12]. Both connexin 43-mediated channel assembly and cell-cell communication were impaired in the disturbed region compared with the undisturbed region confirming regional detectable differences in the levels of gene expression, protein expression and functional communication.

With the relatively recent emergence of high throughput genomic technologies, novel mechanistic insights into the pathobiology of this tissue have been established. The application of high throughput microarray analysis of individual cells of the vascular wall enables a more complete appreciation of the extent and biologic significance of vascular cell activation by biomechanical stimuli, particularly in the context of pathophysiologically relevant phenotypic changes. Available array and SAGE analysis of gene expression of specific vascular wall cellular components will be described.

10.3.1
Endothelial Cells

Endothelial cells reside at the precise interface of the vessel wall and the circulation to rapidly sense changes in blood flow and respond by secreting or metabolizing vasoactive substances to maintain pressure/flow homeostasis. Furthermore, diverse pathophysiologic factors including biomechanical and humoral stimuli all serve to modulate the functional phenotype of the vascular endothelium [12, 19]. Increasing evidence suggests that biomechanical stimulation plays a key role in

---

Fig. 10.1 Schematic of the serial analysis of gene expression (SAGE) method. In the first step, cDNA is synthesized from poly (A)$^+$RNA using the biotin-oligo-dT followed by cleavage of the cDNA (Step 2) by the anchoring enzyme (e.g. *Nla*III), a frequent-cutting (four base-pair recognition sequence) enzyme and expected to cleave most transcripts at least once. The 3'-portion of the cDNA is captured by the streptavidin-coated magnetic beads. Next, two different linkers containing a five base-pair recognition site for a type II restriction enzyme called tagging enzyme such as *Bsm*FI are ligated to two aliquots of the captured cDNA, respectively (Step 3). Short tags plus linkers are then released from the cDNA by tagging enzyme digestion (Step 4), followed by blunting of the ends of the released short tags with the DNA polymerase I, large (Klenow) fragment. In step 6, the linker-tag molecules are ligated tail to tail to form ditags which are amplified by PCR. The ditags are released from linkers by the anchoring enzyme digestion and concatenated by ligase (Step 7) and the concatemers are cloned into a sequencing vector and are subjected to DNA sequencing (Step 8). Finally, in step 9, sequencing of concatemer clones reveals the identity and abundances of each tag. Relative abundance can be calculated by dividing the observed abundance of any tag by the total number of tags analyzed to generate quantitative result. The digital data format of transcripts in both normal and diseased samples can be easily analyzed by SAGE software to identify the genes of interest in specific physiologic or disease states. This figure was modified from Ye *et al.* [9].

the maintenance of vascular integrity as well as in the development of vascular disease [16, 20]. Other analyses of signal cascades, transcriptional factor systems, and the identification of shear-stress-response elements in the promoters of several genes (PDGF-A, VCAM-1, etc.) relevant to the atherosclerotic process [17, 18], have provided insight into the cellular mechanisms linking shear stress stimuli and genetic regulatory events [12, 15]. García-Cardeña G. *et al.* [21] applied transcriptional profiling via cDNA arrays to assess the gene expression in cultured human umbilical vein endothelial cells (HUVEC) exposed to either steady laminar shear stress (10 dynes/cm^2) or turbulent or non-laminar shear stress (spatially and temporally averaged shear stress of 10 dynes/cm^2) for 24 h. Whereas a greater number of genes were downregulated by either laminar or turbulent shear stress than were up-regulated, direct comparison of the two forms of shear stress revealed 100 genes which were differentially regulated (68 up-regulated; 32 down-regulated, turbulent vs. laminar). The application of laminar shear stress to static monolayers resulted in greater than two-fold changes in 205 genes (from a total of 11,397 unique genes) whereas only 86 genes increased greater than two-fold with turbulent shear stress. These studies clearly indicate that cultured endothelial cells can discriminate between these distinct types of fluid mechanical stimulation.

Similar studies in human aortic endothelial cells exposed to laminar shear stress at 12 dynes/cm^2 for 24 hours [22] revealed 125 genes (from 1,185 gene arrays), which exhibited significant difference between the sheared and static samples. Genes related to endothelial cell inflammation including MYD-88 (an adaptor protein essential for signaling via the IL-1 receptor), CD30 ligand (a member of the TNF superfamily and an inducer for T-cell proliferation) and platelet basic protein (a chemokine for neutrophils and monocytes) were down-regulated in response to shear stress as detected by microarray analysis and confirmed by Northern blotting. Since exposure of endothelial cells to proinflammatory cytokines such as IL-1 and TNF results in the modulation of "prothrombotic-proinflammatory" program, shear stress may antagonize inflammatory responses through gene modulation to prevent the induction of the atherogenic "prothrombotic-proinflammatory" program. In contrast, genes such as Tie2 (angiopoeitin-1 and -2 receptor), Flk-1 (VEGF receptor) and MMP-1 (matrix metalloproteinase 1) were significantly upregulated, supporting the concept that shear stress promotes endothelial cell survival, angiogenesis [23], migration and wound healing [24].

Additional studies in human umbilical vein endothelial cells exposed to higher levels of shear stress (25 dynes/cm^2 for 6 or 24 h) also appear to provide evidence to support the concept that physiological levels of shear stress are protective to endothelium [25]. A total of 52 genes were differentially regulated genes (32 up-, 20 down-regulated out of 4,000 genes) in response to shear stress results which were further confirmed by Northern blot analysis. For example, expression of members of the cytochrome P450 (CYP) family of genes (including 1A1 and 1B1) whose gene products are associated with cellular detoxification mechanisms, were dramatically increased. Consistent with this concept, the connective tissue growth factor (CTGF) gene whose product has been postulated to play a role in the develop-

ment of fibrotic disease [26], and is highly expressed in vascular cells in atherosclerotic lesion, was significantly downregulated [27]. These observations are consistent with the hypothesis that physiological arterial shear stress is protective and
retards fibrotic and atherosclerotic disease processes.

Nitric oxide (NO) is a key bioregulatory molecule with expression of endothelial
nitric oxide synthase (NOS) regulated by shear stress. Microarray studies have provided mechanistic information as to how shear stress regulates the synthesis of
NO in endothelial cells [25] with marked increases in arginiosuccinate synthetase
gene expression (2.5 to 3 fold) suggesting that shear stress-induced increase in NO
synthesis may dependent on an increase in L-arginine synthesis from L-citrulline.
The vasoactive intestinal polypeptide (VIP) receptor precursor gene (1.5 to 2 fold)
is also upregulated by shear stress indicating that increased availability of VIP receptors potentially increase VIP-mediated NO production. Similarly, aside from increased NOS expression via shear stress promoter elements, increased elastin
gene expression (2 to 3 folds) by shear stress may increase the binding of elastin
peptides to the elastin-laminin receptor, leading to an increase in intracellular
Ca^{2+} and activation of eNOS with subsequent NO release.

In contrast to shear stress, there is a paucity of studies in which the effect of cyclic strain on endothelial cell gene expression relevant to normal physiologic and
pathophysiologic processes has been examined. We have evaluated the effect of

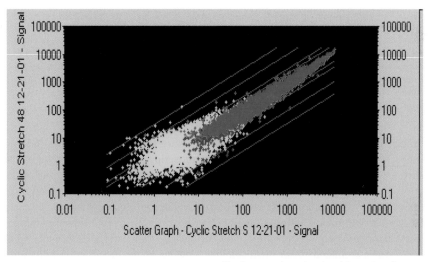

Fig. 10.2 Scatter Plot on Human Lung Endothelium Gene Expression Exposed to Cyclic
Stretch. Expression profiling of HPAEC gene
expression after 18% cyclic stretch for 48 h
was analyzed using the Affymetrix Microarray
Suite software (MAS, ver 5.0). The X axis represents the signal intensity of gene expression in cells exposed to static condition. The
Y axis represents the signal intensity of gene
expression in cells exposed 18% cyclic stretch
for 48 h. Colored dots represent the level of
expression whereas parallel lines represent
fold changes (2, 5, 10 and 30 folds). Expression levels high to low correlate with color
change from red to yellow.

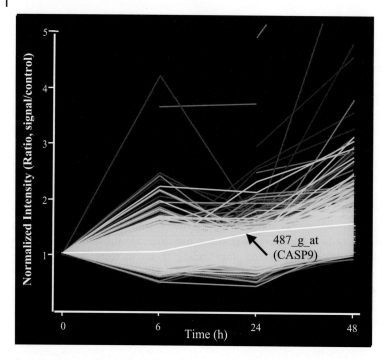

Fig. 10.3 K-mean Clustering Analysis of Gene Expression in Human Lung Endothelium exposed to 18% cyclic stretch. Expression profiling of HPAEC gene expression after 18% cyclic stretch for varied time intervals was analyzed using GeneSpring (ver. 4.2, Silicon Genetics). The representative clusters representing 1,232 genes are displayed. X axis lists the time points and Y axis represent the normalized intensity. Illustrated clusters identify a unique class of genes that are upregulated at 48 h. Selection of a particular gene (such as CASP9 (487_g_ at, white line), an apoptosis regulator gene (listed in Tab. 10.1) illustrates the expression behavior of the individual gene in a time dependent manner.

human pulmonary artery endothelial cell exposure to 18% cyclic stretch (CS) on gene expression profiling (Birukov *et al.* manuscript submitted). Extended cyclic stretch (simulates pathophysiological lung expansion) exposure (48 hr) enhanced thrombin-induced paracellular gap formation and decreased transcellular electrical resistance in conjunction with significant alterations in gene expression detected by Affymetrix human Genechip oligonucleotide arrays. Our preliminary results indicate approximately 60 genes (of $\sim$12,000 genes) were upregulated with $\sim$140 genes down-regulated (more than 2-fold) following cyclic stretch preconditioning (Fig. 10.2). Fig. 10.3 illustrates the expression profiling of human lung endothelium exposed to 18% cyclic stretch for varied time intervals with a representative K-mean cluster (1232 genes) depicted by K-mean analysis which identifies a unique class of genes that are upregulated at 48 h. To assess gene-gene interaction within the cluster, a tree view (dendrogram) analysis was applied and illustrated in Fig. 10.4. Through dendrogram analysis, the related genes within the same cluster, which shows the similar expression pattern, can be identified. Our data reveal that apoptosis-related genes such as DED caspase, DAD-1, CED-3, ICE etc. were among the most up-regulated genes (Tab. 10.1). While experiments utilizing RT-PCR, real-time RT-PCR, Northern and Western blots are currently in progress to validate these data, these data suggest that excessive mechanical cyclic strain promotes programmed cell death and impairs vascular functions.

To identify the repertoire of genes differentially expressed after stimuli associated with the atherosclerotic process, de Waard *et al.* [28] used serial analysis of gene expression (SAGE) to evaluate quiescent human arterial endothelial cells

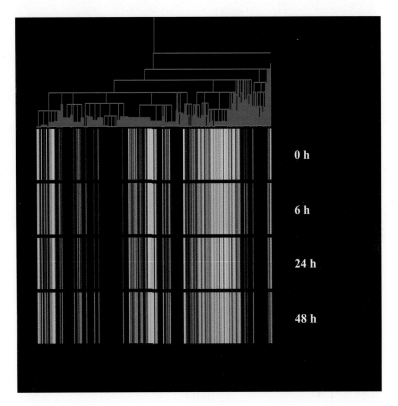

Fig. 10.4 **A** Tree View (Dendrogram) Gene Expression in Human Lung Endothelium Exposed to Cyclic Stretch. Panel A: Expression profiling of HPAEC gene expression after 18% cyclic stretch treatment for 0, 6, 24 and 48 h was analyzed using GeneSpring (ver. 4.2, Silicon Genetics). Approximately 1,232 genes are displayed which allow us to identify the related genes within the same cluster which shows the similar expression pattern. Panel B: Display the adjacent genes of 487_g_ at (Caspase 9) identified in K-mean cluster. The color depicts the raw gene expression levels after CS where the brighter the more expressed. Gray rectangle represents absent of gene. Within the vicinity of CASP9, at least 3 genes (Tissue specific extinguisher, Microtube-associated protein 1A/1B light chain 3, and Quinone oxidoreductase homolog) were known to be involved in apoptosis. Inversin, syntaxin 7 and integrin α V may be implicated to have functions related to apoptosis. The hypothetical protein 37884_f_ at) may be indicated to have a functional role in the apoptosis.

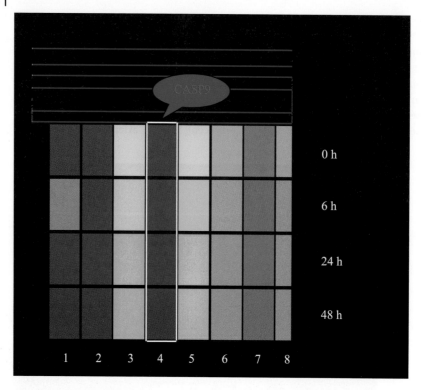

Fig. 10.4 B

Tab. 10.1 HPAEC apoptosis regulatory genes showing upregulation after 18% CS for 48 h

Probe ID	Gene name	48 h
32746_ at	DED caspase	1.66
33774_ at	Death trigger	1.53
34892_ at	DR4 homolog	1.36
35662_ at	New member	1.28
38413_ at	DAD-1	1.50
39436_ at	BNIP 3a	1.56
487_g_ at	CASP 9	1.66
34449_ at	ICE	1.30
33419_ at	Putative	2.88

treated with oxidized LDL, a strong atherogenic stimulus (6 hrs). Among the 12,000 tags analyzed, ∼600 tags (∼5%) were differentially expressed representing 56 differentially expressed genes (42 known genes), including the hallmark endothelial cell activation markers such as interleukin 8 (IL-8), monocyte chemoattractant protein 1 (MCP-1), vascular cell adhesion molecule 1 (VCAM-1), plasminogen

Tab. 10.2 List of genes within the same cluster identified by tree view analysis shows similar expression patterns after 18% CS for 48 h

Gene order	Probe ID	Gene name	Fold of expression
1	37132_ at	Inversin	1.59
2	37884_ f_ at	Hypothetical protein	1.58
3	39370_ at	Microtube-associated protein 1A/1B, light chain 3	1.57
4	487_ g_ at	Caspase 9 (CASP9)	1.56
5	226_ at	CAMP dependent protein kinase	1.48
6	36079_ at	Quinone oxidoreductase homolog	1.45
7	38744_ at	Syntaxin 7	1.41
8	39071_ at	Integrin alpha V	1.38

activator inhibitor 1 (PAI-1), Gro-α, Gro-β and E-selectin. Differential transcription of a selection of the upregulated genes was confirmed by Northern blot analysis.

An important application of the genomic information obtained by microarray analysis is the potential for novel disease-specific therapeutic approaches. To explore aspects of tumor angiogenesis [29, 30], St. Croix *et al.* [31] compared gene expression patterns of endothelial cells derived from blood vessels of normal and malignant colorectal tissues. Of the 170 transcripts predominantly expressed in endothelium, 46 were specifically elevated in tumor-associated endothelium including extracellular matrix proteins (such as several collagens: type IV, $\alpha 2$; type VI, $\alpha 1$ and $\alpha 2$; CD146), while many genes remain with an as yet unknown function. These studies provide molecular information on tumor biological processes, which may have significant implications for the development of anti-angiogenic therapies. In fact, data obtained from St. Croix's group have been further examined by Novatachkova *et al.* [32] concluded that up-regulated transcripts in angiogenesis are involved in extracellular matrix remodeling, cellular migration, adhesion, cell-cell communication rather than in angiogenesis initiation or integrative control. These studies strongly suggest potential application of microarray in therapeutic approaches.

10.3.2
Smooth Muscle Cells

In the artery wall, vascular smooth muscle cells participate in regulating vascular tone and extracellular matrix synthesis and degradation. Under pathophysiological conditions, smooth muscle cells respond to growth factors and cytokines secreted by endothelial cells and surrounding macrophages and de-differentiate into a synthetic, proliferative phenotype and migrate from the media into the intima to produce large amounts of extracellular matrix constituents. Cumulative data (*http://www.ncbi.nlm.nih.gov/UniGene*) identified genes (ICAM-1, GM-CSF, IL-8, NF-κB,

FGF-5 etc.) that are specifically induced in cultured smooth muscle cells upon exposure to atherosclerotic stimuli. The overall alterations in smooth muscle gene expression are largely unknown.

In addition to growth factors and cytokines, the mechanical environment is a major stimulus to alter smooth muscle cell function. Cheng *et al.* [33] have shown that large deformations *in vitro* (10% deformation) lead to transient cell injury, release of FGF-2) and cell proliferation. Similar observations were reported by Lindner *et al.* [34] using an *in vivo* balloon injury model and suggest that mechanical forces can directly regulate vascular smooth muscle cell function. To explore the relationship between mechanical forces and vascular diseases, Feng *et al.* [35] utilized a biaxial cyclic device that generates uniform deformation on vascular smooth muscle cell and monitored gene expression ($\sim 5,000$ gene microarrays) in cells exposed to 12 and 24 hours of continuous uniform strain. Only three genes (cyclooxygenase-1, tenascin-C and plasminogen activator inhibitor-1) were upregulated and 13 genes such as matrix metalloproteinase-1 and thrombomodulin were downregulated by strain, with the results verified by Northern and Western blot analyses. As smooth muscle cells that reside in the artery wall are constantly exposed to a mechanically active environment with variable mechanical loads, smooth muscle cells play a pivotal role in the inappropriate growth or remodeling of the vascular system during arteriosclerosis and hypertension. While incomplete, these studies highlight the possibility of assessing the biological effects of mechanical strain in the vessel wall via gene expression analysis with microarrays.

DNA microarray analysis was utilized to explore the synthesis of proteoglycans, a major component of arterial extracellular matrix, by vascular smooth muscle cells in response to precisely controlled mechanical strains [36]. Large aggregates formed by proteoglycans also contribute to tissue mechanical properties, providing a hydrated sponge-like matrix that resists or cushions against deformation of arterial structure. Since proteoglycans are matrix molecules that participate in tissue mechanics, mechanical deformation (4% cyclic stretch) was applied to arterial smooth muscle cells grown on a thin and transparent membrane to generate a nearly homogeneous biaxial strain profile. While this magnitude of strain does not lead to cell injury, microarray data revealed increased expression of versican (3.2-fold), biglycan (2.0-fold), and perlecan (2.0-fold), whereas decorin mRNA levels decreased to a third of control levels [35] with protein expression correlating well with gene expression. Low-level cyclic stretch did not alter the hydrodynamic size of proteoglycans as evidenced by molecular sieve chromatography but increased sulfate incorporation in both chondroitin/dermatan sulfate proteoglycans and heparan sulfate proteoglycans [36]. These and other data demonstrate that mechanical deformation increases specific vascular smooth muscle cell proteoglycan synthesis and aggregation, suggesting that arterial smooth muscle cells may modify their biomechanical environment in a manner that limits potential biomechanical injury.

10.3.3
Fibroblasts

High throughput microarray analyses have been utilized to explore the temporal profile of gene expression in fibroblasts exposed to serum, a suitable model for growth control and cell cycle progression studies. Iyer V. R. *et al.* [37] utilized cDNA microarrays ($\sim$8,600 human genes) to assess the transcriptional program in primary cultured foreskin fibroblasts stimulated with serum. Clustering analysis of the temporal patterns of expression revealed, that transcriptional alterations were related to the physiology of wound repair. These data suggest that fibroblasts play a larger and richer role in this complex multi-cellular response.

Extracellular interactions of plasma clotting factor VIIa (FVIIa) with tissue factor (TF) on cell surfaces trigger the intracellular signaling events. Pendurthi *et al.* [38] performed limited microarray studies in human fibroblasts treated with factor VIIa (90 min), and found five upregulated genes and one downregulated gene. Two of the upregulated genes (Cyr61 and CIGF) are known to affect cell adhesion, as well as mitogenesis and migration in some cell types. Thus, a potential signaling pathway relevant to the known effects of factor VIIa has emerged from these studies.

10.3.4
Monocytes/Macrophages

Monocytes play a pivotal role in various human infectious and inflammatory diseases and participate as a key effector in the formation of the atherosclerotic plaque. SAGE analysis in lipopolysaccharide (LPS)-stimulated human monocytes [39] yielded total of 35,874 tags corresponding to more than 12,000 different transcripts. LPS-inducible gene products are involved in cell activation and migration, angiogenesis, tissue remodeling, metabolism, and inflammatory reactions (including IL-6, IL-1a, IL-1β, TNF-a, COX2, macrophage inflammatory protein (MIP) MMP-9, PAI-2 RANTES, growth-regulated oncogene (GRO)-a and IL-8) [40]. Exposure of the monocyte/macrophage cell line (THP-1) to Ox-LDL (30 min to 4 days) identified 268 genes (10,000 human gene cDNA microarray) containing known and novel molecular components of the cellular response implicated in the growth, survival, migratory, inflammatory, and matrix remodeling activity of vessel wall macrophages [41]. The observed induction of the orphan receptor LXR-alpha and retinoid X receptor (RXR) and the previously reported induction of peroxisome-proliferator-activated receptor (PPAR)-gamma under these conditions points to the potential involvement of nuclear receptor in the macrophage response to ox-LDL loading.

Although macrophages are critically involved in both atherogenesis and plaque instability where macrophages are subjected to excess mechanical stress, the mechanism in which biomechanical forces affect macrophage function remains incompletely defined. DNA microarray analysis (1,056 genes) was used to assess the transcriptional profile of mechanically induced genes in a human acute monocytic leukemia cell line (THP-1) [42]. The mechanical deformation (1 Hz) was applied to a thin and transparent membrane on which THP-1 cells were cultured. Among the 1,056 well-characterized genes with putative functions, only three genes were induced more than 2.5-fold at 3 and 6 h, and no genes were mechanically induced at 1 h. The increased expression of prostate apoptosis response-4 (PAR-4), interleukin-8 and the NF-κB inducible immediate-early response gene, IEX-1 in an amplitude-dependent induction pattern were confirmed by RT-PCR [42]. Induction of IEX-1 by mechanical deformation may participate in differentiation of monocytes/macrophages and promotion of atherogenesis. Although the number of genes on the array platform was limited, these findings still suggest that mechanical stress in vivo, such as that associated with hypertension, may have an important role in atherogenesis and instability of coronary artery plaques.

Finally, the pathogenesis of unruptured intracranial aneurysms, a common vascular abnormality with a strong genetic component, is poorly understood. Peters *et al.* [43] used a global gene expression analysis approach (SAGE-Lite) in combination with a novel data-mining approach to perform a high-resolution transcript analysis of a single intracranial aneurysm obtained from a 3-year-old girl. Aneurysmal dilation results in a highly dynamic cellular environment in which extensive wound healing and tissue/extracellular matrix remodeling are taking place. Specifically, significant over-expression of genes encoding extracellular matrix components (collagen and elastin), genes involved in extracellular matrix turnover (TIMP-3, OSF-2), cell adhesion and anti-adhesion (SPARC, hevin), cytokinesis (PNUTL2), and cell migration (tetraspanin-5) was observed. Although these data represent the analysis of only one individual, the study highlights the potential for this technique to provide unique insights into the molecular basis of aneurysmal disease and to define numerous candidate markers for future biochemical, physiological, and genetic studies of intracranial aneurysm.

10.4
Future Perspectives

Analyses of vascular cell gene expression patterns, particularly when adequately validated (RT-PCR, Northern and Western blot analysis), may result in direct improvement of disease diagnosis and therapy, and ultimately reduce morbidity and mortality for these devastating illnesses. Unfortunately, vascular cells are not a readily accessible population and represent a heterogenous population of cells whose cell-specific gene expression may be lost when cellular mixtures are analyzed. The increasingly utilization of laser-assisted micro-dissection to provide a more homogenous sample may ultimately lead to the generation of gene expression profiles in disease tissues, tissue- or pathologic stage-specific markers, identify treatment responders and prove useful in identifying new targets for therapeutic agents. Further advances in functional proteomics analysis is vascular components (2D gel electrophoresis coupled to mass spectrometry) will not only confirm array results at the protein level, but together, provide basic information integral to biologic and clinical investigation for years to come.

10.5
References

1 BERTELSEN A. H., VELCULESCU V. E. High-throughput gene expression analysis using SAGE. *DDT.* **1998**, *3*, 152–159.

2 HEDRICK S. M., COHEN D. I., NIELSEN E. A., DAVIS M. M. Isolation of cDNA clones encoding T cell-specific membrane-associated proteins. *Nature.* **1984**, *308*, 149–153.

3 LIANG P., PARDEE A. B. Differential display of eukaryotic messenger RNA by means of the polymerase chain reaction. *Science.* **1992**, *257*, 967–971.

4 VELCULESCU V. E., ZHANG L., VOGELSTEIN B., KINZLER K. W. Serial analysis of gene expression. *Science.* **1995**, *270*, 484–487.

5 LOCKHART D. J., DONG H., BYRNE M. C., FOLLETTIE M. T., GALLO M. V., CHEE M. S., MITTMANN M., WANG C., KOBAYASHI M., HORTON H., BROWN, E. L. Expression monitoring by hybridization to high-density oligonucleotide arrays. *Nat. Biotechnol.* **1996**, *14*,1675–1680.

6 SCHENA M., SHALON D., DAVIS R. W., BROWN P. O. Quantitative monitoring of gene expression patterns with a complementary DNA microarray. *Science.* **1995**, *270*, 467–470.

7 SOUTHERN E. M. et al. Arrays of complementary oligonucleotides for analysis the hybridization behavior of nucleotide acids. *Nucleic Acids Res.* **1994**, *22*, 1368–1373.

8 MADDEN, S. L., WANG, C. J., LANDES, G. Serial analysis of gene expression: from gene discovery to target identification. *DDT* **2000**, *5*, 415–425.

9 YE S. Q., ZHANG L. Q., ZHENG F., VIRGIL D., KWITEROVICH P. O. miniSAGE: gene expression profiling using serial analysis of gene expression from 1 microgram total RNA. *Anal Biochem.* **2000**, *287* (1), 144–152.

10 VIRCHOW R. Der Atermatose Prozess der Arterien. Wien. Med. *Wochenshr.* **1856**, *6*, 825–841.

11 KU D. N., GIDDENS D. P., ZARINS C. K., GLAGOV S. Pulsatile flow and atherosclerosis in the human carotid bifurcation. Positive correlation between plaque location and low and oscillating shear stress. *Arteriosclerosis.* **1985**, *5*, 293–301.

12 DAVIES P. F. Flow-mediated endothelial mechanotransduction. *Physiol. Rev.* **1995**, *75*, 519–560.

13 DEWEY C. F. JR., BUSSOLARI S. R., GIMBRONE M. A. JR., DAVIES P. F. The dynamic response of vascular endothelial cells to fluid shear stress. *J Biomech Eng.* **1981**, *103*(3), 177–185.

14 NEREM R. M., LEVESQUE M. J., CORNHILL J. F. Vascular endothelial morphology as an indicator of blood flow. *J. Biochem. Eng.* **1981**, *103*, 172–178.

15 RESNICK N., GIMBRONE M. A. JR. Hemodynamic forces are complex regulators of endothelial gene expression. *FASEB J.* **1995**, *9*(10), 874–882.

16 GIMBRONE M. A. JR., TOPPER J. N., NAGEL T., ANDERSON K. R., GARCIA-GARDENA G. Endothelial dysfunction, hemodynamic forces, and atherogenesis. *Ann. N. Y. Acad. Sci.* **2000**, *902*, 230–239.

17 KHACHIGIAN L. M., ANDERSON K. R., HALNON N. J., GIMBRONE M. A. JR., RESNICK N., COLLINS T. Egr-1 is activated in endothelial cells exposed to fluid shear stress and interacts with a novel shear-stress response element in PDGF A-chain promoter. *Arteriscler. Thromb. Vasc. Biol.* **1997**, *17*, 2280–2286.

18 KHACHIGIAN L. M, RESNICK N, GIMBRONE M. A JR, COLLINS T. Nuclear factor-kappa B interacts functionally with the platelet-derived growth factor B-chain shear-stress response element in vascular endothelial cells exposed to fluid shear stress. *J Clin Invest.* **1995**, *96*(2), 1169–1175.

19 GIMBRONE M. A. JR., NAGEL T., TOPPER J. N. Biomechanical activation: an emerging paradigm in endothelial adhesion biology. *J. Clin. Invest.* **1997**, *100*, S61–S65.

20 TRAUB O., BERK B. C. Laminar shear stress: mechanisms by which endothelial cells transduce an atheroprotective force. *Arterioscler. Thromb. Vasc. Biol.* **1998**, *18*, 677–685.

21 GARCÍA-CARDEA G., COMANDER J., ANDERSON K. R., BLACKMAN B. R., GIMBRONE M. A. JR. Biomechanical activation of vascular endothelium as a determinant of its functional phenotype. *Proc. Natl. Acad. Sci. USA* **2001**, *98* (8), 4478–4485.

22 CHEN B.P.C., LI Y.S., ZHAO Y., CHEN K.D., LI S., LAO J., YUAN S., SHYY J.Y.J., CHIEN S. DNA microarray analysis of gene expression in endothelial cells in response to 24-h shear stress. *Physiol. Genomics* **2001**, *7*, 55–63.

23 PARTANEN J., DUMONT D.J. Function of Tie1 and Tie2 receptor tyrosine kinases in vascular development. *Current Top Microbiol Immunol* **1999**, *237*, 159–172.

24 GOUPILLE P., JAYSON M.I., VALAT J.P., FREEMONT A.J. Matrix metalloproteinases: the clue to intervertebral disc degeneration? *Spine* **1998**, *23*, 1612–1626.

25 McCORMICK S.M., ESKIN S.G., McINTIRE L.V., TENG C.L., LU C-M., RUSSELL C.G., CHITTUR K.K. DNA microarray reveals changes in gene expression of shear stressed human umbilical vein endothelial cells. *Proc. Natl. Acad. Sci. USA* **2001**, *98*(16), 8955–8960.

26 HAHN A., HEUSINGER-RIBEIRO J., LANZ T., ZENKEL S., GOPPELT-STRUEBE M. Induction of connective tissue growth factor by activation of heptahelical receptors. Modulation by Rho proteins and the actin cytoskeleton. *J. Biol. Chem.* **2000**, *275*, 37429–37435.

27 OEMAR B.S, WERNER A, GARNIER J.M, DO D.D, GODOY N, NAUCK M, MARZ W, RUPP J, PECH M, LUSCHER T.F. Human connective tissue growth factor is expressed in advanced atherosclerotic lesions. *Circulation.* **1997**, *95*(4), 831–839.

28 DE WAARD V., VAN DEN BERG B.M., VEKEN J., SCHULTZ-HEIENBROK R., PANNEKOEK H., VAN ZONNEVELD A.J. Serial analysis of gene expression to assess the endothelial cell response to an atherogenic stimulus. *Gene.* **1999**, *226*, 1–8.

29 FOLKMAN J. Tumor angiogenesis: therapeutic implications. *N. Engl. J. Med.* **1971**, *285*, 1182–1186.

30 KERBEL R.S. Tumor angiogenesis: past, present and the near future. *Carcinogenesis.* **2000**, *21*, 505–515.

31 ST. CROIX B., RAGO C., VELCULESCU V., TRAVERSO G., ROMANS K.E., MONTGOMERY E., LAL A., RIGGINS G.J., LENGAUER C., VOGELSTEIN B., KINZLER K.W. Gene expressed in human tumor endothelium. *Science.* **2000**, *289*, 1191–1202.

32 NOVATCHKOVA M., EISENHABER F. Can molecular mechanisms of biological processes be extracted from expression profiles? Case study: endothelial contribution to tumor-induced angiogenesis. *BioEssays* **2001**, *23*(12), 1159–1175.

33 CHENG G.C., BRIGGS W.H., GERSON D.S., LIBBY P., GRODZINSKY A.J., GARY M.L., LEE R.T. Mechanical strain tightly controls fibroblast growth factot-2 release from cultured human vascular smooth muscle cells. *Circ Research* **1997**, *80*, 28–36.

34 LINDNER V., REIDY M.A. Proliferation of smooth muscle cells after vascular injury is inhibited by an antibody against basic fibroblast growth factor. *Proc. Natl. Acad. Sci. USA.* **1991**, *88*, 3739–3743.

35 FENG Y., YANG J. HUANG H., KENNEDY S., TURI T., THOMPSON J., LIBBY P., LEE R. Transcriptional profile of mechanically induced genes in human vascular smooth muscle cells. *Circ Res* **1999**, *85*, 1118–1123.

36 LEE R.T., YAMAMOTO C., FENG Y., POTTER-PERIGO S., BRIGGS W.H., LANDSCHULZ K.T., TURI T.G., THOMPSON J.F., LIBBY P., WIGHT T.N. Mechanical strain induces specific changes in the synthesis and organization of proteoglycans by vascular smooth muscle cells. *J Biol Chem.* **2001**, *276*(17), 13847–13851.

37 IYER V.R., EISEN M.B., ROSS D.T., SCHULER G., MOORE T., LEE J.C.F., TRENT J.M., STAUDT L.M., HUDSON J. JR., BOGUSKI M.S., LASHKARI D., SHALON D., BOTSTEIN D., BROWN P.O. The transcriptional program in the response of human fibroblasts to serum. *Science.* **1999**, *283*, 83–87.

38 PENDURTHI U., ALLEN K., EZBAN M., RAO L. Factor VIIa and thrombin induce the expression of Cyr61 and connective tissue growth factor, extracellular matrix signaling proteins that could act as possible downstream mediators in factor VIIa x tissue factor-induced signal transduction. *J Biol Chem* **2000**, *275*, 14632–14641.

39 SUZUKI T, HASHIMOTO S, TOYODA N, NAGAI S, YAMAZAKI N, DONG H.Y, SAKAI J,

YAMASHITA T, NUKIWA T, MATSUSHIMA K. Comprehensive gene expression profile of LPS-stimulated human monocytes by SAGE. *Blood*. **2000**, *96*(7), 2584–2591.

40 Ross R. Atherosclerosis – an inflammatory disease. *N Engl J Med*. **1999**, *340*(2), 115–126.

41 SHIFFMAN D., MIKITA T., TAI J.T., WADE D.P. PORTER J.G., SEIHAMER J.J., SOMOGYI R., LIANG S., LAWN R.M. Large scale gene expression analysis of cholesterol loaded macrophages. *J Biol Chem* **2000**, *275*(48), 37 324–37 332.

42 OHKI R., YAMAMOTO K., MANO H., LEE R.T., IKEDA U., SHIMADA K. Identification of mechanically induced genes in human monocytic cells by DNA microarrays. *J Hypertens*. **2002**, *20*(4), 685–691.

43 PETERS D.G, KASSAM A.B, FEINGOLD E, HEIDRICH-O'HARE E, YONAS H, FERRELL R.E, BRUFSKY A. Molecular anatomy of an intracranial aneurysm: coordinated expression of genes involved in wound healing and tissue remodeling. *Stroke*. **2001**,*32*(4),1036–1042.

11
Proteomics, a Step Beyond Genomics: Applications to Cardiovascular Disease

Emma McGregor and Michael J. Dunn

11.1
Introduction

The concept of mapping the human complement of protein expression was first proposed more than twenty years ago [1] following the development of a technique where complex protein mixtures could be separated at high resolution by two-dimensional polyacrylamide gel electrophoresis (2DE) [2, 3]. However, it was not until 1995, that the term 'proteome', defined as the PROTEin complement of a genOME, was first coined by Wilkins working as part of a collaborative team at Macquarie (Australia) and Sydney Universities (Australia) [4, 5].

With the advent of rapid gene sequencing techniques, the genomes from nearly 100 species have been completed at the time of writing and a total of more than 600 genome sequencing projects are in progress (GOLD, Genomics Online Database, http://igweb.integratedgenomics.com/GOLD/). However, genomic information, though a powerful resource, does not attribute function to genes. Although genomic approaches provide information on all of the possible ways in which an organism may express its genes, it does not provide insights into the ways in which an organism may modify its pattern of gene expression in response to particular conditions. In addition it is now apparent that one gene does not encode a single protein, because of processes such as alternative mRNA splicing, RNA editing and post-translational protein modification. Therefore the functional complexity of an organism far exceeds that indicated by its genome sequence alone.

To overcome this problem, gene expression can be studied directly either at the mRNA level or at the protein level. Powerful techniques such as cDNA and oligonucleotide micro-arrays and serial analysis of gene expression (SAGE) make rapid screening of mRNA expression possible. However, there is often a poor correlation between mRNA abundance and the quantity of the corresponding functional protein present within a cell [6, 7]. In addition concomitant co- and post-translational modification (PTM) events can result in a diversity of protein products from a single open reading frame. These modifications can include phosphorylation, sulphation, glycosylation, hydroxylation, N-methylation, carboxymethylation, acetylation, prenylation and N-myristolation.

Whilst there are many new technologies that have been developed to investigate changes in protein expression and PTM's such as LC-MS/MS, discussed in Chapter 11 and ICAT™, discussed in Chapter 12, here we will focus on protein separation, visualization, and quantitation using the traditional 2DE approach.

11.2
Protein Solubilization

2DE is a powerful technique which can be used to analyse the proteome and has the advantage of being able to resolve and reveal post-translational modifications of proteins, as these generally result in changes in protein charge and/or mass. It has become the method of choice for the analysis of protein expression in complex biological systems such as whole cells, tissues and organisms. In order to exploit this technique to the full, sample preparation is of paramount importance. Maximising protein solubilization is key to resolving any proteome. To this end, the correct choice of solubilization buffer is vital. Samples are generally prepared using denaturing lysis buffers containing a high urea concentration (7–9 M) with the addition of a non-ionic (e.g. NP-40) or zwitterionic (e.g. CHAPS) detergent and a reducing agent (e.g. DTT). These conditions are intended to break all the non-covalent interactions between the sample proteins [8]. Preventing any artifactual or chemical modifications once sample proteins are solubilized is imperative. Solubilization buffer should maintain all extracted polypeptides in their intact state (amino acid composition and PTM's), without introducing further artifactual modifications, prior to solubilization. Therefore all enzymes that are able to modify proteins, such as proteases, must be quickly and irreversibly inactivated. It is important to understand that any one solubilization buffer, e.g. the 'standard' buffer described above, is not suitable for all proteins. Currently the major challenge in proteomics is the development of appropriate methods for the solubilization of particular classes of proteins such as hydrophobic membrane and membrane-associated proteins. Variations in solubilization buffer constituents using newly developed detergents, e.g. sulfobetaine, additional denaturing agents such as thiourea, and alternative reducing agents, e.g. trubutyl phosphine, can help to improve protein solubilization and hence the concentration of extracted protein for certain sample types [9, 10]. The choice of solubilisation buffer must be optimized for each sample type to be analysed by 2DE.

11.3
Protein Separation

Protein separation by 2DE involves two steps. The first step or dimension subjects solubilized proteins to isoelectric focusing (IEF) under denaturing conditions. On the application of a current, the charged polypeptide subunits migrate along a polyacrylamide gel strip that contains a pH gradient of an appropriate range, until they reach the pH at which their net charge is zero (isoelectric focusing point or p*I*). The gel strip thus contains discrete protein bands along its length, separated upon the basis of their charge. This is combined with a second-dimension separation on sodium dodecyl sulfate (SDS) polyacrylamide gels where the gel strip is applied to the edge of the second-dimension SDS gel and the focussed polypeptides migrate in an electric current into the second gel, proteins now separating on the basis of their relative molecular mass. This orthogonal combination of charge separation (isoelectric point, p*I*) with size separation (relative molecular mass, M_r) results in the sample proteins being distributed across the two-dimensional gel profile (Fig. 11.1). Resulting protein spot patterns are conventionally orientated with acidic isoelectric points to the left of the gel and low molecular weight proteins at the bottom of the gel (Fig. 11.1).

To be effective and accurate 2DE must produce highly reproducible protein separations and have high resolution. This has been achieved in recent years by the use of immobilized pH gradients (IPG) (Amersham Biosciences) [11]. Previously synthetic carrier ampholytes (SCA) were used to generate the pH gradients required for IEF. Electroendosmotic flow of water (migration of H_3O^+ towards the

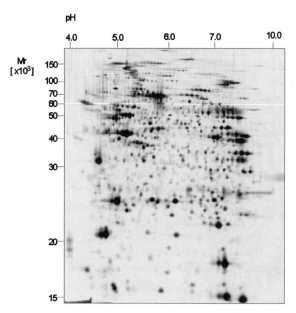

Fig. 11.1 A two-dimensional electrophoresis (2DE) separation of 80 µg of heart (ventricle) proteins. The first dimension comprised an 18 cm non-linear pH 3–10 immobilised pH gradient (IPG) subjected to isoelectric focusing. The second dimension was a 21 cm 12% SDS-PAGE (sodium dodecylsulphate polyacrylamide gel electrophoresis) gel. Proteins were detected by silver staining. The non-linear pH range of the first-dimension IPG strip is indicated along the top of the gel, acidic pH to the left. The apparent isoelectric points (p*I*) of the separated proteins can be estimated from these values. The M_r (relative molecular mass) scale can be used to estimate the molecular weights of the separated proteins.

cathode during electrophoresis) occurring during IEF results in migration of the small SCA molecules towards the cathode. This is known as cathodic drift and results in pH gradient instability and the loss of the more basic proteins from the gel. The application of IPG technology to 2DE overcame this problem. The immobiline reagents consist of a series of eight acrylamide derivatives with the structure CH2=CH-CO-NH-R, where R contains either a carboxyl or tertiary amino group, giving a series of buffers with different pK values distributed throughout the pH 3 to 10 range. The appropriate IPG reagents can be added to the gel polymerization mixture according to published recipes. During the polymerization process the buffering groups that form the pH gradient are covalently attached *via* vinyl bonds to the polyacrylamide backbone. IPG generated in this way are immune to the effects of electroendosmosis. IEF separations are therefore extremely stable. As with anything new, implementing the IPG technology to 2DE separations proved problematic. However, these were overcome and now IPG IEF is the current method of choice for the first dimension of 2DE [12]. IPG IEF is performed on individual gel strips, 3–5 mm wide, cast on a plastic support. A range of pre-cast IPG strips carrying various pH ranges are available commercially (IPG DryStrips, Amersham Biosciences; IPG Ready Strips, Bio-Rad Laboratories). IEF is carried out generally using a horizontal flat-bed IEF system (e.g. IPGphor, Amersham Biosciences; Multiphor, Amersham Biosciences; Protean II XL Cell, Bio-Rad). Following steady-state IEF, strips are equilibrated and then applied to the surface of either vertical or horizontal slab SDS-PAGE gels as described above [12]. Using this method, inter-laboratory studies of heart, barley and yeast proteins have demonstrated that excellent reproducibility can be achieved both in proteins migrating to the same position on a gel and quantitative data [13, 14].

Recently, improvements in protein separation in the first dimension have been reported, where a combined IPGphor/Multiphor approach has been used [15]. The main difference between the two systems is that in-gel rehydration using the IPGphor is carried out under low voltage overnight whereas with the Multiphor it is carried out using rehydration trays at room temperature. An additional advantage, in theory, is that IEF can be performed much faster at higher voltages (up to 8 kV) using the IPGphor. The advantages associated with both pieces of equipment when performing IEF are controversial. By applying a low voltage during in-gel rehydration, improved protein entry, especially of high molecular weight proteins has been reported [16]. However, an increased number of detectable protein spots in 2DE gels has been reported following IEF using the Multiphor compared to the same samples subjected to IEF using the IPGphor [17]. By using the IPGphor for sample loading under a low voltage, then transferring strips to the Multiphor for focusing, an increased sample entry compared to using the Multiphor alone was visualized in the resulting gel pattern [15].

Using standard format SDS-gels (20×20 cm) with 18 cm IPG strips it is possible to routinely separate 2000 proteins from whole-cell and tissue extracts. Resolution can be significantly enhanced (separation of 5000–10,000 proteins) using large format (40×30 cm) 2DE gels [18]. However, gels of this size are very rarely used due to the handling problems associated with such large gels. In contrast,

IGP 3-10 L ⟶

12%
SDS-PAGE
↓

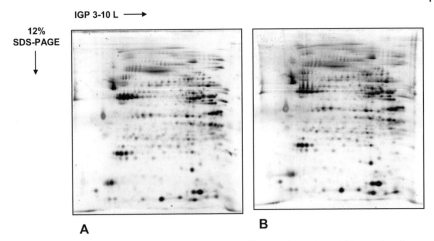

A   B

Fig. 11.2 Small format (7×7 cm) 2DE gel separations of human heart (ventricle) proteins. In the first dimension proteins were separated by isoelectric focusing along a linear pH 3-10 gradient and the second dimension was a 12% SDS-PAGE (sodium dodecylsulphate polyacrylamide gel electrophoresis) gel. Proteins were detected by silver staining (A) 4 µg, (B) 6 µg protein loading.

mini-gels (7×7 cm) can be run using 7 cm IPG strips. Whilst these gels will only separate a few hundred proteins they can be very useful for rapid screening purposes or for studies where the amount of protein available is limited (Fig. 11.2).

Sodium dodecyl sulfate polyacrylamide gel electrophoresis (SDS-PAGE) in the second dimension is performed using apparatus dedicated to running multiple large-format 2DE SDS-gels. Currently available commercial gel-tanks have the capacity to run batches of up to twelve gels at one time (e.g. DALT 2 Ettan DALT, Amersham Biosciences; Protean Plus Dodeca Cell, Bio-Rad). The main advantage that these gel-tanks have over older models is that gels can be run in only 5–6 hours.

Separation and resolution can be further maximised using IPG strips covering a variety of pH ranges. Wide pH 3–10 gradients are used to give an overview of the protein profile of a sample. Narrow range IPG's (e.g pH 4–7, 6–9, 4.0–5.0, 4.5–5.5, 5.0–6.0, 5.5–6.7) are now available and have the capability of 'pulling apart' this protein profile and increasing the resolution, and thereby increasing proteomic coverage, in particular regions (Figs. 11.3a & 3b) [19]. Thus, the task of resolving more and more proteins and increasing the amount of information about a particular proteome becomes easier but perhaps a little interminable. A recent example, looking at 2DE electrophoresis of *Saccharomyces cerevisiae*, revealed an increase in the number of protein spots detected in a particular sample when narrow-range pH gradient strips were used. Using a pH 3–10 gradient, 755 protein spots were detected in the second dimension which rose to a total of 2286 protein spots being detected in the same sample across a number of 2DE gels when a selection of narrow-range pH gradients were employed in the first dimen-

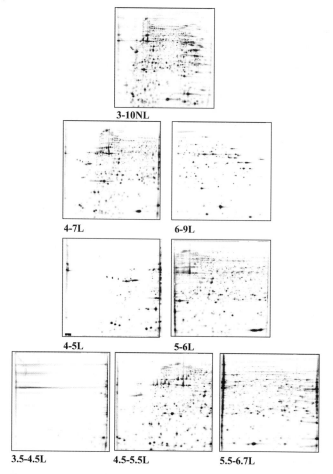

3-10NL

4-7L   **6-9L**

4-5L   **5-6L**

3.5-4.5L   **4.5-5.5L**   **5.5-6.7L**

Fig. 11.3 a, b Zoom gels: increasing proteomic coverage using narrow-range pH gradients. Fig. 11.3a shows human heart (ventricle) proteins which were initially separated in the first dimension using an 18 cm non-linear pH 3–10 immobilised pH gradient (IPG). In order to separate this sample still further, very narrow-range pH gradients were used (e.g. pH 4–7L, 6-9L, 4–5L, 5–6L, 3.5–4.5L, 5.5–6.7L). The second dimension for all gels was a 21 cm 12% SDS-PAGE (sodium dodecylsulphate polyacrylamide gel electrophoresis) gel. Proteins were detected by silver staining. The advantages of using this approach are illustrated in Fig. 11.3 b. One protein spot is detected by silver staining on a 2DE gel after using a pH 3–10 IPG in the first dimension. Analysis by MS (mass spectrometry) identified this spot as either enoyl-CoA-hydratase (EH) or HSP27 (Fig. 11.3 bi). If the same sample is separated further in the first dimension using a narrower pH 4–7 IPG, this same spot now begins to separate out into two protein spots. Spot 1 was identified by MS as EH and spot 2 as HSP27 (Fig. 11.3bii). Increasing the resolution even further by employing a pH 5.5–6.7 IPG in the first dimension allows three protein spots to now be detected in this region by silver staining. Subsequent MS analysis identified spot 1 and 2 as EH and spot 3 as HSP27 (Fig. 11.3biii). (Figures 11.3a and 11.3b taken from Westbrook *et al*, 2001, 22 2865–2871).

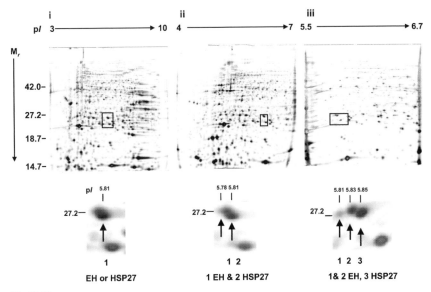

Fig. 11.3b

sion [20]. An important additional advantage of narrow range or so-called zoom IPG's is that they can tolerate higher protein loading for micro-preparative purposes. By generating 2DE gels using IPG strips with overlapping pH ranges it is possible to view the gels side-by-side to form the equivalent of proteomic 'contigs' [21]. As well as narrow pH gradients, very basic gradients up to pH 12 are now available, and are useful for separating very basic cellular components e.g. nuclear and ribosomal proteins [22].

An additional/alternative approach to increasing proteomic coverage, termed subproteomics, is subcellular fractionation. This is based upon sample prefractionation into specific cellular compartments to determine protein location. Combining this with immobilized pH gradients spanning single pH units can be a powerful tool to use to visualize poorly abundant proteins within a cellular system [23].

11.4
Protein Detection/Visualisation

In order to visualize proteins after electrophoresis they have to be detected at high sensitivity. Therefore the stain used should ideally possess a high dynamic range (i.e. it can detect proteins over a wide range of concentrations and linearity to allow rigorous quantitative analysis). Also, it is an advantage if the staining method used is compatible with subsequent protein identification by mass spectrometry. Visualisation of proteins fixed in a gel is usually achieved by staining the whole gel. Traditionally protein staining following gel electrophoresis was performed using Coomassie Brilliant Blue (CBB), however this has very limited sensitivity.

While increased sensitivity can be achieved using CBB in a colloidal form [24], a preferred alternative is silver staining of polyacrylamide gels. This was first introduced in 1979 by Switzer *et al.* [25] and quickly became very popular due to its high sensitivity (approximately 0.1 ng protein per spot, some 100 times more sensitive than Coomassie blue). However, silver staining is not without its own problems. Using this method as a quantitative procedure can be challenging because it is known to be non-stoichiometric and prone to saturation and negative staining effects, where regions of very high protein concentration do not stain and appear as 'holes' in the pattern of stained spots. Not only does silver staining pose a problem when trying to quantify differential protein expression but it can also interfere with subsequent identification by mass spectrometry (MS). The reason for this is the inclusion of gluteraldehyde in the sensitization step of silver staining. Gluteraldehyde reacts with the amino groups of proteins, both ε-amino (lysine side chain) and a-amino, causing extensive cross-linking of proteins. To achieve optimal MS analysis of silver stained proteins, gluteraldehyde must be eliminated from the sensitization step [26–28].

Detection methods based on the use of fluorescent compounds are claimed to have a much larger linear dynamic range, and hence sensitivity, than silver staining. Over recent years we have seen the introduction of fluorescent post-electrophoretic stains such as Sypro Ruby and Sypro Orange which are said to offer improved sensitivity, higher dynamic range and easy handling [29]. Images of fluorescently stained gels can be easily visualized/captured using UV tables and multiwavelength fluorescent scanners.

Extensive studies have been carried out to evaluate the sensitivity of these stains in comparison to silver staining [30, 31]. In our laboratory we have compared three commercially available fluorescent stains, SYPRO Ruby, SYPRO Orange and SYPRO Red (Molecular Probes, Eugene, OR, USA), and compared their sensitivity and dynamic range on 2DE gels with silver staining [32]. We found that SYPRO Ruby had a similar sensitivity to silver staining, but SYPRO Orange and SYPRO Red were considerably less sensitive, requiring respectively four and eight times more protein to be loaded to achieve equivalent 2DE protein profiles. Not unexpectedly, silver staining provided pour linearity even for the less abundant proteins where saturation was not a problem. In contrast the SYPRO dyes provided much greater linearity over their extended dynamic range [32].

Recently much effort has been put in to investigate the compatibility of fluorescent stains, in particular Sypro Ruby, with mass spectrometry. The general finding is that in comparison with conventional silver staining, Sypro Ruby demonstrates a broad linear dynamic range and enhanced recovery of peptides from in-gel digests for matrix assisted laser desorption/ionization-time of flight (MALDI-TOF) mass spectrometry [30, 33]. In addition to increased sensitivity, ease of use is another advantage to the fluorescent stain and the user, since staining a gel using Sypro Ruby follows a much simpler and far less labour intensive protocol than silver staining.

Whilst fluorescent post-electrophoretic stains have increased the sensitivity of detection and quantitation of proteins on 2-D gels, many pairs of gels are still required in order to establish biologically statistically significant differences between for example a control and disease state. However, the development of fluorescent 2DE differential gel electrophoresis (DIGE) by Unlu *et al.* [34] has now enabled us to be able to detect and quantitate differences between experimentally paired samples resolved on the same 2DE gel. DIGE uses two mass- and charge-matched *N*-hydroxy succinimidyl ester derivatives of the fluorescent cyanine dyes Cy3 and Cy5, which possess distinct excitation and emission spectra. Cy3 and Cy5 covalently bind to lysine residues. Two samples that are to be compared are both labeled with each of the dyes. Once both samples are labeled, equivalent aliquots of each are mixed and the resulting mixture of both samples is then subjected to 2DE (Fig. 11.4). The resulting gel is fluorescently imaged twice (once for the Cy3 dye and once for the Cy5 dye) and then both images are superimposed. The labeling process for DIGE takes only 45 minutes which is much faster than staining

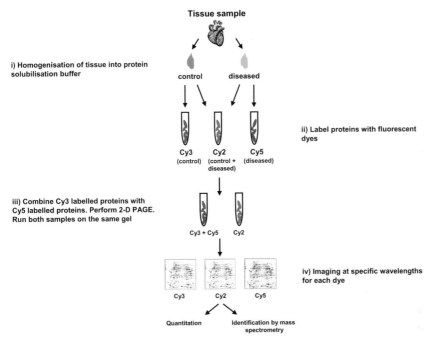

Fig. 11.4   DIGE (differential gel electrophoresis) analysis. A flow diagram illustrating the use of Cy3, Cy5 and Cy2 dyes in the analysis of control versus diseased samples. One protein sample is labeled with Cy3 (e.g. control sample), whilst the other is labeled with Cy5 (e.g. diseased/experimental sample). After the labeling reaction is terminated, equal amounts of each labeled sample are combined. In parallel, equal concentrations of both control and diseased samples are pooled in one tube and labeled with Cy2. Both the mixture of Cy3 and Cy5 labeled proteins and the Cy2 labeled pooled sample are separated on the same gel. The Cy2 labeled pooled sample is used for normalisation purposes. Each 2DE protein profile associated with each individual dye is visualized at a specific wavelength.

methods with similar detection sensitivities such as silver (>2 hours) and Sypro Ruby (>3 hours). The main advantage to DIGE is the ability to compare two samples on the same gel. This reduces inter-gel variability, decreases the number of gels required by 50% and allows for more accurate and rapid analysis of differences. Improved accuracy of quantitation, for comparison of multiple samples, can be achieved using a pooled internal standard containing all experimental samples labeled with a third dye Cy2. This internal standard is included with experimental samples run on each gel (i.e. those labeled with Cy3 and Cy5 dyes) (Fig. 11.4) [35, 36].

DIGE does not interfere with subsequent analysis by mass spectrometry, however there are some disadvantages associated with DIGE. DIGE relies on the fluorescence of the dye for quantitation. Therefore carefully controlled labeling is critical for accurate quantitation. The number of lysine residues contained within a particular protein will dictate the labeling efficiency of that protein. Proteins with a high percentage of lysine residues could be labeled more efficiently than proteins with little or no lysine. Hence, DIGE cannot detect proteins without lysine. Therefore a protein spot can appear to be highly abundant when silver stained but appear significantly reduced when using DIGE.

DIGE can affect protein spot patterns compared to patterns obtained with conventional systems. Proteins are physically labeled i.e. Cy3 and Cy5 molecules are covalently linked to lysine residues. This can alter the location of a protein spot within a gel slightly. The dyes used do not change the p*I* values of proteins because the dye molecules are positively charged which negates the loss of the charge on the lysine group. However, the dye molecule has a molecular weight of 0.5 kDa. An additional 0.5 kDa will not significantly affect the migration of large proteins because only a small percentage of the protein becomes labeled. However this small increase in molecular weight becomes more evident as the molecular weight of proteins decreases.

The labeling reaction is carried out at low efficiency (protein:dye, 5:1) such that at most only one lysine residue per protein molecule is conjugated with dye. This minimizes any change in molecular weight. Since only a small percentage (typically 2–5%) of a protein becomes labeled with Cy3 or Cy5, most of the protein molecules cannot be visualised in the Cy3 and Cy5 images. To locate unlabelled protein the 2DE gel can be stained post-DIGE using Sypro Ruby [35, 37]. The protein of interest is identified by comparing the Sypro Ruby stained image and the Cy3 and Cy5 images.

Whilst the various staining methods described are common in most proteomics laboratories, another technique used to detect proteins is radiolabelling. This too can also achieve very high sensitivity, although not all proteins can be readily radiolabelled. The use of phosphorimaging screens has increased the dynamic range of radiolabelling and has overcome the problem of non-linearity associated with X-ray film images of radiolabelled 2DE separations. The surface of phosphorimaging screens contain a thin layer of $BaFbr:Eu^{2+}$ in a plastic support. A dried 2-D gel is placed in contact with the screen. During this exposure step the β-particles emitted by the radiolabelled proteins pass through the layer converting Eu^{2+}

to Eu^{3+}. After a suitable exposure time the screen is transferred to a phosphorimaging scanner where light from a high intensity HeNe laser (633 nm) is absorbed causing the excited state Eu^{3+} ions to decay back to the Eu^{2+} ground state by the emission of blue (390 nm) luminescence proportional to the amount of radiation incident on the screen. This approach requires relatively short exposure times compared to conventional autoradiography, has a high dynamic range and good linearity of response. The major disadvantage is the high cost of the phosphorimaging screens and the dedicated phosphorimaging device required. The high cost aspect also applies to using fluorescently labeled proteins since a dedicated fluorescent imager is required for their imaging.

11.5
Protein Identification

In this section we aim to provide the reader with a brief overview of methods and techniques used to identify and characterize proteins from 2-D gels. For a more comprehensive review we refer the reader to Chapters 12 and 13.

Although 2DE can effectively separate all the component proteins of a proteome, providing quantitative data, protein identification and function will remain unknown. Conventional methods of identifying proteins from 2-D gels have relied on Western blotting, microsequencing by automated Edman degradation [38] and amino acid compositional analysis [39]. The disadvantages of these techniques is that they are very time consuming with limited sensitivity.

Over the past few years a variety of sensitive methods, centered around mass spectrometry (MS), have become available for the identification and characterization of proteins and peptides (Fig. 11.5). The development of techniques such as matrix-assisted laser desorption/ionization (MALDI) and electrospray (ESI) ionization methods have brought with them very high sensitivity requiring small amounts of sample (a sensitivity of femtomole to attomole) and the capacity for high sample throughput [40, 41]. Both MALDI and ESI are capable of ionizing very large molecules with little or no fragmentation. The type of analyzer used with each can vary. MALDI sources are usually coupled to time-of-flight (TOF) analyzers, whilst ESI sources can be coupled to analyzers such as quadropole, ion-trap and hybrid quadropole time-of-flight (Q-TOF).

Peptide mass fingerprinting (PMF) is the primary tool for MS identification of proteins in proteomic studies. Prior to analysis, an excised protein spot is digested with a protease, typically trypsin, which cleaves proteins at basic amino acids (arginine/lysine residues), if present, breaking proteins down into a mixture of peptides. The masses of the peptides present can then be measured by mass spectrometry to produce a characteristic mass profile or 'fingerprint' of that protein (Fig. 11.6). The mass profile is then compared with peptide masses predicted from theoretical digestion of known protein sequences contained within current databases or predicted from nucleotide sequence databases. This approach proves very effective when trying to identify proteins from species whose genomes are

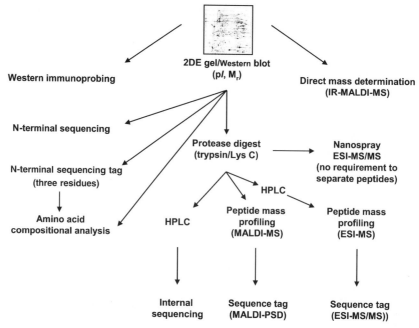

Fig. 11.5 Methods used for the identification and characterization of proteins separted by 2DE. The following abbreviations have been used: 2DE, two-dimensional electrophoresis; ESI, electrospray ionization; HPLC, high performance liquid chromatography; IR, infrared; Mr, relative molecular mass; MALDI, matrix-assisted laser desorption ionization; MS, mass spectrometry; MS/MS tandem mass spectrometry; p*I*, isoelectric point; PSD, post-source decay.

completely sequenced, but is not so reliable for organisms whose genomes have not been completed. This has been a problem in the past for proteomic studies of some animal models of heart disease, for example those involving rats, dogs, pigs and cows. This problem has been shown to be overcome effectively by improving PMF by adopting an orthogonal approach combined with amino acid compositional analysis [42].

However, in some instances it is impossible to assign an unequivocal identity to a protein based on PMF. Amino acid sequence information is then required to confirm an identity. This can be generated by conventional automated chemical Edman microsequencing but is now most readily accomplished using tandem mass spectrometry (MS/MS). MS/MS is a two-stage process, either by means of MALDI-MS with post-source decay (PSD) or ESI-MS/MS triple-quadropole, ion-trap or Q-TOF machines, to induce fragmentation of peptide bonds. One approach, termed peptide sequence tagging, is based on the interpretation of a portion of the ESI-MS/MS or PSD-MALDI-MS fragmentation data to generate a short partial sequence or 'tag'. Using the 'tag' in combination with the mass of the intact parent peptide ion provides significant additional information for the

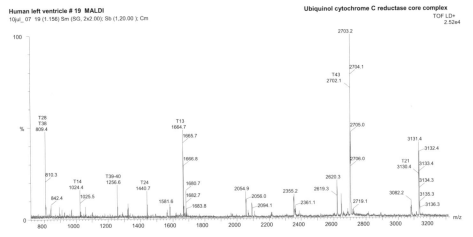

Fig. 11.6 Peptide mass profiling of a silver stained spot from a 2DE separation of human heart (ventricle) proteins. A MALDI-TOF mass spectrum of tryptic peptides is shown, analyzed using a Micromass Tofspec 2E spectrometer (Manchester, UK) operated in the positive ion reflectron mode at 20 kV acceler- ating voltage with "time-lag focusing" enabled. The protein spot of interest was identified as human ubiquinol cytochrome *c* reductase core complex. 2DE gels were silver stained using a modified Amersham Biosciences kit.

homology search [43]. A second approach uses a database searching algorithm SE- QUEST [44]. This matches uninterpreted experimental MS/MS spectra with predicted fragment patterns generated *in silico* from sequences in protein and nucleotide databases. These techniques are highly sensitive as stated above. A nano-electrospray ion source that has a spraying time of more than 30 minutes from only 1 μl of sample, can achieve sensitivities of low femtomole range. This allows the sequencing of protein spots from silver stained 2DE gels containing down to 5 ng protein [45].

11.6
Bioinformatics

Bioinformatics plays a central role in proteomics. At present the greatest bottleneck faced by all proteomic labs is image analysis of 2DE gels. This is a very time consuming process involving a certain amount of technical expertise if accurate conclusions are to be derived from the data. There are a variety of specialized software packages commercially available for quantitative analysis and databasing of electrophoretic separations and a range of bioinformatic tools for identifying proteins based on data from micro-chemical analyses. In addition, special bioinformatic tools have been developed for the further characterization of proteins, ranging from the calculation of their physicochemical properties (e.g. pI, M_r) to the prediction of their potential post-translational modifications and their three-dimensional structure. Most of these tools with their associated databases are available on the Internet through the World Wide Web (WWW), and can be accessed through the ExPASy proteomics server (http://www.expasy.ch/tools/).

Once image analysis is complete and proteins have been identified and characterized it is essential that annotated and curated databases are constructed to store all the data. These databases must allow effective interrogation by the user and, where possible, make data available to other scientists worldwide. Currently this is best achieved using the WWW. To maintain optimal inter-connectivity between these 2DE gel protein databases and other databases of related information, it has been suggested that such 2DE databases are constructed according to a set of fundamental rules [46]. Databases conforming to these rules are said to be 'federated 2-DE databases', while many of the other databases conform to at least some of the rules. An index to these 2DE protein databases can be accessed via WORLD-2DPAGE (http://www.expasy.ch/ch2d/2d-index.html).

11.7
Cardiovascular Proteomics

Heart diseases resulting in heart failure are among the leading causes of morbidity and mortality in the developed world. They can result from either systemic diseases (such as hypertensive heart disease and ischaemic heart disease) or specific heart muscle disease (such as dilated cardiomyopathy). The present generation of therapeutic treatments available for treating most cases of heart failure, do little more than ameliorate symptoms so that, in the majority of cases, the only option is heart transplantation. The causes of cardiac dysfunction in heart failure are still largely unknown. However the pathogenic mechanisms underlying the disease processes are likely to involve significant alterations in myocardial gene and protein expression which will in turn determine the disease progression and outcome. Detailed characterization of these changes should further our understanding of cardiac dysfunction in heart disease and heart failure providing new diagnostic and therapeutic markers.

A relatively recent review describing alterations in the expression of contractile proteins, calcium homeostasis and signal transduction, highlights the fact that most molecular studies of cardiac dysfunction have been carried out on specific cellular systems [47]. However, as with many other areas of research, scientists in this area have quickly acknowledged the potential of proteomics in being able to characterize differential protein expression in heart disease and heart failure. The use of proteomics should provide new insights into the cellular mechanisms involved in cardiac dysfunction.

11.7.1
Heart 2DE Protein Databases

To facilitate proteomic research into heart diseases three groups have established 2DE gel protein databases of human cardiac proteins (Tab. 11.1). These databases known as HSC-2DPAGE [48], HEART-2DPAGE [49], and HP-2DPAGE [50], are accessible through the WWW and conform to the rules for federated 2DE protein databases [46]. The databases contain information on several hundred cardiac proteins that have been identified by protein chemical methods. In addition, 2DE protein databases for other mammals, such as the mouse, rat [51], dog [52], pig and cow, are also under construction to support work on animal models of heart disease and heart failure.

11.7.2
Dilated Cardiomyopathy

Dilated cardiomyopathy (DCM) is a disease of unknown aetiology. It is a severe disease characterized by impaired systolic function resulting in heart failure. So far, proteomic investigations into human heart disease have centered around DCM. Known contributory factors of DCM are viral infections, cardiac-specific autoantibodies, toxic agents, genetic factors and sustained alcohol abuse. As many as 100 cardiac proteins have been observed to significantly alter in their expression in DCM, with the majority of these proteins being less abundant in the diseased heart. This has been reported in numerous studies [51–57]. Many of these proteins have been identified using chemical methods such as mass spectrometry [53, 58–60] and have been classified into three broad functional classes:

1. Cytoskeletal and myofibrillar proteins
2. Proteins associated with mitochondria and energy production
3. Proteins associated with stress responses

Tab. 11.1 2DE heart protein databases accessible via the World Wide Web

Database	Web address	Organ
HEART-2DPAGE	http://userpage.chemie.fu-berlin.de/~pleiss/dh-zb.html	Human heart (ventricle) Human heart (atrium)
HP-2DPAGE	http://www.mdc-berlin.de/~emu/heart/	Human heart (ventricle)
HSC-2DPAGE	http://www.harefield.nthames.nhs.uk/nhli/protein/	Human heart (ventricle) Rat heart (ventricle) Dog heart (ventricle)
RAT HEART-2DPAGE	http://www.mpiib-berlin.mpg.de/2D-PAGE/RAT-HEART/2d/	Rat heart

The next major challenge is to now investigate the contribution of these changes to altered cellular function underlying cardiac dysfunction. This has already begun. For example fifty-nine isoelectric isoforms of HSP27 have been observed to be present in human myocardium using traditional 2DE large format gels. Twelve of these protein spots in the pI range of 4.9–6.2 and mass range of 27,000–28,000 Da were significantly altered in intensity in myocardium taken from patients with DCM. Ten of these protein spots were significantly changed in myocardium taken from patients with ischaemic heart failure [57].

11.7.3
Animal Models of Heart Disease

Investigations of human diseased tissue samples can be compromised by factors such as the disease stage, tissue heterogeneity, genetic variability and the patient's medical history/therapy. It is extremely difficult to avoid any of the above complications when working with human samples. An alternative approach is to apply proteomics to appropriate models of human disease. Animal models are an attractive alternative.

There are several models of cardiac hypertrophy, heart disease and heart failure in small animals, particularly the rat. Examples of proteomic analysis of these models are studies of changes in cardiac proteins in response to alcohol [61, 62] and lead [63] toxicity. Unfortunately the cardiac physiology of these small animal models and their normal pattern of gene expression (e.g. isoforms of the major cardiac contractile proteins) differ from that in larger mammals such as humans. Researchers have thus moved their investigations into higher mammals and recently two proteomic studies of heart failure in large animals have been published. One study investigated pacing-induced heat failure in the dog [64, 65], whilst the second, based in our laboratory, investigated bovine DCM [66]. Both studies demonstrated shared similarities with the proteome analysis of human DCM, with the majority of changes involving reduced protein abundance in the diseased heart.

Identifying altered canine and bovine proteins proved to be particularly challenging since these species are poorly represented in current genomic databases. As a result of this, new bioinformatic tools (MultiIdent) have had to be developed to facilitate cross-species protein identification [67]. For bovine DCM, the most significant change was a seven-fold increase in the enzyme ubiquitin carboxyl-terminal hydrolase (UCH) [66]. This could facilitate increased protein ubiquitination in the diseased state, leading to proteolysis via the 26S proteosome pathway. Interestingly there is evidence to suggest that inappropriate ubiquitination of proteins could contribute to the development of heart failure [68].

More recently we have investigated whether the ubiquitin-proteosome system is perturbed in the heart of human DCM patients [69]. As in bovine DCM, expression of the enzyme UCH was more than 8-fold elevated at the protein level and more than 5-fold elevated at the mRNA level in human DCM. Moreover, this increased expression of UCH was shown by immunocytochemistry to be associated with the myocytes which do not exhibit detectable staining in control hearts. Overall protein ubiquitination was increased 5-fold in DCM relative to control hearts and using a selective affinity purification method we were able to demonstrate enhanced ubiquitination of a number of distinct proteins in DCM hearts. We have identified a number of these proteins by mass spectrometry. Interestingly many of these proteins were the same proteins that we have previously found to be present at reduced abundance in DCM hearts [53]. This new evidence strengthens our hypothesis that inappropriate ubiquitin conjugation leads to proteolysis and depletion of certain proteins in the DCM heart and may contribute to loss of normal cellular function in the diseased heart.

11.7.4
Proteomic Characterization of Cardiac Antigens in Heart Disease and Transplantation

Proteomics can be utilized to identify cardiac-specific antigens that elicit antibody responses in heart disease and following cardiac transplantation. This approach makes use of Western blot transfers of 2DE separations of cardiac proteins. These are probed with patient serum samples and developed using appropriately conjugated anti-human immunoglobulins. This strategy has revealed several cardiac antigens that are reactive with autoantibodies in DCM [70, 71] and myocarditis [72]. Cardiac antigens that are associated with antibody responses following cardiac transplantation have also been characterized and may be involved in acute [73] and chronic [74] rejection.

11.8
Summary

The use of proteomics in cardiovascular research has enabled the identification and characterization of differential protein expression in heart disease. By complementing genomic-based approaches, new insights into complex cellular processes will improve our understanding of cardiac dysfunction. Proteomics should allow an understanding of disease at the molecular level and provide the means for identifying novel diagnostic disease markers. This in turn will accelerate the development of new therapeutic approaches.

11.9
References

1 O'Farrell, P. H. High resolution two-dimensional electrophoresis of proteins. *J. Biol. Chem.* **1975**, *250*, 4007–4021.

2 ANDERSON, N. G., ANDERSON, L. The Human Protein Index. *Clin.Chem.* **1982**, *28*, 739–748.

3 ANDERSON, N. G., MATHESON, A., ANDERSON, N. L. Back to the future: the human protein index (HPI) and the agenda for post-proteomic biology. *Proteomics.* **2001**, *1*, 3–12.

4 WASINGER, V. C., CORDWELL, S. J., CERPA-POLJAK, A., YAN, J. X., *et al.* Progress with gene-product mapping of the Mollicutes: Mycoplasma genitalium. *Electrophoresis.* **1995**, *16*, 1090–1094.

5 WILKINS, M. R., SANCHEZ, J. C., GOOLEY, A. A., APPEL, R. D., *et al.* Progress with proteome projects: why all proteins expressed by a genome should be identified and how to do it. *Biotechnol. Genet. Eng Rev.* **1996**, *13*, 19–50.

6 ANDERSON, L., SEILHAMER, J. A comparison of selected mRNA and protein abundances in human liver. *Electrophoresis.* **1997**, *18*, 533–537.

7 HAYNES, P. A., GYGI, S. P., FIGEYS, D., AEBERSOLD, R. Proteome analysis: biological assay or data archive? *Electrophoresis* **1998**, *19*, 1862–1871.

8 DUNN, M. J. Two-dimensional polyacrylamide gel electrophoresis. *Adv. Electrophoresis.* **1987**, *1*, 1–109.

9 RABILLOUD, T., GIANAZZA, E., CATTO, N., RIGHETTI, P. G. Amidosulfobetaines, a family of detergents with improved solubilization properties: application for isoelectric focusing under denaturing conditions. *Anal. Biochem.* **1990**, *185*, 94–102.

10 RABILLOUD, T. Use of thiourea to increase the solubility of membrane proteins in two- dimensional electrophoresis. *Electrophoresis.* **1998**, *19*, 758-760.

11 BJELLQVIST, B., EK, K., RIGHETTI, P. G., GIANAZZA, E., *et al.* Isoelectric focusing in immobilized pH gradients: principle, methodology and some applications. *J. Biochem. Biophys. Methods.* **1982**, *6*, 317–339.

12 GORG, A., BOGUTH, G., OBERMAIER, C., POSCH, A., *et al.* Two-dimensional polyacrylamide gel electrophoresis with immobilized pH gradients in the first dimension (IPG-Dalt): the state of the art and the controversy of vertical versus horizontal systems. *Electrophoresis.* **1995**, *16*, 1079–1086.

13 CORBETT, J. M., DUNN, M. J., POSCH, A., GORG, A. Positional reproducibility of protein spots in two-dimensional polyacrylamide gel electrophoresis using immobilised pH gradient isoelectric focusing in the first dimension: an interlaboratory comparison. *Electrophoresis.* **1994**, *15*, 1205–1211.

14 BLOMBERG, A., BLOMBERG, L., NORBECK, J., FEY, S. J., *et al.* Interlaboratory reproducibility of yeast protein patterns analyzed by immobilized pH gradient two-dimensional gel electrophoresis. *Electrophoresis.* **1995**, *16*, 1935–1945.

15 CRAVEN, R. A., JACKSON, D. H., SELBY, P. J., BANKS, R. E. Increased protein entry together with improved focussing using a combined IPGphor/Multiphor approach. *Proteomics.* **2002**, *2*, 1061–1063.

16 GORG, A., OBERMAIER, C., BOGUTH, G., HARDER, A., *et al.* The current state of two-dimensional electrophoresis with immobilized pH gradients. *Electrophoresis.* **2000**, *21*, 1037–1053.

17 CHOE, L. H., LEE, K. H. A comparison of three commercially available isoelectric focusing units for proteome analysis: the multiphor, the IPGphor and the protean IEF cell. *Electrophoresis.* **2000**, *21*, 993–1000.

18 KLOSE, J., KOBALZ, U. Two-dimensional electrophoresis of proteins: an updated protocol and implications for a functional analysis of the genome. *Electrophoresis.* **1995**, *16*, 1034–1059.

19 WESTBROOK, J. A., YAN, J. X., WAIT, R., WELSON, S. Y., *et al.* Zooming-in on the proteome: very narrow-range immobilised pH gradients reveal more protein species and isoforms. *Electrophoresis.* **2001**, *22*, 2865–2871.

20 WILDGRUBER, R., HARDER, A., OBER-
MAIER, C., BOGUTH, G., *et al.* Towards
higher resolution: two-dimensional electrophoresis of Saccharomyces cerevisiae
proteins using overlapping narrow immobilized pH gradients. *Electrophoresis.*
2000, *21*, 2610–2616.

21 WASINGER, V. C., BJELLQVIST, B., HUM-
PHERY-SMITH, I. Proteomic 'contigs' of
Ochrobactrum anthropi, application of
extensive pH gradients. *Electrophoresis.*
1997, *18*, 1373–1383.

22 GORG, A., OBERMAIER, C., BOGUTH, G.,
CSORDAS, A., *et al.* Very alkaline immobilized pH gradients for two-dimensional
electrophoresis of ribosomal and nuclear
proteins. *Electrophoresis.* **1997**, *18*, 328–
337.

23 CORDWELL, S. J., NOUWENS, A. S., VER-
RILLS, N. M., BASSEAL, D. J., *et al.* Subproteomics based upon protein cellular location and relative solubilities in conjunction with composite two-dimensional
electrophoresis gels. *Electrophoresis.* **2000**,
21, 1094–1103.

24 NEUHOFF, V., AROLD, N., TAUBE, D., EHR-
HARDT, W. Improved staining of proteins
in polyacrylamide gels including isoelectric focusing gels with clear background
at nanogram sensitivity using Coomassie
Brilliant Blue G-250 and R-250. *Electrophoresis.* **1988**, *9*, 255–262.

25 SWITZER, R. C., III, MERRIL, C. R., SHIF-
RIN, S. A highly sensitive silver stain for
detecting proteins and peptides in polyacrylamide gels. *Anal. Biochem.* **1979**, *98*,
231–237.

26 SHEVCHENKO, A., WILM, M., VORM, O.,
MANN, M. Mass spectrometric sequencing of proteins silver-stained polyacrylamide gels. *Anal. Chem.* **1996**, *68*, 850–
858.

27 YAN, J. X., WAIT, R., BERKELMAN, T., HAR-
RY, R., *et al.* A modified silver staining
protocol for visualization of proteins
compatible with matrix-assisted laser de-
sorption/ionization and electrospray ioni-
zation-mass spectrometry. *Electrophoresis.*
2002, *21*, 3666–3672.

28 Sinha, P., Poland, J., Schnolzer, M., Rabilloud, T. A new silver staining apparatus and procedure for matrix-assisted laser desorption/ionization-time of flight

analysis of proteins after two-dimensional electrophoresis. *Proteomics.* **2001**,
1, 835–840.

29 PATTON, W. F. A thousand points of light:
the application of fluorescence detection
technologies to two-dimensional gel electrophoresis and proteomics. *Electrophoresis.* **2000**, *21*, 1123–1144.

30 LOPEZ, M. F., BERGGREN, K., CHERNO-
KALSKAYA, E., LAZAREV, A., *et al.* A comparison of silver stain and SYPRO Ruby
Protein Gel Stain with respect to protein
detection in two-dimensional gels and
identification by peptide mass profiling.
Electrophoresis. **2000**, *21*, 3673–3683.

31 BERGGREN, K. N., SCHULENBERG, B., LO-
PEZ, M. F., STEINBERG, T. H., *et al.* An improved formulation of SYPRO Ruby protein gel stain: Comparison with the original formulation and with a ruthenium II
tris (bathophenanthroline disulfonate)
formulation. *Proteomics.* **2002**, *2*, 486–
498.

32 YAN, J. X., HARRY, R. A., SPIBEY, C.,
DUNN, M. J. Postelectrophoretic staining
of proteins separated by two-dimensional
gel electrophoresis using SYPRO dyes.
Electrophoresis. **2000**, *21*, 3657–3665.

33 LAUBER, W. M., CARROLL, J. A., DUFIELD,
D. R., KIESEL, J. R., *et al.* Mass spectrometry compatibility of two-dimensional gel
protein stains. *Electrophoresis.* **2001**, *22*,
906–918.

34 UNLU, M., MORGAN, M. E., MINDEN, J. S.
Difference gel electrophoresis: a single
gel method for detecting changes in protein extracts. *Electrophoresis.* **1997**, *18*,
2071–2077.

35 GHARBI, S., GAFFNEY, P., YANG, A., ZVE-
LEBIL, M. J., *et al.* Evaluation of two-dimensional differential gel electrophoresis
for proteomic expression analysis of a
model breast cancer cell system. *Mol. Cell
Proteomics.* **2002**, *1*, 91–98.

36 ALBAN, A., OLU DAVID, S., BJORKESTEN,
L., ANDERSSON, C., *et al.* A novel experimental design for comparative two-dimensional gel analysis: two-dimensional
difference gel electrophoresis incorporating a pooled internal standard. *Proteomics.* **2003**, *3*, in press.

37 Zhou, G., Li, H., DeCamp, D., Chen, S.,
et al. 2D differential in-gel electropho-

resis for the identification of esophageal scans cell cancer-specific protein markers. *Mol.Cell Proteomics.* **2002**, *1*, 117–124.

38 PATTERSON, S. D. From electrophoretically separated protein to identification: strategies for sequence and mass analysis. *Anal. Biochem.* **1994**, *221*, 1–15.

39 YAN, J.X., WILKINS, M.R., OU, K., GOOLEY, A.A., *et al.* Large-scale amino-acid analysis for proteome studies. *J. Chromatogr. A.* **1996**, *736*, 291–302.

40 PATTERSON, S. D., AEBERSOLD, R. Mass spectrometric approaches for the identification of gel-separated proteins. *Electrophoresis.* **1995**, *16*, 1791–1814.

41 YATES, J.R., III Database searching using mass spectrometry data. *Electrophoresis.* **1998**, *19*, 893–900.

42 WHEELER, C.H., BERRY, S.L., WILKINS, M.R., CORBETT, J.M., *et al.* Characterisation of proteins from two-dimensional electrophoresis gels by matrix-assisted laser desorption mass spectrometry and amino acid compositional analysis. *Electrophoresis.* **1996**, *17*, 580–587.

43 MANN, M., WILM, M. Error-tolerant identification of peptides in sequence databases by peptide sequence tags. *Anal. Chem.* **1994**, *66*, 4390–4399.

44 ENG, J.K., MCCORMACK, A.L., and YATES, J.R., III. An approach to correlate tandem mass spectral data of peptides with amino acid sequenecs in a protein database. *J. Am. Soc. Mass Spec.* **1994**, *5*, 976–989.

45 WILM, M., SHEVCHENKO, A., HOUTHAEVE, T., BREIT, S., *et al.* Femtomole sequencing of proteins from polyacrylamide gels by nano-electrospray mass spectrometry. *Nature.* **1996**, *379*, 466–469.

46 APPEL, R.D., BAIROCH, A., SANCHEZ, J.C., VARGAS, J.R., *et al.* Federated two-dimensional electrophoresis database: a simple means of publishing two-dimensional electrophoresis data. *Electrophoresis.* **1996**, *17*, 540-546.

47 MITTMANN, C., ESCHENHAGEN, T., SCHOLZ, H. Cellular and molecular aspects of contractile dysfunction in heart failure. *Cardiovasc. Res.* **1998**, *39*, 267–275.

48 EVANS, G., WHEELER, C.H., CORBETT, J.M., DUNN, M.J. Construction of HSC-

2DPAGE: a two-dimensional gel electrophoresis database of heart proteins. *Electrophoresis.* **1997**, *18*, 471–479.

49 PLEISSNER, K.P., SANDER, S., OSWALD, H., REGITZ-ZAGROSEK, V., *et al.* The construction of the World Wide Web-accessible myocardial two-dimensional gel electrophoresis protein database "HEART-2DPAGE": a practical approach. *Electrophoresis.* **1996**, *17*, 1386–1392.

50 MULLER, E.C., THIEDE, B., ZIMNY-ARNDT, U., SCHELER, C., *et al.* High-performance human myocardial two-dimensional electrophoresis database: edition 1996. *Electrophoresis.* **1996**, *17*, 1700–1712.

51 LI, X.P., PLEISSNER, K.P., SCHELER, C., REGITZ-ZAGROSEK, V., *et al.* A two-dimensional electrophoresis database of rat heart proteins. *Electrophoresis.* **1999**, *20*, 891–897.

52 DUNN, M.J., CORBETT, J.M., WHEELER, C.H. HSC-2DPAGE and the two-dimensional gel electrophoresis database of dog heart proteins. *Electrophoresis.* **1997**, *18*, 2795–2802.

53 CORBETT, J.M., WHY, H.J., WHEELER, C.H., RICHARDSON, P.J., *et al.* Cardiac protein abnormalities in dilated cardiomyopathy detected by two-dimensional polyacrylamide gel electrophoresis. *Electrophoresis.* **1998**, *19*, 2031–2042.

54 KNECHT, M., REGITZ-ZAGROSEK, V., PLEISSNER, K.P., EMIG, S., *et al.* Dilated cardiomyopathy: computer-assisted analysis of endomyocardial biopsy protein patterns by two-dimensional gel electrophoresis. *Eur. J. Clin. Chem. Clin. Biochem.* **1994**, *32*, 615–624.

55 KNECHT, M., REGITZ-ZAGROSEK, V., PLEISSNER, K.P., JUNGBLUT, P., *et al.* Characterization of myocardial protein composition in dilated cardiomyopathy by two-dimensional gel electrophoresis. *Eur. Heart J.* **1994**, 15 Suppl D, 37–44.

56 PLEISSNER, K.P., REGITZ-ZAGROSEK, V., WEISE, C., NEUSS, M., *et al.* Chamber-specific expression of human myocardial proteins detected by two-dimensional gel electrophoresis. *Electrophoresis.* **1995**, *16*, 841–850.

57 SCHELER, C., LI, X.P., SALNIKOW, J., DUNN, M.J., *et al.* Comparison of two-dimensional electrophoresis patterns of heat shock protein Hsp27 species in normal and cardiomyopathic hearts. *Electrophoresis*. **1999**, *20*, 3623–3628.

58 PLEISSNER, K. P., SODING, P., SANDER, S., OSWALD, H., *et al*. Dilated cardiomyopathy-associated proteins and their presentation in a WWW-accessible two-dimensional gel protein database. *Electrophoresis*. **1997**, *18*, 802–808.

59 THIEDE, B., OTTO, A., ZIMNY-ARNDT, U., MULLER, E. C., *et al*. Identification of human myocardial proteins separated by two-dimensional electrophoresis with matrix-assisted laser desorption/ionization mass spectrometry. *Electrophoresis*. **1996**, *17*, 588–599.

60 OTTO, A., THIEDE, B., MULLER, E. C., SCHELER, C., *et al*. Identification of human myocardial proteins separated by two-dimensional electrophoresis using an effective sample preparation for mass spectrometry. *Electrophoresis*. **1996**, *17*, 1643–1650.

61 PATEL, V. B., CORBETT, J. M., DUNN, M. J., WINROW, V. R., *et al*. Protein profiling in cardiac tissue in response to the chronic effects of alcohol. *Electrophoresis*. **1997**, *18*, 2788–2794.

62 PATEL, V. B., SANDHU, G., CORBETT, J. M., DUNN, M. J., *et al*. A comparative investigation into the effect of chronic alcohol feeding on the myocardium of normotensive and hypertensive rats: an electrophoretic and biochemical study. *Electrophoresis*. **2000**, *21*, 2454–2462.

63 TORAASON, M., MOORMAN, W., MATHIAS, P. I., FULTZ, C., *et al*. Two-dimensional electrophoretic analysis of myocardial proteins from lead-exposed rabbits. *Electrophoresis*. **1997**, *18*, 2978–2982.

64 HEINKE, M. Y., WHEELER, C. H., CHANG, D., EINSTEIN, R., *et al*. Protein changes observed in pacing-induced heart failure using two-dimensional electrophoresis. *Electrophoresis*. **1998**, *19*, 2021–2030.

65 HEINKE, M. Y., WHEELER, C. H., YAN, J. X., AMIN, V., *et al*. Changes in myocardial protein expression in pacing-induced canine heart failure. *Electrophoresis*. **1999**, *20*, 2086–2093.

66 WEEKES, J., WHEELER, C. H., YAN, J. X., WEIL, J., *et al*. Bovine dilated cardiomyopathy: proteomic analysis of an animal model of human dilated cardiomyopathy. *Electrophoresis*. **1999**, *20*, 898–906.

67 WILKINS, M. R., GASTEIGER, E., WHEELER, C.aH., LINDSKOG, I., *et al*. Multiple parameter cross-species protein identification using MultiIdent–a world-wide web accessible tool. *Electrophoresis*. **1998**, *19*, 3199–3206.

68 FIELD, M. L., CLARK, J. F. Inappropriate ubiquitin conjugation: a proposed mechanism contributing to heart failure. *Cardiovasc.Res*. **1997**, *33*, 8–12.

69 WEEKES, J., MORRISON, K., MULLEN, A., WAIT, R., *et al*. Hyper-ubiquitination of proteins in dilated cardiomyopathy. *Proteomics*. **2003**, in press.

70 LATIF, N., BAKER, C. S., DUNN, M. J., ROSE, M. L., *et al*. Frequency and specificity of antiheart antibodies in patients with dilated cardiomyopathy detected using SDS-PAGE and western blotting. *J. Am. Coll. Cardiol*. **1993**, *22*, 1378–1384.

71 POHLNER, K., PORTIG, I., PANKUWEIT, S., LOTTSPEICH, F., *et al*. Identification of mitochondrial antigens recognized by antibodies in sera of patients with idiopathic dilated cardiomyopathy by two-dimensional gel electrophoresis and protein sequencing. *Am. J. Cardiol*. **1997**, *80*, 1040–1045.

72 PANKUWEIT, S., PORTIG, I., LOTTSPEICH, F., MAISCH, B. Autoantibodies in sera of patients with myocarditis: characterization of the corresponding proteins by isoelectric focusing and N-terminal sequence analysis. *J. Mol. Cell Cardiol*. **1997**, *29*, 77–84.

73 LATIF, N., ROSE, M. L., YACOUB, M. H., DUNN, M. J. Association of pretransplantation antiheart antibodies with clinical course after heart transplantation. *J. Heart Lung Transplant*. **1995**, *14*, 119–126.

74 WHEELER, C. H., COLLINS, A., DUNN, M. J., CRISP, S. J., *et al*. Characterization of endothelial antigens associated with transplant-associated coronary artery disease. *J. Heart Lung Transplant*. **1995**, *14*, S188–S197.

12
Mass Spectrometry – A Powerful Analytical Tool

HELEN L. BYERS and MALCOLM A. WARD

Molecular weight is an intrinsic property of every substance and mass spectrometry enables this property to be investigated by separating charged particles, so called ions, according to their mass to charge (m/z) ratio. The first stage of a mass spectrometric analysis involves the introduction of the sample into the source region of the instrument where the formation of ions occurs. Then, once formed, the ionised particles are accelerated into the analyser, where they are separated and mass analysed before being detected electronically at the detector. This chapter will provide the reader with an overview of the different ionisation methods and introduce various mass analysers and instrument configurations. Strategies for protein identification and characterisation will be explained, illustrated with examples to show how these powerful analytical tools can be applied to the study of proteins and peptides.

12.1
Ionisation Processes

In order that ions can be analysed they have to be created and introduced into the gaseous phase. There are various methods of ionisation including electron impact (EI), chemical ionisation (CI), plasma desorption (PD) [1] and fast atom bombardment (FAB) [2]. EI and CI are most suitable for small volatile substances, however, the latter two processes were widely used in bioanalytical laboratories during the 1980's and early 90's [3, 4]. This was the era preceding the advent of the now perhaps more familiar processes of MALDI and ESI. These two techniques now dominate scientific literature concerned with the analysis of large biomolecules.

Karas and Hillencamp are credited with the development of MALDI [5]. Here, analyte solutions are allowed to dry onto stainless steel target plates in the presence of a matrix material. This causes the formation of co-crystals. The matrix is generally a low molecular weight organic acid chosen for it's characteristic absorption spectra. Hence, by firing laser light at the target, a gaseous plume of ions is created because, in simple terms, the energy of the beam is absorbed by the matrix and transferred to the analyte molecules, which become charged and can therefore be accelerated into the analyser.

ESI is an atmospheric ionisation technique, which can be used to analyse a wide range of polar molecules. Pioneered by Fenn et al. [6] in 1985, its first application to protein analysis was published four years later [7]. The mechanism of ESI involves the emission of ions from a droplet into the gas phase at atmospheric pressure. Typically the analyte-containing solution is passed directly into a stainless steel capillary, which carries a high potential, normally set at 3 to 5kV. The strong electric field and action of nebulising nitrogen gas causes the formation of a fine spray containing highly charged droplets at the tip of the capillary. These charged droplets are desolvated in a heated source causing the emission of ions. The ions are then transferred from atmospheric pressure into the mass analyser. ESI is a "soft" ionisation technique and typically gives rise to mass spectra dominated by protonated molecules for positive ion analysis and deprotonated molecules for negative ion analysis. Higher molecular weight compounds such as proteins can produce a series of multiply charged ions, which can be mathematically deconvoluted to determine a pseudo-molecular ion species [8]. The presence of contamination, such as ammonium or sodium salts, may cause some compounds to form adducts. In addition to the $[M+H]^+$ molecular ion, resulting spectra show adduct ions such as $[M+NH_4]^+$, $[M+Na]^+$ and $[M+K]^+$. These adducts are observed 17, 22 and 38 Da, respectively up from the protonated molecular ion.

Typical flow rates for ESI are in the range 5 μl/min to 200 μl/min although sources can be operated easily up to 1 ml/min. However, when attempting to analyse low abundant proteins and peptides there are key advantages of using much lower flowrates. Results are dependent on the concentration of the sample in the solvent rather than the total amount of sample entering the source. Wilm and Mann built a miniaturised electrospray device allowing analytes to be introduced into the instrument at nanolitre flowrates [9]. This development meant that peptide samples could be analysed more efficiently because the droplets produced are ∼100 times smaller in volume than those in conventional electrospray sources. The resulting yield of ions is far more efficient therefore reducing the total amount of protein required. Initially fitted onto a triple quadrupole instrument, the nanospray device soon became popular. Typically the peptide mixtures were desalted and pre-concentrated using a small glass capillary packed with Poros R2 resins [10–12]. Additionally, increased sensitivity can be achieved by coupling ESI mass spectrometry with liquid chromatography (LC) and this will be discussed in detail later in the chapter.

12.2
Mass Analysers

Once the ions have been created they enter the analyser where they are separated and mass analysed. Time of flight tubes, quadrupoles and iontraps are all types of analysers and these can be arranged in a number of configurations to allow complicated scan functions to be performed. For example MS/MS experiments where two steps of mass spectrometry are performed in tandem. This generally involves selection of a particular precursor ion, which enters a collision cell to be fragmented by colliding under low pressure with argon or nitrogen. This process is called Collision Induced Dissociation (CID). Resulting fragment ions are subsequently measured using a second mass analyser.

Perhaps the simplest way to separate ions is via their time of flight. TOF analysers are commonly used in conjunction with MALDI and rely on digitisation frequencies of 20 Hz or greater to allow very accurate timing of the ions. Smaller ions travel faster and arrive at the detector before the heavier ions. The Kinetic Energy (KE) of each ion is proportional to the square route of it's mass (M) as given by the formula $KE = \frac{1}{2} MV^2$, where ion velocity is V. Simple calibration routines using known compounds are used to convert time to m/z to give a mass spectral representation. Reflectron devices are simple electronic mirrors which improve resolution by increasing the flight time of the ions.

A quadrupole analyser consists of a set of four steel rods. Opposing rods have a radio-frequency (rf) and a dc voltage applied to them. The two sets of rods are 180° out of phase. All ions entering the analyser take a helical path through the rods with the dc function only allowing ions of a particular mass to pass through and be detected. A triple quadrupole analyser is essentially three sets of quadrupole rods arranged one after another. This arrangement allows Tandem MS/MS experiments to be performed.

The mid 90's saw the development of hybrid instruments such as the Q-Tof (Micromass) and QStar (Applied Biosystems). These instruments offer advantages in terms of sensitivity and resolution. The use of a reflectron time of flight analyser replaces the third quadrupole enabling more ions to be sampled in a shorter timeframe (duty cycle) thus greater sensitivity is achieved. The reflectron device also separates ions more effectively than a quadrupole and thus it is possible to obtain resolution of 10,000 or better [13]. MS/MS spectra of peptides can therefore be acquired with each fragment ion resolved into individual isotopes. This is extremely important if peptide sequencing is to be undertaken accurately. The mass difference between amides and carboxylic acid groups is easily recognised so that asparagines and aspartic acids, glutamines and glutamic acids can be readily assigned. It is even possible to differentiate glutamine and lysine residues, which differ in mass by 0.03638 Da [14]. Isobaric residues leucine and isoleucine are not normally distinguished although this can be achieved by high energy CID using a Tof-Tof analyser [15].

Iontrap mass analysers are similar to quadrupoles and consist of a ring electrode and two endcaps. Ions, produced in the ion source, are effectively stored in the trap. In MS mode these ions are sequentially ejected and scanned onto the detector to produce a full mass spectrum. In MS/MS mode the ions produced by the source are again stored in the ion-trap. At this stage, all ions, except the preselected precursor ion, are ejected from the trap. The remaining precursor ions are then excited, causing them to collide with background helium atoms, and to undergo fragmentation. The resultant fragment ions are then sequentially ejected and scanned onto the detector to produce an MS/MS spectrum. Further fragmentation of the fragment ions, to give MS^3 (MS/MS/MS) spectrum, is also possible.

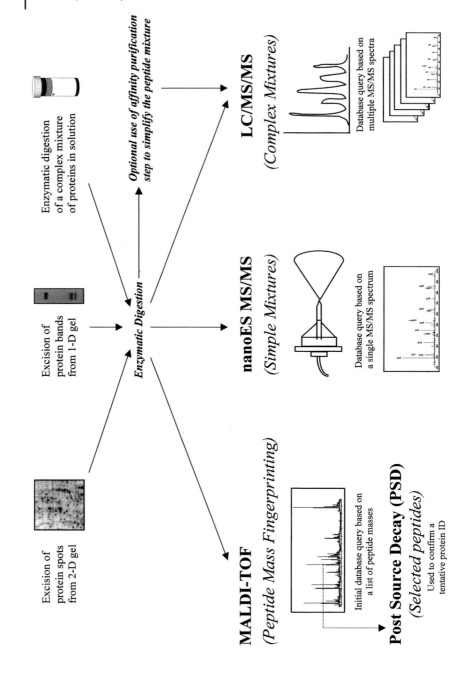

There are two key factors, which drive the development of new instruments. Improvements in overall sensitivity and resolution are always welcome, although this typically impacts on the overall cost of the machinery. Many proteomics laboratories are geared towards high throughput analyses and so the speed at which measurements can be made is also important.

12.3
Strategies for Protein and Characterisation

Mass spectrometry may be applied to answer many questions and there are several ways to address protein characterisation using mass spectrometry (Fig. 12.1). The choice of technology primarily depends on the complexity of the problem and the amount of protein available. The emergence of proteomics has seen a vast increase in the number of proteins requiring analysis. In 1993 several groups reported on the use of mass spectrometry-derived data to interrogate protein sequence databases [16–19]. Since this time methodologies for the identification of low levels of proteins separated by 1-D or 2-D gel electrophoresis have improved [20]. Nowadays it is routinely possible to analyse as little as 5 ng of starting material (equating to 100 femtomoles for a 50 kDa protein), from CBB or silver-stained gels. Typically, proteins of interest are excised from the gel, washed, reduced and alkylated and enzymatically digested using trypsin. An advantage of this method is that detergents, salts and reducing agents are eliminated during the process. It is important to minimise contamination from keratin as this gives rise to peptides which can dominate the resulting mass spectrum. Other strategies involve digesting the proteins in solution or on blots [21, 22]. Following digestion a variety of analytical routes are available as indicated in the schematic. The following section will describe each technique in more detail illustrated with examples.

12.3.1
Protein Identification by MALDI-TOF

MALDI-TOF mass spectrometry can be used as a "first pass" to generate Peptide Mass Fingerprints (PMF's) for enzymatically digested protein samples. This peptide mass mapping approach was first reported by Henzel and co-workers [16]. Each PMF represents a collection of accurately determined peptide molecular weights, which can be submitted as a database search query (Fig. 12.2). To improve sensitivity, the sample supernatants can be desalted and concentrated using commercially available ZipTips [23] prior to target loading. These miniature columns are packed with a small amount of C^{18} reversed phase resin.

Fig. 12.1 Schematic to show a general overview of strategies for protein identification and characterisation using mass spectrometry.

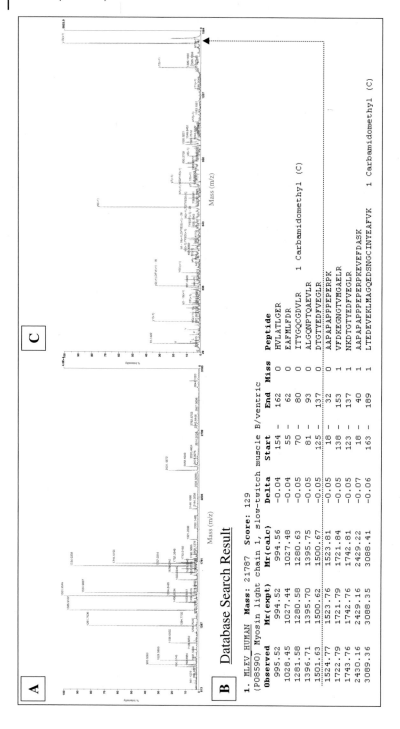

A

B Database Search Result

1. MLEV_HUMAN **Mass:** 21787 **Score:** 129
(P08590) Myosin light chain 1, slow-twitch muscle B/ventric

Observed	Mr(expt)	Mr(calc)	Delta	Start	End	Miss	Peptide		
995.52	994.52	994.56	-0.04	154 –	162	0	HVLATLGER		
1028.45	1027.44	1027.48	-0.04	55 –	62	0	EAFMLFDR		
1281.58	1280.58	1280.63	-0.05	70 –	80	0	ITYGQCGDVLR	1	Carbamidomethyl (C)
1396.71	1395.70	1395.75	-0.05	81 –	93	0	ALGQNPTQAEVLR		
1501.63	1500.62	1500.67	-0.05	125 –	137	0	DTGTYEDFVEGLR		
1524.77	1523.76	1523.81	-0.05	18 –	32	0	AAPAPAPPPEPERPK		
1722.79	1721.79	1721.84	-0.05	138 –	153	1	VFDKEGNGTVMGAELR		
1743.76	1742.76	1742.81	-0.05	123 –	137	1	NKDTGTYEDFVEGLR		
2430.16	2429.16	2429.22	-0.07	18 –	40	1	AAPAPAPPPEPERPKEVEFDASK		
3089.36	3088.35	3088.41	-0.06	163 –	189	1	LTEDEVEKLMAGQEDSNGCINYEAFVK	1	Carbamidomethyl (C)

C

Not all proteins can be identified directly by PMF's. For example, if the protein sequence is not present in the database or if the spectrum only contains a limited number of peptides. In the latter cases, PSD may be used to confirm ambiguous protein identifications. Here, a particular peptide molecular ion may be selected using an iongate and the fragments produced as a consequence of the high energy ionisation process, can be mass analysed by focusing them at the detector (Fig. 12.2). Typically PSD spectra contain complex patterns of fragment ions, which are difficult to interpret. Although such data may yield the identity of a putative protein hit, PSD is not generally used to sequence peptides. Additionally, MALDI-TOF is less effective for the analysis of complex protein mixtures as generally only the most abundant proteins are identified. Small proteins (20 kDa or less) may also prove difficult to analyse as these tend to generate fewer appropriately sized tryptic peptides therefore decreasing the chances of an unambiguous identification. In such cases further analytical techniques must be employed.

12.3.2
Protein Identification by Tandem MS/MS

Tandem MS/MS of peptides may be used to provide sequence information by virtue of the fragment ions produced (Fig. 12.3). Fragmentation occurs generally across the peptide bond leading to a ladder of sequence ions that are diagnostic of the amino acid sequence. The difference between consecutive ions in a series indicates the mass of the amino acid at that position in the peptide. The most common ion types are b and y ions. The C-terminal containing fragments are designated y-ions and the N-terminal containing fragments are designated b-ions [24]. Peptides created by trypsin proteolysis and ionised by electrospray generally form ions that are doubly charged. This stems from the presence of basic groups within the peptide, namely, the alpha amino group at the N-terminus and the side chain of the C-terminal lysine or arginine. MS/MS spectra of such peptides generally yield a prominent y-type ion series in the high mass end of the spectrum [25]. Ideally, for *de novo* sequencing purposes, a complete set of complementary b and y ions will ensure a high confidence level for the proposed peptide sequence.

Peptides derived from gel separated proteins can be delivered into the mass spectrometer either by using a static nanoelectrospray device or, more dynamical-

---

Fig. 12.2 A – MALDI-TOF PMF of a 21 kDa spot from human heart left ventricle. The spectrum shows isotopically resolved molecular ions for the peptides resulting from tryptically digested material. Following internal lock mass calibration, the spectrum can be processed to give a monoisotopic masslist, which is then submitted as a database search query. B – The results of the database search displayed via a web browser. In this instance the protein is clearly identified as myosin light chain 1 with 10 peptides matched within a mass tolerance of 30 ppm. The peptides cover 62% of the total sequence. C – PSD spectrum of m/z 1501.85 showing the fragment ions matching the sequence DTGTYEDFVEGLR. PSD data can be used to verify an ambiguous protein identity. Data courtesy of Nicola Leeds.

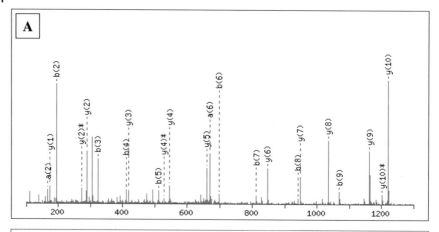

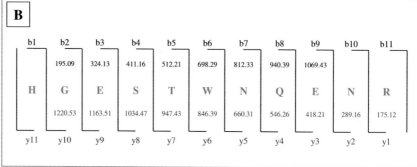

Fig. 12.3 A – MS/MS spectrum of a peptide originating from phosphoglycerate mutase (P15 259; PGM), PGM was isolated from a 1-D gel band reduced, alkylated and digested with trypsin. Peptides were separated using a 75 micron ID PepMap RP column and eluted at a flowrate of 200 nl/min into an ESI source fitted to a Q-Tof instrument. The doubly charged peptide ion at m/z 679.26, relating to Mr 1356.58 Da, was selected for MS/MS. B – The fragment ions produced matched the sequence HGESTWNQENR. A strong series of b and y ions are matched to this sequence. The presence of the histidine at the N-terminus helps to stabilise the b ions.

ly, by liquid chromatography. The on-line coupling of LC to Tandem MS/MS instrumentation (LC/MS/MS) is presently unparalleled in terms of sensitivity and dynamic range and is particularly powerful when used in conjunction with mass spectrometers capable of data dependent scanning. Automatic switching, for example, from survey mode to MS/MS mode, enables a great deal of data to be generated in a short space of time.

LC/MS/MS datasets can contain hundreds of MS/MS spectra and these are typically processed into a format to allow direct database query submissions.

LC/MS/MS can be used alone or in combination with 1-D or 2-D electrophoresis. RP chromatography is the method of choice due to the solvents being compatible with the ionisation process. Sample cleanup, separation and concentration

A) Injection of sample

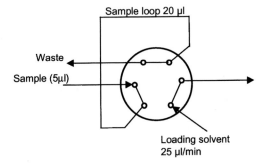

B) Sample loading to pre-column

C) Separation with the analytical column in-line

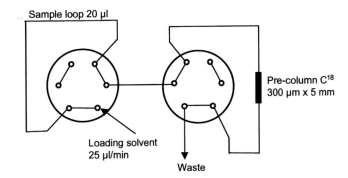

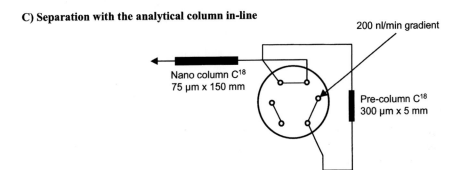

Fig. 12.4 Schematic showing the valve configurations for an LC system set up to pre-concentrate the peptide sample. This system can be linked to an autosampler capable of injecting µl amounts of sample. Valve positions are shown for A) sample injection B) sample loading onto the pre-column and C) switching the analytical column in line with the pre-column.

are all achieved in a single experiment with the temporal separation making it possible for complex mixtures of peptides to be analysed effectively. Nowadays, 50–100 μm ID columns, packed with polymeric or silica based C^{18} coated stationary phase, are preferred as these minimise the elution volume of peptides and hence increase the concentration of the peptides entering the mass spectrometer.

To improve sensitivity, pre-concentration techniques can be employed. For example, the sample can be concentrated on a pre-column of RP media, prior to separation on the analytical column. This is made possible due to an elegant valve switching system (Fig. 12.4). The pre-column not only concentrates the sample but also removes salts, buffers and other low molecular contaminants. This helps to preserve the life time of the analytical column. Other advantages of such a system include the ability to automate sample introduction, via autosamplers capable of injecting microlitre volumes.

To enhance the MS/MS information obtained for low level peptides the ability to selectively extend the analysis time of interesting peaks has been demonstrated [26]. When a peak of interest is eluted the flow rate is reduced so as the time spent acquiring data on the peptides is increased. This is referred to as "peak parking" and flow rates between 5–200 nl/min are typically employed. CZE can also be coupled to Tandem MS/MS instrumentation [27], although this technique is less robust than LC/MS/MS and consequently less popular.

Multidimensional LC is beginning to emerge as an alternative approach to 2-DE [28, 29]. Generally, this combines strong cation exchange (SCX) chromatography and RP chromatography. Discrete fractions of peptides are displaced from the SCX column onto the RP column using a salt gradient. Peptides are retained on the RP column and contaminating salts and buffers are washed to waste. The peptides are then eluted from the RP column into the mass spectrometer. Coupling direct-identification approaches with quantitative methods, such as ICAT [30] and MCAT [31], to measure relative protein expression will greatly increase the value of the data produced. Even when analysing complex mixtures using multidimensional LC, pre-fractionation is still beneficial because, as with any direct analysis of crude protein mixtures, the major proteins present are likely to mask the less abundant proteins. Multidimensional LC also has the ability to detect and identify a wide variety of protein classes, including proteins with extreme pI's, Mr abundance and hydrophobicity (e.g. Transmembrane proteins).

12.3.3
Database Searching

The recent advances in mass spectrometry technologies have been closely followed by the development of bioinformatics tools to extract the relevant information contained within information-rich datasets. Programmes such as MASCOT, Protein Prospector, Sequest and Global server [32–35] are invaluable and by accepting uninterpreted data, make the task of analysing large datasets much more amenable. *De novo* sequencing algorithms hold promise however, in our laboratory, this task still remains largely a manual process. Typically, MS/MS data is initially searched against protein sequence databases and subsequently against nucleotide databases, such as expressed sequence tag (EST), and more recently raw genomic sequences [36, 37]. There are several advantages of searching genomic databases directly. The MS/MS data may help to define the structure of the gene by locating start and stop codons and also intron-exon boundaries. Unfortunately, nucleotide databases cannot be interrogated effectively by MALDI alone. Modifications of proteins, that are not apparent from the DNA sequence, can also be identified using proteomic methodologies.

12.3.4
Post Translational Modifications

Once translated, protein molecules are transported through the cell and many become modified to suit particular tasks or functions. Such modifications include disulphide bridge formation, the addition of lipids, glycosylation and phosphorylation. It is these modification events which greatly diversify the proteome. Although there are only 30000 gene products encoded within the human genome [38] each gene could potentially produce a heterogeneous population of related protein molecules, perhaps hundreds of unique species. Whilst minimal sequence information is sufficient to identify a protein in a sequence database, perhaps by matching two or three peptides, the analysis of post translational events can be far more challenging. Ideally, the goal is to obtain complete sequence coverage of the protein and hence more intricate sample preparation and analyses are required.

It is possible to locate modified peptides by virtue of an alteration to an expected molecular weight (Tab. 12.1). Modified peptides can be chemically/enzymatically treated to remove the modification. A comparison of mass spectra recorded before and after such treatments can be compared. For example, alkaline phosphatase can be used to remove phosphate groups [39] and PNGase F can be used to remove complete glycan structures from asparagine residues [40]. Sequential removal of sugar residues from N-linked oligosaccharides using a set of endoproteinase enzymes has also been demonstrated [41]. The detection of phosphopeptides and glycopeptides can be enhanced by the use of product ion scanning [42]. These scans are based on the formation of specific signature ions during CID. Precursor ions that give rise to the signature ion of interest are selectively recorded. For example, selection of m/z 204 as the product ion of N-acetylhexosamine, $[HexNAc]^+$, is indicative of glycopeptides [43], whilst selection of m/z 79, the phosphate ion PO_3^-, may be used to locate phosphopeptides [44]. Alternatively, the loss of phosphoric acid, H_3PO_4, can be induced from phosphorylated species, which facilitates their detection via a Constant Neutral Loss (CNL) experiment [45]. An MS/MS spectrum of a phospho-peptide is shown (Fig. 12.5). Here, the phospho-peptide originates from Neurofilament triplet H (NFH) protein and contains the sequence motif KSP, which is recognised by a family of proline-dependent protein kinases. A comprehensive analysis of the phosphorylation sites of NFH and other neurofilament proteins has been reported [46].

Tab. 12.1 Mass changes of common post-translational modifications of peptides and proteins

Modification	Monoisotopic mass change	Average mass change
Homoserine formed from Met by CNBr	−29.99281	−30.0935
Pyroglutamic acid from Gln	−17.02655	−17.0306
Disulphide bond formation	−2.01565	−2.0159
C-terminal amide from Gly	−0.98402	−0.9847
Deamidation of Asn and Gln	0.98402	0.9847
Methylation	14.01565	14.0269
Hydroxylation	15.99491	15.9994
Oxidation of Met	15.99491	15.9994
Formylation	27.99491	28.0104
Acetylation	42.01056	42.0373
Carboxylation of Asp and Glu	43.98983	44.0098
Carboxyamidomethyl (CAM) Cys	57.02146	57.0520
Carboxymethyl (Cmc) Cys	58.00548	58.0367
Phosphorylation	79.96633	79.9799
Sulphation	79.95682	80.0642
Pyridylethyl (PE-Cys)	105.05785	105.1393
Cysteinylation	119.00410	119.1442
Pentose	132.04226	132.1161
Deoxyhexose	146.05791	146.1430
Hexosamine	161.06881	161.1577
Hexose	162.05282	162.1424
Lipoic acid (amide bond to Lys)	188.03296	188.3147
N-acetylhexosamine	203.07937	203.1950
Farnesylation	204.18780	204.3556
Myristoylation	210.19836	210.3598
Biotinylation (amide bond to Lys)	226.07760	226.2994
Pyridoxal phosphate (Schiff base on Lys)	231.02966	231.1449
Palmitoylation	238.22966	238.4136
Stearoylation	266.26096	266.4674
Geranylgeranlylation	272.25040	272.4741
N-acetylneuraminic acid (Sialic acid)	291.09542	291.2579
Glutathionlyation	305.06816	305.3117
N-glycolylneuraminic acid	307.09033	307.2573
5'-Adenosylation	329.05252	329.2091
4'-Phosphopantetheine	339.07797	339.3294
ADP-ribosylation	541.06111	541.3052

The emergence of proteomics has caused a growing demand for high sensitivity protein identification and characterisation and the use of mass spectrometry has become a central component of many proteomic research programmes. Already the emphasis of proteomics is changing from an unfocused high-throughput approach, where the results are of limited value, to a more focused approach involving detailed analyses of the protein samples. For example, specific proteins of interest can be enriched either by cell fractionation methods and/or affinity based

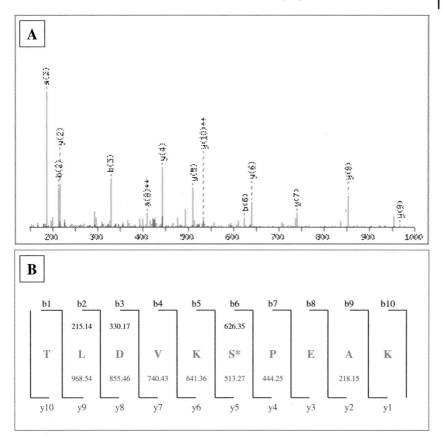

Fig. 12.5 A – MS/MS spectrum of a phosphopeptide originating from Neurofilament triplet H protein (P12,036; NFH). NFH was isolated from a 1D gel band reduced, alkylated and digested with trypsin. Peptides were separated using a 75 micron ID PepMap RP column and eluted at a flowrate of 200 nl/min into a Z-spray source fitted to a Q-Tof instrument. The doubly charged peptide ion at m/z 584.29, relating to Mr 1166.56 Da, was selected for MS/MS. B – The fragment ions produced matched the sequence TLDVKS*PEAK were S* represents a phosphorylated serine residue. Sample courtesy of Diane Hanger.

protein purification strategies. Typically, the enriched protein sample is then separated by SDS-PAGE and the protein bands characterised by mass spectrometry. In this way organelles can be selectively purified and their protein complement determined by mass spectrometry. Two recent studies on the golgi and nucleolus exemplify this approach [47, 48]. In the case of affinity-based purification methods the target protein is either "pulled-down" using a specific antibody or via a molecular tag [21]. Recently, Eaton et al. have targeted proteins which become S-thiolated during oxidative stress following ischemia and repurfusion [49]. Here reactive cysteines were labelled with a biotin reagent and purified using streptavidin. Additionally, if the purification process is carried out under non-denaturing conditions, other proteins associated to the target protein may be pulled down. The N-methyl-D-aspartate (NMDA) neurotransmitter receptor complex was characterised in this manner [50]. Here 77 proteins were found to be associated with the complex.

Successful purification by affinity based strategies relies on sufficient affinity of the protein complex to the bait and on optimised purification steps. To this end, Rigaut and co workers have developed an elegant tandem affinity purification (TAP) strategy [51]. This double tagging approach improves complex recovery and reduces non-specific binding. Pre-enrichment of a specific protein will also provide more comprehensive sequence coverage, which is necessary for the complete characterisation of PTM's or indeed any changes to the published sequences. For example, Gatlin et al. recently demonstrated the identification of amino acid sequence variations resulting from single nucleotide polymorphisms (SNPs) by obtaining 99% sequence coverage of human hemoglobin in a single LC/MS/MS experiment [52].

At this point we hope the reader will have gained an appreciation of the power and versatility of mass spectrometry. The exponential growth in the number of scientific publications in the field of biological mass spectrometry is clear evidence of the tremendous impact the technologies described have had during the last decade. We are convinced that the next few years will see many advances in the technology and also, perhaps more importantly, that mass spectrometry will enable new discoveries and insights into cellular biology and protein function.

To end, it is important to note that the results from any mass spectrometry experiment will ultimately depend on the quality of the sample initially. Proteome research requires many scientific disciplines and we believe that the challenges posed can be best met by closely integrating good separation science with state-of-the-art mass spectrometry and bioinformatics. Finally, we would like to thank all our colleagues at Proteome Sciences for their assistance and support during the preparation of this manuscript.

12.4
References

1 MacFarlane R. D., Togerson D. F. Californium-252 plasma desorption mass spectroscopy. *Science*, **1976**, *191*, 920.

2 Morris, H. R., Panico, M., Barber, M., Bordoli, R. S., et al. Fast atom bombardment: a new mass spectrometric method for peptide sequence analysis. *Biochem. Biophys. Res. Commun.* **1981**, *101*, 623–631.

3 Chen, L., Cotter, R. J., Stults, J. T. Plasma desorption mapping of the tryptic digest of 23-kDa recombinant human growth hormone. *Anal Biochem*, **1989**, *183*, 190–194.

4 Barber, M., Bordoli, R. S., Elliott, G. J., Horoch, N. J., Green, B. N., Fast atom bombardment mass spectrometry of human proinsulin. *Biochem. Biophys. Res. Commun.* **1983**, *110*, 753–757.

5 Karas, M., Hillenkamp, F. Laser desorption ionization of proteins with molecular masses exceeding 10,000 daltons. *Anal. Chem.* **1988**, *60*, 2299–2301.

6 Whitehouse, C. M., Dreyer, R. N., Yamashita, M., Fenn, J. B. Electrospray interface for liquid chromatographs and

mass spectrometers. *Anal. Chem.* **1985**, *57*, 675–679.

7   FENN, J. B., MANN, M., MENG, C. K., WONG, S. F., WHITEHOUSE, C. M. Electrospray ionization for mass spectrometry of large biomolecules. *Science.* **1989**, *246*, 64–71.

8   VAN DORSSELAER, A., BITSCH, F., GREEN, B., JARVIS, S., et al. Application of electrospray mass spectrometry to the characterization of recombinant proteins up to 44 kDa. *Biomed. Environ. Mass Spectrom.* **1990**, *19*, 692–704.

9   WILM, M., MANN, M. Analytical properties of the nanoelectrospray ion source. *Anal. Chem.* **1996**, *68*, 1–8.

10 OTTO, A., THIEDE, B., MÜLLER, E. C., SCHELER, C., WITTMAN-LIEBOLD, B., JUNGBLUT, P. Identification of human myocardial proteins separated by two-dimensional electrophoresis using an effective sample preparation for mass spectrometry. *Electrophoresis.* **1996**, *17*, 1643.

11 ERDJUMENT-BROMAGE, H., LUI, M., LACOMIS, L., TEMPST, P. Characterizing proteins from 2-DE gels by internal sequence analysis of peptide fragments. Strategies for microsample handling. *Methods Mol Biol* **1999**, *112*, 467–472.

12 JENSEN O. N., WILM M., SHEVCHENKO A., MANN M. Sample preparation methods for mass spectrometric peptide mapping directly from 2-DE gels. *Methods Mol Biol* **1999**, *112*, 513–530.

13 KOVTOUN, S. V., ENGLISH, R. D., COTTER, R. J. Mass correlated acceleration in a reflectron MALDI TOF mass spectrometer: an approach for enhanced resolution over a broad mass range. *J. Am. Soc. Mass Spectrom.* **2002**, *13*, 135–143.

14 TAYLOR, J. A., JOHNSON, R. S. Implementation and uses of automated de novo peptide sequencing by tandem mass spectrometry. *Anal. Chem.* **2001**, *73*, 2594–2604.

15 MEDZIHRADSZKY, K. F., CAMPBELL, J. M., BALDWIN, M. A., FALICK, A. M., JUHASZ, P., VESTAL, M. L., BURLINGAME, A. L. The characteristics of peptide collision-induced dissociation using a high-performance MALDI-TOF/TOF tandem mass spectrometer. *Anal. Chem,* **2000**, *72*, 552–558.

16 HENZEL, W. J., BILLECI, T. M., STULTS, J. T., WONG, S. C. Identifying proteins from two-dimensional gels by molecular mass searching of peptide fragments in protein sequence databases. *Proc. Natl. Acad. Sci. USA.* **1993**, *90*, 5011–5015.

17 MANN, M., HOJRUP, P., ROEPSTORFF, P. Use of mass spectrometric molecular weight information to identify proteins in sequence databases. *Biol. Mass Spectrom.* **1993**, *22*, 338–345.

18 PAPPIN, D. J. C., HOJRUP, P., BLEASBY, A. J. Rapid identification of proteins by peptide-mass fingerprinting. *Curr. Biol.* **1993**, *3*, 327–332.

19 YATES, J. R., SPEICHER, S., GRIFFIN, P. R., HUNKAPILLER, T. Peptide mass maps: a highly informative approach to protein identification. *Anal. Biochem.* **1993**, *214*, 397–408.

20 SHEVCHENKO, A., WILM, M., VORM, O., MANN, M. Mass spectrometric sequencing of proteins silver-stained polyacrylamide gels. *Anal. Chem.* **1996**, *68*, 850–858.

21 McCORMACK, A. L., SCHIELTZ, D. M., GOODE, B., YANG, S., et al. Direct analysis and identification of proteins in mixtures by LC/MS/MS and database searching at the low-femtomole level. *Anal. Chem.* **1997**, *69*, 767–776.

22 LEE, C. H., McCOMB, M. E., BROMIRSKI, M., JILKINE, A., ENS, W., et al. On-membrane digestion of beta-casein for determination of phosphorylation sites by matrix-assisted laser desorption/ionization quadrupole/time-of-flight mass spectrometry. *Rapid Commun. Mass Spectrom.* **2001**, *15*, 191–202.

23 HUMPHERY-SMITH, I., WARD, M. A. Proteome research: methods for protein characterization, S. P. HUNT, F. J. LIVESEY (Eds.), Oxford University Press, New York 2000, pp. 217–218.

24 ROEPSTORFF, P., FOHLMAN, J. Proposal for a common nomenclature for sequence ions in mass spectra of peptides. *J. Biomed. Mass Spectrom.* **1984**, *11*, 601.

25. BONNER, R., SHUSHAN, B. The characterization of proteins and peptides by automated methods. *Rapid Commun. Mass Spectrom.* **1995**, *9*, 1067–1076.

26 DAVIES, M. T., LEE, T. D. Rapid protein identification using a microscale electro-

spray LC/MS system on an ion trap mass spectrometer. *J. Am. Soc. Mass Spectrom.* **1998**, *9*, 194–201.

27 FIGEYS, D., DUCRET, A., YATES III, J.R., ABERSOLD R. Protein identification by solid phase microextraction-capillary zone electrophoresis-microelectrospray-tandem mass spectrometry. *Nat. Biotechnol.* **1996**, *14*, 1579–1583.

28 LINK, A.J., ENG, J., SCHIELTZ, D.M., CARMACK, E., et al. Direct analysis of protein complexes using mass spectrometry. *Nat. Biotechnol.* **1999**, *17*, 676–682.

29 WOLTERS, D.A., WASHBURN, M.P., YATES, J.R. An automated multidimensional protein identification technology for shotgun proteomics. *Anal. Chem.* **2001**, *73*, 5683–5690.

30 GYGI, S.P., RIST, B., GERBER, S.A., TURECEK, F., et al. Quantitative analysis of complex protein mixtures using isotopecoded affinity tags. *Nat. Biotechnol.* **1999**, *17*, 994–999.

31 CAGNEY, G., EMILI, A. De novo peptide sequencing and quantitative profiling of complex protein mixtures using masscoded abundance tagging. *Nat. Biotechnol.* **2002**, *20*, 163–170.

32 PERKINS, D.N., PAPPIN, D.J., CREASY, D.M., COTTRELL, J.S. Probability-based protein identification by searching sequence databases using mass spectrometry data. *Electrophoresis.* **1999**, *18*, 3551–3567.

33 ENJ, J.K., McCORMACK, A.L., YATES, J.R. An approach to correlate tandem mass spectral data of peptides with amino acid sequences in protein databases. *J. Am. Soc. Mass Spectrom.* **1994**, *5*, 976–989.

34 CLAUSER, K.R., BAKER, P.R., BURLINGAME, A.L. Role of Accurate Mass Measurement (±10 ppm) in Protein Identification Strategies Employing MS or MS/MS and Database Searching. *Anal. Chem.* **1999**, *71*, 2871–2882.

35 YOUNG, P., KAPPE, E., SWAINSTON, N., YAO, T., et al. Search algorithm implemented in a client server architecture. 15th International Mass Spectrometry Conference, Barcelona, Spain 2000.

36 CHOUDHARY, J.S., BLACKSTOCK, W.P., CREASY, D.M., COTTRELL, J.S. Interrogating the human genome using uninter-

preted mass spectrometry data. *Proteomics.* **2001**,*1*, 651–667.

37 KUSTER, B., MORTENSEN, P., ANDERSEN, J.S., MANN, M. Mass spectrometry allows direct identification of proteins in large genomes. *Proteomics.* **2001**, *1*, 641–650.

38 CLAVERIE, J.M. Gene number. What if there are only 30,000 human genes? *Science.* **2001**, *291*, 1255–1257.

39 LARSEN, M.R., SORENSEN, G.L., FEY, S.J., LARSEN, P.M., ROEPSTORFF, P. Phosphoproteomics: evaluation of the use of enzymatic de-phosphorylation and differential mass spectrometric peptide mass mapping for site specific phosphorylation assignment in proteins separated by gel electrophoresis. *Proteomics.* **2001**, *1*, 223–238.

40 MILLS, P.B., MILLS, K., JOHNSON, A.W., CLAYTON, P.T., WINCHESTER, B.G. Analysis by matrix assisted laser desorption/ionisation-time of flight mass spectrometry of the post-translational modifications of alpha 1-antitrypsin isoforms separated by two-dimensional polyacrylamide gel electrophoresis. *Proteomics.* **2001**, *1*, 778–786.

41 HARVEY, D.J. Identification of protein-bound carbohydrates by mass spectrometry. *Proteomics.* **2001**, *1*, 311–328.

42 WILM, M., NEUBAUER, G., MANN, M. Parent ion scans of unseparated peptide mixtures. *Anal. Chem.* **1996**, *68*, 527–533.

43 CARR, S.A., HUDDLESTON, M.J., BEAN, M.F. Selective identification and differentiation of N- and O-linked oligosaccharides in glycoproteins by liquid chromatography-mass spectrometry. *Protein Sci.* **1993**, *2*, 183–196.

44 NEUBAUER, G., MANN, M. Mapping of phosphorylation sites of gel-isolated proteins by nanoelectrospray tandem mass spectrometry: potentials and limitations. *Anal. Chem.* **1999**, *71*, 235–242.

45 CARR, S.A., HUDDLESTON, M.J., ANNAN, R.S. Selective detection and sequencing of phosphopeptides at the femtomole level by mass spectrometry. *Anal. Biochem.* **1996**, *239*, 180–192.

46 BETTS, J.C., BLACKSTOCK, W.P., WARD, M.A., ANDERTON, B.A. Identification of phosphorylation sites on neurofilament proteins by nanoelectrospray mass spectrometry. *J. Biol. Chem.* **1997**, *272*(20), 12922–12927.

47 BELL, A. W., WARD, M. A., BLACKSTOCK, W. P., FREEMAN, H. N. M. et al. Proteomics characterization of abundant Golgi membrane proteins. *J. Biol. Chem.* **2001**, *276*, 5152–5165.

48 ANDERSON, J. S., LYON, C. E., FOX, A. H., LEUNG A. K,, et al. Directed Proteomic Analysis of the Human Nucleolus. *Curr. Biol.* **2002**, *12*(1), 1–11.

49 EATON, P., BYERS, H. L., LEEDS, N., WARD, M. A., SHATTOCK, M. J. Detection, quantitation, purification, and identification of cardiac proteins S-thiolated during ischemia and reperfusion. *J. Biol. Chem.* **2002**, Jan 2 [epub ahead of print].

50 HUSI, H., WARD, M. A., CHOUDHARY, J. S., BLACKSTOCK, W. P., GRANT, G. N. Proteomic analysis of NMDA receptor-adhesion protein signaling complexes. *Nat. Neurosci.* **2000**, *3*(7), 661–669.

51 RIGAUT, G., SHEVCHENKO, A., RUTZ, B., WILM, M., et al. A generic protein purification method for protein complex characterization and proteome exploration. *Nature.* **1999**, *17*, 1030–1032.

52 GATLIN, C. L., ENG J. K., CROSS, S. T., DETTER, J. C., YATES, J. R. Automated identification of amino acid sequence variations in proteins by HPLC/microspray tandem mass spectrometry. *Anal. Chem.* **2000**, *7*, 757–763.

13

Differential Expression Proteomic Analysis Using Isotope Coded Affinity Tags

Michael E. Wright and Ruedi Aebersold

13.1
Introduction

One of the major goals of biological research is to define all the elements of a biological system and to determine how their collective activities and interactions carry out complex biological functions. It is expected that defining the activity of each individual element and its relationship to other elements in that system will lead to a more integrated and global understanding of how those elements function together in a biological network. The genomics field has provided examples that support this concept through the application of cDNA microarray technology, e.g. [1, 2]. Large-scale gene expression profiling (via cDNA microarray technology) has revealed that co-regulated genes tend to cluster into distinct functional categories, facilitating the classification of uncharacterized genes to discrete functions and specific pathways [1, 3–6]. This technology has also revealed that distinct cellular phenotypes share similar mRNA expression profiles, again suggesting that uncovering gene expression patterns will lead to a better understanding of the genes that give rise or contribute to the maintenance of a particular cellular state (e.g. [7–16]. However, the underlying premise behind the application of DNA microarray technology is that changes in mRNA abundance directly translate to changes in the abundance of the proteins encoded by those mRNAs. Proteins are the molecular effectors that carry out essentially all biochemical activities of the cell. Therefore, mRNA profiling represent an indirect assay from which protein abundance and activity are inferred. Proteomics is the discipline that attempts to systematically and quantitatively analyze the proteins expressed in a cell or tissue and to therefore provide a more direct indication of the functional state of the cell. Mass spectrometry (MS) is a critical component of essentially any current proteome analysis technology and has had a significant impact on the growth and development of the proteomic field. In particular, protein identification and quantification are most commonly achieved by the application of one of a variety of MS technologies [17]. In this chapter we will highlight new developments in protein reactive chemistries (e.g. isotope coded affinity tags ICAT^{TM} reagents) that when used in combined with MS allow for the global identification and quantification of the proteins present in two or more complex samples such as those representing two different cellular states. We will also discuss how differential expression proteomic analysis via ICAT reagent labeling can lead to the identification of critical regulators within defined signaling pathways, and highlight its applications in biology and medicine.

13.2
Traditional Methodologies for Measuring Protein Abundance

Over the past 25 years two-dimensional polyacrylamide-gel electrophoresis (2DE) has been the cornerstone proteomic technology. The method is capable of separating complex protein mixtures that are composed of thousands of proteins into their individual components and of indicating the quantity of each detected feature [18–21]. During 2DE, proteins are extracted from cells or tissues are separated in the isoelectric focusing (IEF) first-dimension based upon their overall intrinsic charge. Each protein will migrate to an equilibrium position in the gel that is referred to as its isoelectric point (pI). IEF is typically performed in polyacrylamide gels that can span narrow or broad range of pI ranges (Fig. 13.1). Once focused, the proteins are then separated in the second dimension by their size. This is achieved by sodium dodecyl sulfate polyacrylamide gel electrophoresis (SDS-PAGE). The gels are subsequently fixed and proteins are being visualized using traditional staining methods (e.g. silver-stain and coomasie blue) or more recently, fluorescent stains [22–32]. The staining intensity indicates the quantity of each detected spot. Spots containing the stained proteins are excised from the gel, trypsin-digested and the resulting peptides are subjected to MS analysis for protein identification [33–36]. The 2DE strategy has been the method of choice for identifying proteins and for measuring differences in protein abundance between two experimental samples (e.g. treated cells versus untreated cells) by detecting differences in the staining intensities of spots with identical coordinates present in two or more gel patterns. Conceptually, 2DE is relatively simple. However, in practice the strategy has been plagued with technical difficulties that have prevented it from being a very robust quantitative and comprehensive tool for quantitative proteome profiling. For example, a high level of skill and expertise are required to generate nearly identical 2D gel patterns with identical samples. Therefore it has been extremely difficult to compare the protein spot intensities between two gels run under the same exact conditions, particularly if the gels were run in different laboratories. Second, it is widely known that 2DE does not resolve complex protein mixtures very well, especially proteins that are very acidic, very basic, of higher molecular weight, or very hydrophobic [37–41]. Despite recent advances in IEF technology (e.g. narrow range gels) and pre-fractionations schemes that reduce sample complexity, these methods commonly lead to increases in the amount of work to be done and are also plagued with reproducibility problems [42–44]. A third limitation to using 2DE for quantitating protein abundance is that low abundance proteins go routinely undetected [40]. This is a direct reflection of the low loading capacities inherent to 2D gels [40]. Exceeding the loading capacity of the

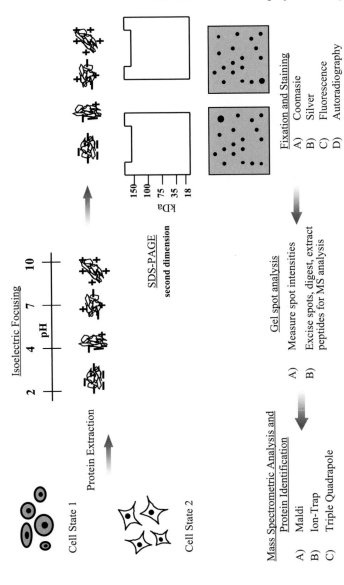

Fig. 13.1 Diagram demonstrating the 2D-PAGE method and its application to quantitative protein profiling. Proteins extracted from cells or tissues are applied to an isoelectric focusing gel strip that separates the proteins according to their isoelectric point. Once focused, the gel strip is applied to a SDS-PAGE gel where the proteins are separated in the second dimension according their size. The gels are fixed and stained with coomasie blue or silver stain. The pattern and quantity of the protein spots are recorded with image analysis software and protein spots of interest are cut out and analyzed by mass spectrometry to facilitate protein identification.

gel will result in poorer gel resolution, which complicates the identification and quantitation of individual proteins within the gel [40]. Fourth, it cannot be assumed that a protein spot contains a single protein species. Frequently two or more proteins co-migrate to the same gel coordinates, making precise protein quantification difficult. Lastly, one of the most significant shortcomings to using 2DE as a quantify proteomic tool is related to the staining methods and image analysis software tools that are used to quantitate the signal intensities of resolved protein spots in the 2D gels [45, 46]. Silver and coomasie staining are the most common staining techniques used for visualizing proteins in 2D gels. Neither method is optimal for quantitative analysis because they display a very limited dynamic range for protein quantification and limited sensitivity [47–50]. The image analysis software tools available for measuring the spot intensities can be compromised by problems associated with accurately defining spot boundaries in the gel, difficulties in normalizing staining between gels, variation in spot intensities between gels and correctly matching a large number of protein spots between gels [45]. New developments in protein detection have tried to make the 2DE method a more viable methodology for quantitative proteomic studies. For example, new fluorescent dyes called propyl-Cy3 and methyl-Cy5 can be covalently bound to proteins and therefore be used for protein detection and quantification (Fig. 13.2) [32]. Two protein samples can be individually labeled with either the Cy3 or Cy5 dye and subsequently combined and run together in one 2D-gel. The ratio of relative signal intensities emitted by the Cy3 and Cy5 dyes for each protein spot will reflect the relative protein abundance of that protein between the two samples. This technology is very similar to the dye technology that commonly used in cDNA microarray analysis [2]. However, labeling proteins with these fluorescent dyes is not without limitations. Covalent modification of the protein may cause alterations in the proteins mobility and solubility during 2DE. Photobleaching, as well as differences in the number of functional groups available for modification with the fluorophore may affect the overall sensitivity of the analysis [51]. Despite these limitations there have been numerous reports of the successful application of 2DE the analysis of clinical or biological samples in the literature. A study that compared differences in the secreted proteins between normal and diseased patients with bladder squamous cell carcinomas found a urinary marker that was subsequently identified as the protein psoriasin [52]. It was suggested that this protein could be used as a marker to track the progression of patients with bladder squamous cell carcinomas. Another study that compared proteins from normal human luminal and myoepithelial breast cancer cells resulted in the identification of 51 proteins out of a total of 170 protein spots that were 2-fold differentially regulated [52]. Generally, 2DE technology has been most successful when used as a differential-display technology to find differences in expression of readily soluble and relatively abundant proteins.

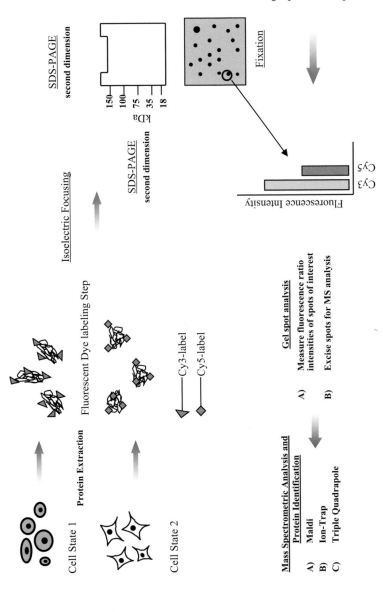

Fig. 13.2 Diagram demonstrating protein quantification via fluorescent dyes in combination with 2D-PAGE. Proteins extracted from cells or tissues are incubated with a fluorescent dye (e.g. Cy3, Cy5) that reacts with functional groups (e.g. amines, carboxyls) in the protein to form covalently linked protein-dye conjugates. The dye-protein conjugates are combined and subjected to 2D-PAGE analysis. The fluorescence intensity ratios of the protein spots are measured using image analysis software tools. Proteins of interested are identified as described in Fig. 13.1.

13.3
Isotopic Methodologies to Quantitative Proteomics

Mass spectrometry (MS) based approaches without the involvement of gel electrophoresis have gained momentum as methods for large-scale protein identification over the last few years [53, 54]. Unfortunately, the intensity of a peptide signal in the mass spectrometer cannot be easily correlated to the amount of analyte present in the sample. To achieve accurate quantification by MS, suitable external or internal standards need to be employed. The optimal internal standard for a particular peptide is the isotopically labeled form of that peptide. Therefore, the newest methods for quantitative proteome analysis are based upon the venerable technique of stable isotope labeling [55]. This method involves incorporation of a stable isotope (e.g. ^{2}H, ^{13}C, ^{15}N) in one of the two samples (Fig. 13.3 A), and the concurrent analysis of the combined samples by mass spectrometry. The ratio of signal intensities for the isotopically heavy and normal form of a peptide indicates the ratio of abundance of the two peptides in the two samples being compared, as it is assumed that the peptide pairs from these samples contain the same physiochemical properties and behave similarly under all conceivable isolation and separation steps. This strategy therefore allows measurements in protein abundance between two samples if one of the protein samples is isotopically different from a second reference sample. One of the first applications of this strategy to quantitate protein expression *in vivo* was performed by Oda et al. (Fig. 13.3 B). In this study one culture of yeast cells was grown on medium containing the natural abundance of the isotopes of nitrogen ^{14}N (99.6%) and ^{15}N (0.4%) and another yeast culture was grown on the same medium enriched in ^{15}N (>96%). In these experiments stable isotope incorporation was therefore achieved by metabolic labeling. After an appropriate growth period the proteins from both yeast samples were extracted and separated by reverse-phase high performance liquid chromatography (RP-HPLC) and then separated by sodium-docecyl sulfate PAGE (SDS-PAGE). In-gel digestion of spots of interest resulted in peptide fragments that were used in protein identification by peptide mass mapping. The ^{15}N incorporation shifted mass of the peptides upwards by one mass unit per incorporated nitrogen. Peptides therefore appeared as doublets in the mass spectrum and the relative peak heights of the two signals indicated the relative abundance of the two forms. The ratio of peak intensities was linear over two orders of magnitude. This study lead to the quantification of 42 highly abundant proteins

---

Fig. 13.3 Demonstration of stable isotope labeling for quantitative protein profiling. A) Digested protein containing an equal amount of the light ^{14}N and heavy ^{15}N isotopes results in isotopic peptide pairs of equivalent ion intensities that are shifted by the mass difference of the isotope in the mass spectrometer. B) Oda et al used *in vivo* stable isotopic labeling in combination with 2-dimensional separation to quantify protein abundance in yeast. Cells grown on regular medium ^{14}N (99.6%) or enriched medium ^{15}N (>96%) were extracted, combined, and separated by HPLC. Resulting HPLC fractions were resolved by SDS-PAGE, protein spots excised, trypsin digested, and isotopic peptides were quantified and identified in the mass spectrometer (MALDI).

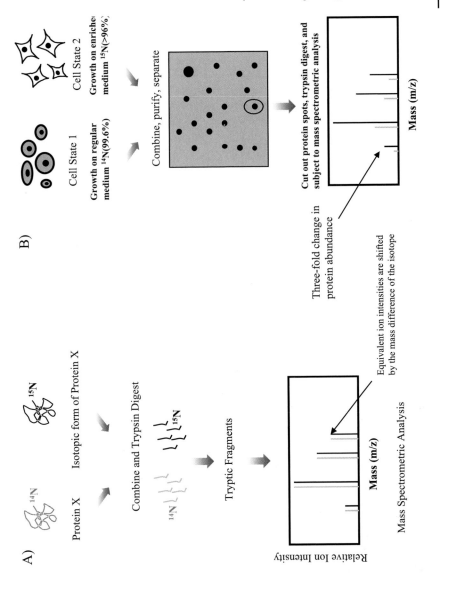

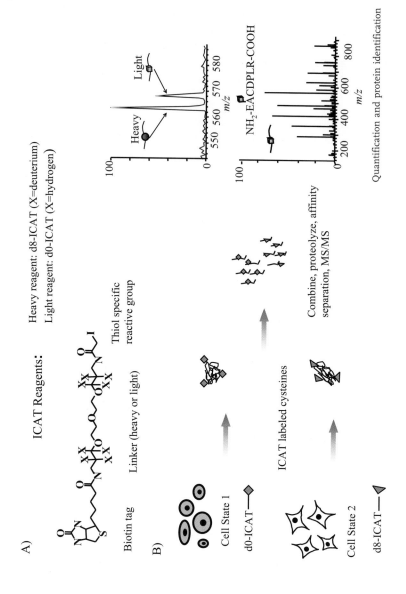

for two pools of yeast cells that were genetically identical but differed in their ability to express the G1 cyclin CLN2. The technique had a percentage error of only ±10%, underscoring the potential for very precise quantification. Metabolic stable isotopic labeling is not without limitations. Its major shortcoming is that it is very impractical and cost prohibitive to grow and label mammalian cells with stable isotopes. This is in contrast to bacterial or yeast samples which are readily metabolically labeled [56]. Furthermore, applying this approach to tissues is impractical if not impossible, even though there is a very strong interest in developing methodologies that can accurately and robustly compare differences in protein abundance in normal and disease tissue [57].

13.4
Isotope-Coded Affinity Tags (ICAT^{TM}) for Quantitative Proteome Analysis

Recently, we have developed a new class of reagent termed ICAT^{TM} [58]. These reagents are designed to incorporate stable isotope tags into proteins post isolation, by chemical reactions. The method based on these reagents is intended to measure quantitative differences in protein abundance between two samples, irrespective of the source of the sample. As shown in Fig. 13.4A, the ICAT reagent introduces stable isotope tags into proteins via the selective alkylation of the side chains of reduced cysteines with either a deuterium-labeled heavy form (d8) or hydrogen-labeled light form (d0) of the reagent (Fig. 13.4B). The ICAT-reagents are composed of three elements: a biotin affinity tag, a linker region in which the 8 hydrogens or 8 deuteriums are located, and a functional group that reacts with sulfhydryl groups within proteins. In ICAT reagent based protein profiling experiments protein mixtures derived from two distinct cell populations are reduced and treated with the isotopically heavy and normal forms of sulfhydryl-specific ICAT-reagent, respectively. The labeled samples are then combined and digested with trypsin, and the resulting complex peptide mixtures are fractionated by ion-exchange chromatography. The biotinylated (stable isotope labeled, cysteine containing peptides) in individual ICAT-labeled ion-exchange peptide fractions are selectively enriched by avidin-biotin chromatography and then subjected to microca-

---

Fig. 13.4 The ICAT reagent method for quantitative protein profiling. A) The ICAT reagent is composed of three elements: an affinity tag (biotin), used for purification of ICAT-labeled peptides, a linker that incorporates an isotope, and a reactive group that has specificity toward thiol containing moieties. The ICAT reagent exist in two forms, a light form referred to as d0 and a heavy form referred to as d8. B) The ICAT method involves extracting protein from cells in two different states, denaturing, reducing, and reacting these proteins with the ICAT reagent. The d0 or d8 labeled protein samples are combined, trypsinized, and subjected to ion-exchange chromatography. The ICAT-labeled peptides are isolated by purification over an aviding affinity matrix and subsequently analyzed by reverse phase HPLC MS/MS. The ratio of the ion intensities of the ICAT pairs is quantified which results in the quantification of protein abundance between the two cell states.

pillary reversed-phase high performance liquid chromatography electrospray ionization tandem mass spectrometry (μHPLC-ESI-MS/MS). Pairs of chemically identical ICAT-labeled peptides essentially co-elute from the RP column (the d8 form of the peptide elutes slightly before the d0 form of the peptide) and are recognized by the mass spectrometer. The relative ion intensities of the two differentially isotopically tagged peptides indicate their relative abundance in the sample. Concurrently, the detected peptides are selected for (CID) and the CID spectra are searched against sequence databases for the purpose of identifying the protein from which the peptide originated. The method therefore allows for the simultaneous identification and quantification of the components of complex protein mixtures.

In the intial study describing the method Gygi et al. identified and quantified 34 yeast proteins that were extracted from yeast cells grown on either galactose or ethanol carbon sources. Many of the proteins identified and quantified were actually enzymes involved in carbon metabolism. For example the ADH2 protein, which is repressed by glucose and galactose sugars, was shown to be induced by more than 200-fold in yeast cells grown on ethanol as a carbon source. These results support the known biochemical activity of the ADH2 enzyme, since this enzyme catalyzes the conversion of ethanol into acetaldehyde, which is ultimately fed back into the tricarboxylic and glyoxylate cycles for energy utilization [59]. The measured ICAT peptide signal intensity ratios were very robust as reflected in the low percent error values (<12%) for ratios of a standard ICAT-labeled peptide mixture *in vitro*. The ICAT method has a number of advantages. First, the method provides quantitative protein profiles of complex samples without the need for prior protein separation by gel electrophoresis or other methods. Second, all classes of proteins with the exception of proteins devoid of cysteine are amenable to the ICAT method. The detection of low abundance proteins and hydrophobic membrane proteins has already been demonstrated [40, 60]. Third, the ICAT labeling strategy is applicable to the extraction of proteins from any biological source, including cells, tissues, or bodily fluids under any number of physiological or non-physiological conditions. Fourth, the alkylation of cysteines by the ICAT reagent is very specific and is not compromised by the presence of non-ionic detergents (NP-40, Triton-X 100) and strong denaturing agents (SDS, urea, guanidine-HCl) [58, 60, 61]. This is critical for the analysis of hydrophobic or very insoluble protein complex. Fifth, the complexity of peptide mixtures is greatly reduced by targeting only those peptides that contain a reactive cysteine residue. This purification step is critical because it reduces the number of peptides that have to be mass spectrometrically analyzed and processed. This is important because the number of peptides eluting from the reverse phase chromatograph has to be low enough to not overwhelm the duty cycle of the mass spectrometer operating in MS/MS mode. Lastly, the ICAT method can be applied to any type of front-end biochemical, immunological, or physical fractionation scheme for the analysis of protein complexes and or low abundance proteins. However, the ICAT method is not without limitations. First, the ICAT label is relatively large and the fragmentation pattern induced after CID can complicate the interpretation of the fragment ion spectra that are used for protein identification via sequence database searching. Second, the ICAT method only works for those proteins that contain at least one cysteine residue. Even though the number of proteins that are cysteine-free is quite small in yeast and other eukaryotic species, some microbial species have a much higher fraction of cysteine-free proteins ($\sim 8\%$). The development of new ICAT reagents with specificities different from SH groups will alleviate this problem. These will also give the ICAT method a broader application to those fields (e.g. neurobiology, cardiovascular biology) where small peptides that are critical for signaling pathways have been difficult to quantitatively analyze to date [62–64].

Since the original publication of the ICAT method there have been several studies that have applied a similar strategy to isolating and quantifying proteins via isotopic labeling [65–68]. These reports either label proteins *in vivo* similar to the approach of Oda et al., or capture proteins post-isolation via their reactive cysteines. These studies were successful in measuring protein abundance through the quantification of isotopic peptide pairs in different mass spectrometers (e.g. ion-trap, fourier transform ion cyclotron resonance (FTICR) mass spectrometers).

13.5
Biological Applications of the ICAT Method

The ICAT method was initially used to characterize protein expression differences in yeast cells grown on either galactose or ethanol [58]. More recently, the method has also been successfully applied to mammalian systems. Han et al. reported the quantitative protein profiling of differentiation-induced microsomal proteins from human myeloid leukemia (HL-60) cells. The report identified and quantified 491 proteins isolated from the microsomal fractions of naive and 12-phorbol 13-myristate acetate (PMA) treated HL-60 cells. Of the 491 proteins $\sim 11\%$ were known cell surface antigens, receptors, and membrane proteins. A large class of previously uncharacterized proteins containing transmembrane domains were identified and quantified as well. This was an important study for several reasons. It successfully identified and quantified membrane and membrane associated proteins from the microsomal fraction of mammalian cells, a class of proteins that has been difficult to analyze by more traditional methods. Proteins located in or associated with the plasma membrane have very important physiological roles in cells. These proteins encompass a number of critical homeostatic and regulatory functions including ion channels, pores, pumps, and transmembrane receptors that can sense the environment and transit signals within or between other cells. Acquiring a better understanding of their localization and abundance during specific signals may help define their role and potential function within the cell. These proteins also contain considerable diagnostic and therapeutic value because of their accessibility [69]. Traditionally, the analysis of membrane proteins has been extremely difficult because of low solubility of many of these proteins in commonly used solvents [70]. A crude microsomal preparation strategy was used to enrich for membrane and membrane associated proteins (Fig. 13.5) [71]. Some of the observed results also suggested new hypotheses on the mechanisms of

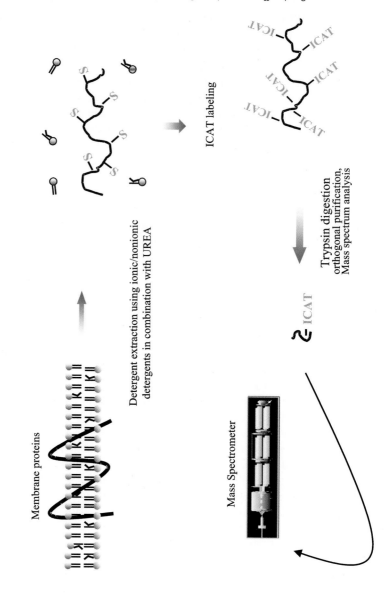

PMA induced cell differentiation. A 20-fold reduction in squalene synthase (SQS) protein in PMA stimulated microsomal fractions in comparison to unstimulated microsomal fractions and the concurrent change in abundance in the microsomal fraction of a number of proteins that depend upon prenylation for membrane association (e.g. Rac1 and RhoG) is consistent with a critical role of SQS in the distribution of farnesyl pyrophosphate (FPP) utilization, a critical component of the prenylation process [72]. SQS is located at a metabolic branch point in the mevalonate pathway for controlling the distribution of the FPP pools for use in either the cholesterol/sterol biosynthetic pathway or for the production of nonsteroidal products such as precursors of prenylation reactions [73]. Proteolysis-induced release of SQS from membranes could therefore lead to the utilization of FPP in cholesterol and sterol biosynthesis and activation of SQS could result in the depletion of FPP for use in the prenylation pathway, explaining the observed PMA-induced downregulation of prenylated protein products in the microsomal fraction [74]. The use of the ICAT method alone, like any discovery science method, is not by itself sufficient to explain an entire biological process like PMA-induced changes in HL-60. However, the ICAT method can reveal new proteins that may participate in a particular process or pathway that can lead to new hypotheses that can be tested by hypothesis-driven research methods. This iterative cycle of applying discovery science and hypothesis-driven approaches can be extended until a more complete understanding of the system comes to form. This study also demonstrated the utility and the application of the ICAT method to a complex mammalian system and illustrated the need for extensive pre-fractionation methods if complex samples are being studied.

We have also used the ICAT method to examine the role of the hormone androgen upon prostate cancer cells. It is well known that androgens are critical for the development and maintenance of normal and cancerous prostate [76, 77]. We have been interested in defining the complete subset of proteins that are responsive to the hormone androgen, with the expectation that this may help us identify the proteins that are involved in androgen receptor signal transduction. This is clinically important since these proteins may act at critical regulatory junctures in the cell and may give us some clues on how prostate cells transit from androgen-dependent to androgen-independent cell growth during prostate cancer [78]. We have been systematically analyzing the cytosolic, membrane, and nuclear protein components of prostate cancer cells and their respective responses to the hormone androgen. For example, we have isolated nuclear proteins from androgen

Fig. 13.5 Strategy for extracting and analyzing membrane proteins via ICAT. Membrane and membrane associated proteins from mammalian cells were isolated using differential centrifugation. Whole cells were incubated in a hypotonic lysis buffer and dounce homogenized to form a microsomal homogenate. This homogenate was spun at low speeds (3K) to remove unbroken cells and nuclei. The supernatant was subjected to a high-speed centrifugation step (100,000 g) and the resulting microsomal pellet was solubilized in a mixture of urea and SDS. These solubilized proteins were subjected to the ICAT method.

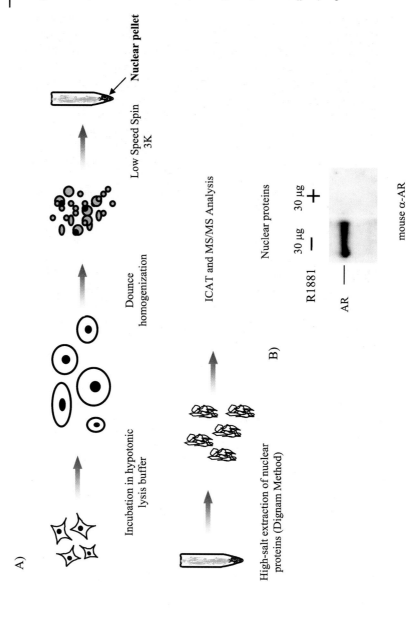

starved and stimulated LNCaP prostate cancer epithelia cells and subjected them to ICAT analysis (Fig. 13.6 A). We have identified and quantified >1,000 proteins located within nuclei or associated with the nuclear membrane of LNCaP cells. These included discrete classes of nuclear proteins associated with nuclear protein transport, architecture, and transcription. Interestingly, we were successful in identifying and quantifying androgen receptor protein expression levels by the ICAT method. ICAT analysis revealed that androgen receptor levels increased by 3-fold in response to androgen starvation. Western-blots on the nuclear proteins isolated from androgen starved and stimulated LNCaP cells confirmed the results obtained by the ICAT method (Fig. 13.6 B), and this conclusively demonstrated that androgen receptor levels increase in the nuclear compartment of LNCaP prostate epithelial cells when starved of the androgen hormone. We are in the process of following up many of the biological observations that are apparent from this large proteomic dataset and are hopeful that these discoveries will help increase our understanding of androgen signaling in prostate cancer.

The ICAT method can resemble a large-scale quantitative western-blot (Fig. 13.7). If used appropriately, the method can provide new insights into protein abundance, localization, and turnover. Application of the ICAT method will grow and undoubtedly lead to new and exciting biological discoveries.

13.6
Conclusions and Future Directions

One of the primary goals of proteomics research is to build functional and physical maps of all the proteins expressed by a cell and understand how these proteins function with each other to form biological networks on a global scale. This scenario has already been taken place at the level of RNA through the application of cDNA microarray technology [2]. Using suitable algorithms, cDNA microarray technology has revealed that many concurrently regulated genes participate in the same biochemical pathway and that cellular states can be ascribed to distinct and unique transcriptional profiles [2]. This phenomenon has yet to be documented at the level of proteins, even though there have been hints that a similar process may be occurring at the protein level as well [79]. Recent advances in the field of proteomics now give scientist the opportunity to address this question more directly. Technologies that can rapidly identify proteins and accurately measure pro-

Fig. 13.6 Strategy for extracting and analyzing nuclear proteins via ICAT. A) Nuclear proteins were extracted using the Dignam method. In short, nuclear pellets isolated using traditional differential centrifugation were incubated in a high-salt (420 mM) solution that extracts most nuclear proteins except the histones. The nuclear proteins were dialyzed to reduce the salt concentrations and subjected to the ICAT method. B) Extracted nuclear proteins (30 µg/lane) from androgen starved/stimulated (10 nM R1881, 72 hours) LNCaP prostate epithelia were subjected to SDS-PAGE and blotted with a monoclonal antibody to the androgen receptor.

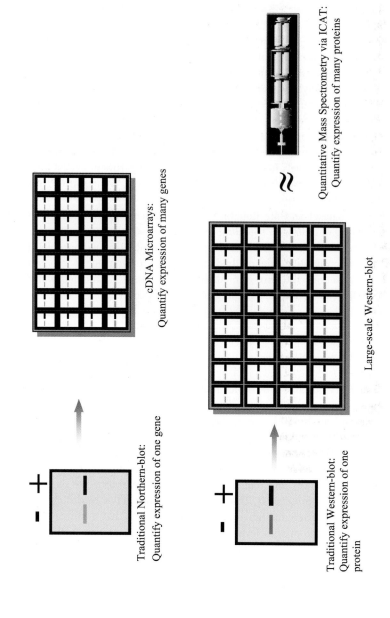

Traditional Northern-blot:
Quantify expression of one gene

cDNA Microarrays:
Quantify expression of many genes

Traditional Western-blot:
Quantify expression of one protein

Large-scale Western-blot

Quantitative Mass Spectrometry via ICAT:
Quantify expression of many proteins

tein expression levels between two cellular states on a global scale will lead the way for this type of analysis. One method that is expected to provide such data is the application of the ICAT reagent method in combination with mass spectrometry [58]. The ICAT reagent technology makes proteomics quantitative, because it effectively measures relative changes in protein abundance between two cellular states. Since proteins are the biological effectors in the cell there is great interest in accurately measuring protein levels in response to a given stimuli. Differential protein profiling can be used to understand differences in protein expression between normal and disease tissue. The method can also provide an added level of information about protein localization and turnover if used appropriately. New chemistries that allow greater flexibility in the labeling of different reactive groups in peptides is forthcoming (e.g. phosphates, amines, carboxyl, sugars, catalytic activity) so that not only the relative abundance but also the level of their post-translantional modifications and, if applicable, their activities will be accurately determined. For example new reagents and methods have recently been developed to capture and purify phosphopeptides from complex peptide mixtures [80, 81]. Minor modifications of these methods will make them applicable for use in differential phosphoprotein analysis on a global scale using a strategy similar to the ICAT method [82–86]. New developments and improvements in mass spectrometers and their corresponding software will considerably speed-up the analysis of ICAT-labeled proteins in the near future, a step that is critically important if proteomic analyses are to reach widespread application. Mass spectrometers are being designed, in particular a matrix assisted laser desorption ionization (MALDI) tandem mass spectrometer that will have the capacity to perform abundance-dependent identification of ICAT reagent labeled samples (manuscript in preparation). This will cut down on the number of protein identifications and quantifications that are required in a particular study since the mass spectrometer will focus on those proteins that show a significant quantitative change under the conditions tested, ignoring the proteins that do not show a change in abundance. These and other incremental changes will increase the sample throughput of quantitative proteomic experiments. It is expected that the widespread application of quantitative proteomics will give us a more comprehensive map of the protein networks and their function in cells and tissues in health and disease.

Fig. 13.7 ICAT method is analogous to a large-scale quantitative western-blot. Traditional methods in molecular biology have used northern-blots to quantify gene expression one gene at a time. New genomic technologies, primarily cDNA microarrays, now allow for measurements in gene expression from hundreds to thousands genes. Similarly, new technologies in the area of proteomics, for example the ICAT method, now allow for measurements in protein abundance on a global scale, which is analogous to a large-scale quantitative western-blot.

13.7
Acknowledgements

M.E.W. was funded by an NIH Postdoctoral Genome Training Grant fellowship. This work was also supported by a grant from the Merck Genome Research Institute (MGRI) and a grant from the National Cancer Institute (1R33CA84698).

13.8
References

1 SCHENA, M., D. SHALON, et al. "Quantitative monitoring of gene expression patterns with a complementary DNA microarray." *Science* **1995**, *270*(5235):467–470.

2 HUGHES, T.R., D.D. SHOEMAKER. "DNA microarrays for expression profiling." *Curr Opin Chem. Biol.* **2001**, *5*(1):21–25.

3 EISEN, M.B., P.T. SPELLMAN, et al. "Cluster analysis and display of genomewide expression patterns." *Proc. Natl. Acad. Sci. USA* **1998**, *95*(25):14863–14868.

4 ROTH, F.P., J.D. HUGHES, et al. "Finding DNA regulatory motifs within unaligned noncoding sequences clustered by whole-genome mRNA quantitation." *Nat. Biotechnol.* **1998**, *16*(10):939–945.

5 BAMMERT, G.F., J.M. FOSTEL. "Genomewide expression patterns in Saccharomyces cerevisiae: comparison of drug treatments and genetic alterations affecting biosynthesis of ergosterol." *Antimicrob. Agents. Chemother.* **2000**, *44*(5):1255–1265.

6 HUGHES, T.R., M.J. MARTON, et al. "Functional discovery via a compendium of expression profiles." *Cell.* **2000**, *102*(1):109–126.

7 KHAN, J., R. SIMON, et al. "Gene expression profiling of alveolar rhabdomyosarcoma with cDNA microarrays." *Cancer. Res.* **1998**, *58*(22):5009–5013.

8 ALON, U., N. BARKAI, et al. "Broad patterns of gene expression revealed by clustering analysis of tumor and normal colon tissues probed by oligonucleotide arrays." *Proc. Natl. Acad. Sci. USA* **1999**, *96*(12):6745–6750.

9 GOLUB, T.R., D.K. SLONIM, et al. "Molecular classification of cancer: class discovery and class prediction by gene expression monitoring." *Science* **1999**, *286*(5439):531–537.

10 PEROU, C.M., S.S. JEFFREY, et al. "Distinctive gene expression patterns in human mammary epithelial cells and breast cancers." *Proc. Natl. Acad. Sci. USA* **1999**, *96*(16):9212–9217.

11 ALIZADEH, A.A., M.B. EISEN, et al. "Distinct types of diffuse large B-cell lymphoma identified by gene expression profiling." *Nature* **2000**, *403*(6769):503–511.

12 BITTNER, M., P. MELTZER, et al. "Molecular classification of cutaneous malignant melanoma by gene expression profiling." *Nature* **2000**, *406*(6795):536–540.

13 CLARK, E.A., T.R. GOLUB, et al. "Genomic analysis of metastasis reveals an essential role for RhoC." *Nature* **2000**, *406*(6795):532–535.

14 PEROU, C.M., T. SORLIE, et al. "Molecular portraits of human breast tumours." *Nature* **2000**, *406*(6797):747–752.

15 ROSS, D.T., U. SCHERF, et al. "Systematic variation in gene expression patterns in human cancer cell lines." *Nat. Genet.* **2000**, *24*(3):227–235.

16 SCHERF, U., D.T. ROSS, et al. "A gene expression database for the molecular pharmacology of cancer." *Nat. Genet.* **2000**, *24*(3):236–244.

17 AEBERSOLD, R., D.R. GOODLETT. "Mass spectrometry in proteomics." *Chem. Rev.* **2001**, *101*(2):269–295.

18 KLOSE, J. "Protein mapping by combined isoelectric focusing and electrophoresis of mouse tissues. A novel approach to testing for induced point mutations in mammals." *Humangenetik* **1975**, *26*(3):231–243.

19 O'FARRELL, P.H. "High resolution two-dimensional electrophoresis of proteins." *J. Biol. Chem.* **1975**, *250*(10):4007–4021.

20 SCHEELE, G. A. "Two-dimensional gel analysis of soluble proteins. Charaterization of guinea pig exocrine pancreatic proteins." *J. Biol. Chem.* **1975**, *250*(14):5375–5385.

21 IBORRA, F., J. M. BUHLER. "Protein subunit mapping. A sensitive high resolution method." *Anal. Biochem.* **1976**, *74*(2):503–511.

22 MEYER, T. S., B. L. LAMBERTS. "Use of coomassie brilliant blue R250 for the electrophoresis of microgram quantities of parotid saliva proteins on acrylamide-gel strips." *Biochim. Biophys. Acta.* **1965**, *107*(1):144–145.

23 TALBOT, D. N., D. A. YPHANTIS. "Fluorescent monitoring of SDS gel electrophoresis." *Anal. Biochem.* **1971**, *44*(1):246–253.

24 KERENYI, L., F. GALLYAS. "A highly sensitive method for demonstrating proteins in electrophoretic, immunoelectrophoretic and immunodiffusion preparations." *Clin. Chim. Acta.* **1972**, *38*(2):465–467.

25 RAGLAND, W. L., J. L. PACE, et al. "Fluorometric scanning of fluorescamine-labeled proteins in polyacrylamide gels." *Anal. Biochem.* **1974**, *59*(1):24–33.

26 SWITZER, R. C., 3RD, C. R. MERRIL, et al. "A highly sensitive silver stain for detecting proteins and peptides in polyacrylamide gels." *Anal. Biochem.* **1979**, *98*(1):231–237.

27 VANDEKERCKHOVE, J., G. BAUW, et al. "Protein-blotting on Polybrene-coated glass-fiber sheets. A basis for acid hydrolysis and gas-phase sequencing of picomole quantities of protein previously separated on sodium dodecyl sulfate/polyacrylamide gel." *Eur. J. Biochem.* **1985**, *152*(1):9–19.

28 SZEWCZYK, B., D. F. SUMMERS. "Fluorescent staining of proteins transferred to nitrocellulose allowing for subsequent probing with antisera." *Anal. Biochem.* **1987**, *164*(2):303–306.

29 JACKSON, P., V. E. URWIN, et al. "Rapid imaging, using a cooled charge-coupled-device, of fluorescent two- dimensional polyacrylamide gels produced by labelling proteins in the first-dimensional isoelectric focusing gel with the fluorophore 2- methoxy-2,4-diphenyl-3(2H)furanone." *Electrophoresis* **1988**, *9*(7):330–339.

30 VERA, J. C., C. RIVAS. "Fluorescent labeling of nitrocellulose-bound proteins at the nanogram level without changes in immunoreactivity." *Anal. Biochem.* **1988**, *173*(2):399–404.

31 HOUSTON, B., D. PEDDIE. "A method for detecting proteins immobilized on nitrocellulose membranes by in situ derivatization with fluorescein isothiocyanate." *Anal. Biochem.* **1989**, *177*(2):263–267.

32 UNLU, M., M. E. MORGAN, et al. "Difference gel electrophoresis: a single gel method for detecting changes in protein extracts." *Electrophoresis* **1997**, *18*(11):2071–2077.

33 HENZEL, W. J., T. M. BILLECI, et al. "Identifying proteins from two-dimensional gels by molecular mass searching of peptide fragments in protein sequence databases." *Proc. Natl. Acad. Sci. USA* **1993**, *90*(11):5011–5015.

34 JAMES, P., M. QUADRONI, et al. "Protein identification by mass profile fingerprinting." *Biochem. Biophys. Res. Commun.* **1993**, *195*(1):58–64.

35 MANN, M., P. HOJRUP, et al. "Use of mass spectrometric molecular weight information to identify proteins in sequence databases." *Biol. Mass. Spectrom.* **1993**, *22*(6):338–345.

36 YATES, J. R., 3RD, S. SPEICHER, et al. "Peptide mass maps: a highly informative approach to protein identification." *Anal. Biochem.* **1993**, *214*(2):397–408.

37 GARRELS, J. I., C. S. MCLAUGHLIN, et al. "Proteome studies of Saccharomyces cerevisiae: identification and characterization of abundant proteins." *Electrophoresis* **1997**, *18*(8):1347–1360.

38 CHEVALLET, M., V. SANTONI, et al. "New zwitterionic detergents improve the analysis of membrane proteins by two-dimensional electrophoresis." *Electrophoresis* **1998**, *19*(11):1901–1909.

39 GORG, A., C. OBERMAIER, et al. "The current state of two-dimensional electrophoresis with immobilized pH gradients." *Electrophoresis* **2000**, *21*(6):1037–53. [pii].

40 GYGI, S. P., G. L. CORTHALS, et al. "Evaluation of two-dimensional gel electrophoresis-based proteome analysis technology." *Proc. Natl. Acad. Sci. USA* **2000**, *97*(17):9390–9395.

41 SANTONI, V., S. KIEFFER, et al. "Membrane proteomics: use of additive main effects with multiplicative interaction model to classify plasma membrane proteins according to their solubility and electrophoretic properties." *Electrophoresis* **2000**, *21*(16):3329–3344.

42 CORTHALS, G.L., V.C. WASINGER, et al. "The dynamic range of protein expression: a challenge for proteomic research." *Electrophoresis* **2000**, *21*(6):1104–1115 [pii].

43 HOVING, S., H. VOSHOL, et al. "Towards high performance two-dimensional gel electrophoresis using ultrazoom gels." *Electrophoresis* **2000**, *21*(13):2617–2621.

44 WILDGRUBER, R., A. HARDER, et al. "Towards higher resolution: two-dimensional electrophoresis of Saccharomyces cerevisiae proteins using overlapping narrow immobilized pH gradients." *Electrophoresis* **2000**, *21*(13):2610–2616.

45 VOSS, T., P. HABERL. "Observations on the reproducibility and matching efficiency of two-dimensional electrophoresis gels: consequences for comprehensive data analysis." *Electrophoresis* **2000**, *21*(16):3345–3350.

46 SMILANSKY, Z. "Automatic registration for images of two-dimensional protein gels." *Electrophoresis* **2001**, *22*(9):1616–1626.

47 DUNN, M.J. "Two-dimensional gel electrophoresis of proteins." *J. Chromatogr.* **1987**, *418*:145–185.

48 LOPEZ, M.F., K. BERGGREN, et al. "A comparison of silver stain and SYPRO Ruby Protein Gel Stain with respect to protein detection in two-dimensional gels and identification by peptide mass profiling." *Electrophoresis* **2000**, *21*(17):3673–3683.

49 LAUBER, W.M., J.A. CARROLL, et al. "Mass spectrometry compatibility of two-dimensional gel protein stains." *Electrophoresis* **2001**, *22*(5):906–918.

50 MALONE, J.P., M.R. RADABAUGH, et al. "Practical aspects of fluorescent staining for proteomic applications." *Electrophoresis* **2001**, *22*(5):919–932.

51 URWIN, V.E., P. JACKSON. "Two-dimensional polyacrylamide gel electrophoresis of proteins labeled with the fluorophore monobromobimane prior to first-dimensional isoelectric focusing: imaging of the fluorescent protein spot patterns

using a cooled charge-coupled device." *Anal. Biochem.* **1993**, *209*(1):57–62.

52 OSTERGAARD, M., H. WOLF, et al. "Psoriasin (S100A7): a putative urinary marker for the follow-up of patients with bladder squamous cell carcinomas." *Electrophoresis* **1999**, *20*(2):349–354.

53 OPITECK, G.J., K.C. LEWIS, et al. "Comprehensive on-line LC/LC/MS of proteins." *Anal. Chem.* **1997**, *69*(8):1518–1524.

54 LINK, A.J., J. ENG, et al. "Direct analysis of protein complexes using mass spectrometry." *Nat. Biotechnol.* **1999**, *17*(7):676–682.

55 DE LEENHEER, A.P., L.M. THIENPONT. "Applications of isotope-dilution-mass spectrometry in clinical chemistry, pharmacokinetics, and toxicology." *Mass. Spectrometry. Reviews* **1992**, *11*:249–307.

56 JENSEN, P.K., L. PASA-TOLIC, et al. "Probing proteomes using capillary isoelectric focusing-electrospray ionization Fourier transform ion cyclotron resonance mass spectrometry." *Anal. Chem.* **1999**, *71*(11):2076–2084.

57 CHAMBERS, G., L. LAWRIE, et al. "Proteomics: a new approach to the study of disease." *J. Pathol.* **2000**, *192*(3):280–288.

58 GYGI, S.P., B. RIST, et al. "Quantitative analysis of complex protein mixtures using isotope-coded affinity tags." *Nat. Biotechnol.* **1999**, *17*(10):994–999.

59 JOHNSTON, M., M. CARLSON. The molecular and cellular biology of the yeast. *The molecular and cellular biology of the yeast.* E.W. JONES, J.R. PRINGLE and J.R. BROACH. New York City, Cold Spring Harbor Press. **1992**.

60 HAN, D.K., J. ENG, et al. "Quantitative profiling of differentiation-induced microsomal proteins using isotope-coded affinity tags and mass spectrometry." *Nat. Biotechnol.* **2001**, *19*(10):946–951.

61 SMOLKA, M.B., H. ZHOU, et al. "Optimization of the isotope-coded affinity tag-labeling procedure for quantitative proteome analysis." *Anal. Biochem.* **2001**, *297*(1):25–31.

62 ARRELL, D.K., I. NEVEROVA, et al. "Cardiovascular proteomics: evolution and potential." *Circ. Res.* **2001**, *88*(8):763–773.

63 GRANT, S.G., W.P. BLACKSTOCK. "Proteomics in neuroscience: from protein to network." *J. Neurosci.* **2001**, *21*(21):8315–8318.

64 MACRI, J., S.T. RAPUNDALO. "Application of proteomics to the study of cardiovascular biology." *Trends Cardiovasc. Med.* **2001**, *11*(2):66–75.

65 JI, J., A. CHAKRABORTY, et al. "Strategy for qualitative and quantitative analysis in proteomics based on signature peptides." *J. Chromatogr. B. Biomed. Sci. Appl.* **2000**, *745*(1):197–210.

66 CONRADS, T.P., K. ALVING, et al. "Quantitative analysis of bacterial and mammalian proteomes using a combination of cysteine affinity tags and 15N-metabolic labeling." *Anal. Chem.* **2001**, *73*(9):2132–2139.

67 SMITH, R.D., L. PASA-TOLIC, et al. "Rapid quantitative measurements of proteomes by Fourier transform ion cyclotron resonance mass spectrometry." *Electrophoresis* **2001**, *22*(9):1652–1668.

68 MARTINOVIC, S., T.D. VEENSTRA, et al. "Selective incorporation of isotopically labeled amino acids for identification of intact proteins on a proteome-wide level." *J. Mass. Spectrom.* **2002**, *37*(1):99–107.

69 DREWS, J. "Research & development. Basic science and pharmaceutical innovation." *Nat. Biotechnol.* **1999**, *17*(5):406.

70 SANTONI, V., M. MOLLOY, et al. "Membrane proteins and proteomics: un amour impossible?" *Electrophoresis* **2000**, *21*(6):1054–1070. [pii].

71 SCOPES, R.K. *Protein Purification.* New York City, Springer-Verlag. **1994**.

72 GOLDSTEIN, J.L., M.S. BROWN. "Regulation of the mevalonate pathway." *Nature* **1990**, *343*(6257):425–430.

73 YOKOYAMA, K., G.W. GOODWIN, et al. "Protein prenyltransferases." *Biochem. Soc. Trans.* **1992**, *20*(2):489–494.

74 SHECHTER, I., E. KLINGER, et al. "Solubilization, purification, and characterization of a truncated form of rat hepatic squalene synthetase." *J. Biol. Chem.* **1992**, *267*(12):8628–8635.

75 TANSEY, T.R., I. SHECHTER. "Structure and regulation of mammalian squalene synthase." *Biochim. Biophys. Acta.* **2000**, *1529*(1–3):49–62.

76 ISAACS, J.T. "Role of androgens in prostatic cancer." *Vitam Horm* **1994**, *49*:433–502.

77 LINDZEY, J., M.V. KUMAR, et al. "Molecular mechanisms of androgen action." *Vitam. Horm.* **1994**, *49*:383–432.

78 ABATE-SHEN, C., M.M. SHEN. "Molecular genetics of prostate cancer." *Genes. Dev.* **2000**, *14*(19):2410–2434.

79 GE, H., Z. LIU, et al. "Correlation between transcriptome and interactome mapping data from Saccharomyces cerevisiae." *Nat. Genet.* **2001**, *29*(4):482–486.

80 ODA, Y., T. NAGASU, et al. "Enrichment analysis of phosphorylated proteins as a tool for probing the phosphoproteome." *Nat. Biotechnol.* **2001**, *19*(4):379–382.

81 ZHOU, H., J.D. WATTS, et al. "A systematic approach to the analysis of protein phosphorylation." *Nat. Biotechnol.* **2001**, *19*(4):375–378

82 ODA, Y., K. HUANG, et al. "Accurate quantitation of protein expression and site-specific phosphorylation." *Proc. Natl. Acad. Sci. USA* **1999**, *96*(12):6591–6596.

83 WECKWERTH, W., L. WILLMITZER, et al. "Comparative quantification and identification of phosphoproteins using stable isotope labeling and liquid chromatography/mass spectrometry." *Rapid. Commun. Mass. Spectrom.* **2000**, *14*(18):1677–1681.

84 GOSHE, M.B., T.P. CONRADS, et al. "Phosphoprotein isotope-coded affinity tag approach for isolating and quantitating phosphopeptides in proteome-wide analyses." *Anal. Chem.* **2001**, *73*(11):2578–2586.

85 FICARRO, S.B., M.L. MCCLELAND, et al. "Phosphoproteome analysis by mass spectrometry and its application to Saccharomyces cerevisiae." *Nat. Biotechnol.* **2002**, *20*(3):301–305.

86 GOSHE, M.B., T.D. VEENSTRA, et al. "Phosphoprotein isotope-coded affinity tags: application to the enrichment and identification of low-abundance phosphoproteins." *Anal. Chem.* **2002**, *74*(3):607–616.

14

Protein Chip Technology in Proteomics Analysis

ROSALIND E. JENKINS and STEPHEN R. PENNINGTON

14.1
Introduction

The progress in genome sequencing projects has been dramatic and widely publicised: a complete draft of the human genome was published in 2001 [1–3] and as of June 2002, 63% of the sequence had been 'finished' to a high standard (see *www.ornl.gov/hgmis/project/progress.html*). A total of 73 genomes had been completed and there were a further 416 genome projects in progress (*ergo.integratedgenomics. com/GOLD/*). These projects have already spawned new approaches to mRNA expression analysis [4–6] and techniques to undertake comprehensive analysis of protein:protein interactions using the two-hybrid assay [7–9]. Recent advances in microarray technology have increased the speed at which differential gene expression data may be gathered [4, 10–13] and have generated data that in their own right have increased our understanding of health and disease. However, attempts to analyse changes in gene expression at the nucleic acid level associated with, for example, disease states, knockout of individual genes, drug treatments and changing extracellular conditions, may be inappropriate since the expression level of individual mRNAs often does not reflect the level of expression of the corresponding protein product(s) [14, 15]. Nonetheless, genomics based studies have provided a vital framework on which to base the analysis of the proteome of cells and tissues.

Proteomics has come to be broadly defined as the global analysis of the expression and activity of proteins within a biological system, but it also includes protein turnover and localisation, post-translational modification, and interaction with other biological moieties [16–21]. Achieving global coverage of such complex systems in a high throughput manner is thus both highly desirable, and extremely demanding. Significant advances are being made in the development of high-throughput techniques for protein localisation [22], protein structure analysis [23–29], prediction of protein function [30–33] and in mapping protein:protein interactions both by practical methods such as the two-hybrid [7–9, 34] and tandem affinity purification (TAP) [35–37] approaches, and by computational methods [38]. Other impressive approaches to genome-wide analysis of gene function have also been reported [39, 40]. This chapter will briefly review the platform technologies currently dominating proteomics, and will then go on to describe the progress made to date in the development of protein and antibody arrays.

14.2
Two-dimensional Gel Electrophoresis

The realisation that mRNA expression does not always reflect the level of expression or activity of the corresponding protein [14, 15], and the fact that not all biological samples are amenable to analysis at the level of nucleic acids, has lead to a resurgence of interest in measuring proteins directly. The most frequently adopted approach both for determining changes in protein expression and for separating proteins prior to characterisation is 2DE [16, 41–49]. This approach, although 25 years old, has been developed and optimised so that it remains unrivalled in terms of the number of proteins that may be 'displayed' on a single gel, up to 10,000 [50], the utility for generating protein maps [17, 51–56], and the analysis of relative levels of protein expression (see Proteomics 1 (10) and [57–62] for examples). There have been developments in acrylamide and buffer formulations and in electrophoresis apparatus that improve the reproducibility of 2DE gels. New stains have also been developed that allow for highly sensitive detection of proteins over a wider dynamic range and that are compatible with subsequent identification by mass spectrometry [45, 63, 64]. For instance, modified silver stains are available that exhibit sensitivity comparable to the standard silver stain but that are more compatible with subsequent protein characterisation by matrix-assisted laser desorption-ionisation (MALDI) mass spectrometry (MS) [65]. Similarly, Sypro Ruby (Molecular Probes) has been reported to match silver staining in terms of sensitivity but it exhibits a more linear correlation between signal intensity and protein level, and interferes minimally with mass spectrometric analysis [63, 66]. Non-fluorescent dyes that do not modify the target proteins and that are compatible with MS, such as India Ink staining of proteins electroblotted onto nitrocellulose, are also being evaluated [67].

There are, of course, many disadvantages with employing 2DE as a platform technology for proteomics. The method is notorious for problems with reproducibility, it is labour intensive and it is relatively demanding with regards to the amount of material that is required for each analysis [45, 61, 68]. The quantitative comparison of multiple gels is also somewhat of a bottleneck in the process from sample to protein identification. The development of image analysis software has eased the problem and semi-automated gel analysis, such as the Progenesis system from Non Linear Dynamics, is being explored as an option. The introduction of dual sample labelling, such as difference in-gel electrophoresis (DIGE, Amersham Biosciences), where a control and test sample are labelled with different fluorescent dyes prior to mixing and electrophoresis on the same 2DE gel, minimises the problems with gel to gel inconsistency and enables quantitative analysis based on the relative amounts of the two dye colours within each protein spot [60, 69, 70]. The combination of laser capture microdissection with 2DE, especially in selective analysis of human tumour cells, has also shown much promise [71, 72], and sequential detergent extraction techniques have improved the ability to study membrane-associated proteins by 2DE [42, 73]. Many of the subsequent steps in protein characterisation have been automated so that spots may be robotically excised, digested and loaded onto MALDI targets and the acquisition of peptide mass fingerprints and searching of protein databases may be performed with minimal user input [74-78]. Despite these advances, the initial separation of proteins on 2D gels is the rate-limiting step in terms of sample throughput so that the development of alternatives to 2DE has become an increasingly high priority.

14.3
Alternative Approaches to Protein Separation

Orthogonal approaches to sample separation, such as multi-dimensional liquid chromatography (MDLC), capillary electrochromatography (CEC), and isoelectric focussing size exclusion chromatography (IEF-SEC) have proven to be useful when coupled with tandem mass spectrometry [46, 47, 79–83] and are the preferred methods of many research groups. However, they suffer from distinct drawbacks, such as the relatively large quantity of starting material required, particularly if low abundance proteins are of interest, and the lack of quantitative data generated. However, new methods are being developed that will allow quantitative analysis of samples by MS. Isotope coded affinity tagging (ICAT) involves chemical modification of cysteine residues in the samples with either heavy or light reagents coupled to a biotin tag [84–87]. The samples are then mixed, digested with protease and the biotin-tagged peptides isolated by affinity chromatography. The simplified peptide mixture may then be separated by gel electrophoresis or column chromatography and the fractions analysed by MS. The ratio of heavy peptide to light peptide provides a measure of the relative quantities of the native proteins in the original samples [84–87]. The process is relatively high throughput, open to automation and shows great promise as a complementary technology to 2DE.

Miniaturisation and automation of sample delivery systems and multi-dimensional separation procedures reduces sample consumption, enables greater sample throughput and may be linked directly to tandem MS for protein detection and identification [88–96]. Thus LC-MS/MS has been performed on complex mixtures of proteins by detecting and separating components on the basis of mass (by precursor ion scanning) and subsequently identifying each component (by high energy collision-induced dissociation) [89, 96–101]. Microfabricated modules that enable one- or two-dimensional separation by liquid chromatography or electrophoresis are being developed [88, 91, 94, 95, 102–106], initially based on etched silicon devices but more recently on soft-lithography of elastomers, a simpler and more reproducible option [104]. These 'lab on a chip' devices have been used for serial and parallel fragmentation of samples prior to MS analysis.

On-chip sample simplification may be performed using the SELDI chip that comprises a capture reagent such as cation exchange resin immobilised to a specialised MALDI target [107]. A subset of the proteins in a mixture bind to the capture surface under a given set of conditions and can be assayed directly by MALDI-MS. This process enables rapid sample profiling and is particularly suited to determining biomarkers for disease where sample volumes are limited but where sample numbers need to be high, thus ruling out 2DE as an appropriate method [108–111]. The recent introduction of an adaptor for the Q-TOF that will enable potential biomarkers to be identified by tandem MS has increased the value of the Ciphergen system, particularly to the clinical research environment.

The above is not an exhaustive review of the methods available for proteome analysis but it provides an overview of separation-dependent approaches. These are providing a wealth of information on comparative protein expression profiles, protein-ligand interactions and differential post-translational modification, but they are still largely dependent on a degree of sample fractionation and the serial analysis of the protein components of complex mixtures. In order to achieve highly parallel analyses of protein mixtures, protein arrays are being developed.

14.4
Protein Arrays

The separation-dependent methods described above have developed rapidly to become automated and miniaturised, but they still require relatively large quantities of protein sample and operate in a sequential rather than a parallel fashion. Interest in protein and antibody arrays as tools to complement and perhaps to replace separation-dependent methods of protein expression analysis has increased dramatically over the past two or three years. The ability to perform highly parallel proteomics studies in a way that is analogous to those made possible by DNA microarray technology would revolutionise the study of dynamic cellular and metabolic processes [14, 61, 112]. These analyses exploit molecular recognition to isolate, detect and/or identify multiple target molecules simultaneously. Nucleic acid molecules are remarkably well-suited to such analyses as the specific base-pairing between an immobilised probe and its target is predictable and occurs with approximately the same kinetics regardless of the specific DNA/mRNA sequence. Also, labelling of the target sequences with radioactive or fluorescent tags is both simple and stoichiometric, and does not affect the interaction of the target with the probe. 'DNA chip' technology has successfully been applied to a wide variety of biological systems and has resulted in significant insights into human diseases such as cancer [22, 113–118].

Arrays for proteome analysis may be separated into two groups, protein arrays and antibody arrays. Protein arrays comprise a capture molecule other than an antibody whose physiological interaction with other molecules is the basis of the scientific interest. Thus protein arrays might be designed to study enzyme-ligand binding, to investigate the activity of proteins of interest in the presence of other biological compounds or drugs, to determine functionally important protein-protein interactions, or to form the basis of a serodiagnostic chip [119–124]. They may also be used for high-throughput screening of antibody specificity, involving the arraying of potentially genome-wide sets of recombinant proteins [120, 125, 126]. Antibody arrays, on the other hand, comprise a series of immobilised antibodies used primarily for relative protein expression analysis and thus may be seen as an alternative to 2DE [61, 120], although they may also be used to evaluate antibody specificity in a high-throughput manner [127].

The basic requirements of antibody and protein arrays are essentially the same: i) specific recognition molecules must be generated and isolated; ii) the recognition molecules must retain their biological activity when immobilised in an addressable format; and iii) bound target must be detected. As described above, these requirements are readily satisfied for DNA arrays but protein capture molecules are considerably more difficult to generate in quantity and at a high level of purity, plus immobilisation may affect their ability to capture the target antigen with high specificity and affinity. Stoichiometric modification of the protein sample with fluorescent or other tags is difficult to achieve and may influence the interaction of the labelled protein with the capture molecule. Optimal conditions for binding of target protein to the array may differ widely for each capture molecule because of their widely divergent physicochemical properties, introducing a further level of complexity to the interpretation of binding patterns. Thus it is difficult to envisage protein or antibody arrays that are analogous to DNA arrays with current technology, but developments that are occurring across many disciplines may eventually lead to more limited but nonetheless useful protein and antibody arrays.

14.4.1
Protein Recognition Molecules

The range of molecules available to be exploited as protein recognition molecules (PRMs) is expanding as new technologies are being developed to replace or complement older, more established strategies, although few methods for generating capture reagents specifically for incorporation into arrays for protein expression profiling have been described [128, 129]. Whatever their nature, the PRMs to be incorporated into protein or antibody arrays need to meet various criteria: firstly, one has to have access to a process that can support the generation and selection of antibodies to individual proteins and possibly each of their post-translationally modified forms; the antibodies must be highly specific, that is they must be capable of recognising and distinguishing the individual protein when presented in a complex mixture; the binding of the individual proteins within the mixture to their cognate proteins must occur under similar conditions, that is with similar affinity, association constant etc; and the array must be amenable to prolonged storage without reduction in the reactivity of some or all of the capture reagents.

The biological function of polyclonal and monoclonal antibodies is molecular recognition, so that they are perhaps the most obvious choice as PRMs. They are capable of exquisite distinction between closely related proteins, such as those differing by phosphorylation at a single site, and there are well-established procedures for the generation of polyclonal and monoclonal antibodies *in vivo* [130, 131]. However, the generation and screening of antibodies is very demanding in terms of the requirement for pure protein and cross-reactivity with molecules other than the target protein cannot be ruled out.

The production of synthetic phage-displayed antibodies provides considerable scope for the generation of protein recognition molecules to support the development of 'protein chips'. Recombinant phage technology was established in the early 1980's and has developed into a rapid and simple method to link phenotype to genotype by inducing recombinant phage to express a protein of interest, such as an antibody fragment, on the surface coat where it is available for functional and binding assays. The advantages of antibody phage display over conventional monoclonal antibodies include the ability to humanise the antibodies for use in immunotherapy and the generation of antibodies against self antigens as a source of potential anti-cancer reagents [132–139]. Problems associated with the technology include the low affinity expressed by many of the antibodies selected from phage displayed libraries, although there is the opportunity to 'mature' the antibodies in order to increase their affinity by processes such as *in vitro* mutagenesis and error-prone PCR [140–142]. Other problems include the genetic instability of some of the libraries [142], and the fact that the selection procedure requires multiple rounds of panning against the cognate protein. As with monoclonal antibodies, selection of phage displayed antibodies requires significant quantities of pure target protein, and unique specificity for the target cannot be assured.

Various methods have been developed that employ peptides as either the capture reagent, or as a readily available source of target antigen for screening PRMs. Thus, random peptide aptamers may be introduced into cells in order to inhibit intracellular functions [143, 144], or peptides may be synthesized *in situ* to generate arrays for screening for interaction partners [145, 146], and the latter are likely to be more flexible in terms of the range of specificities that may be achieved than an array based on, for instance, monoclonal antibodies. However, their usefulness may be restricted when functional interactions are based on conformational rather than linear epitopes. Polysome display involves the *in vitro* expression of a peptide and the stabilisation of the interaction between the ribosome, the mRNA and the peptide being expressed, providing a link between the peptide and its encoding gene. This allows multiple rounds of screening to be performed without the prior availability of large quantities of pure protein [147, 148]. The potential also exists to immobilise polysomes in an array format via molecular modification of the complexes with a DNA linker.

Proteins are not the only PRMs that might be used for 'protein chips'. Any group of molecules that can specifically recognise individual proteins with appropriate affinity and avidity and can do so in an immobilised form would be suitable. Indeed, there are disadvantages to the use of proteins as recognition molecules, the most obvious ones being that the chips might suffer from poor storage properties and that the presence of any protease activity within the protein mixtures to be analysed may 'degrade' the chip. Nucleic acid aptamers are single-stranded oligonucleotides possessing high affinity for conformational biomolecules, such as proteins. Libraries of aptamers up to a few hundred nucleotides in length may be immobilised to an appropriate surface for screening and may provide information on conformation, position of hydrogen bonds and other data which help to build up a three-dimensional model of the target protein [149–153]. Nucleic acid aptamers also have advantages for 'protein chips' because the target protein does not need to be modified prior to capture, the aptamer itself acting as the capture and detection reagent [149].

Molecular imprinting involves the synthesis of artificial recognition sites on a surface by mimicking the shape of the template molecule in a polymeric film, thereby forming a molecularly imprinted polymer (MIP). It is proposed that any biological situation in which shape plays a part, such as antibody-antigen interactions, substrate-enzyme binding and receptor-ligand binding, may be mimicked by MIPs [154–158]. Most of the studies performed to date with MIPs involve the imprinting of small molecules such as drugs and pollutants [155, 159, 160], but there is great potential to create artificial capture surfaces that mimic antibodies and receptors [157]. As with most of the PRMs described, a clear disadvantage of MIPs is that their manufacture requires access to large quantities of the molecule to be imprinted, plus the capture surfaces are in effect polyclonal and therefore there may be problems with specificity.

14.4.2
Immobilisation of PRMs

The specific method employed to array and attach the PRMs to the array surface will obviously depend upon their nature. The basic requirements of the arraying procedure are spatial definition, reproducibility, stability and retention of high specificity and affinity. There are many different methods for arraying proteins (including antibodies) [161] and developments are continuously being made, many of them driven by the advances being made in biosensor technology [61, 162–166].

14.4.2.1 Arraying of Proteinaceous PRMs

Proteinaceous PRMs, be they monoclonal antibodies, phage displayed antibodies or other polypeptides, will need to be attached to a suitable surface in a spatially defined manner. In the case of monoclonal antibodies, some means to attach the molecules via their Fc portions, without affecting the conformation of other regions of the molecule, is likely to be required. If allowed to attach randomly to the support, a percentage of the antibodies to be immobilised may fortuitously align themselves correctly, and in some cases this has been reported to have little effect on overall binding capacity [167]. Sensitivity of target protein detection is likely to be an issue with such antibody arraying conditions indicating that orientated binding may be required. Orientation of immunoglobulins is often achieved through the use of Protein A/G that has the required specificity for Fc regions, but such an approach to generating an antibody array does not seem very elegant.

However, several modifications of this standard immobilisation procedure are being investigated. For instance, Protein A has been engineered to express 5 copies of the immunoglobulin G binding domain plus a cysteine residue at the C-terminus, the latter to allow strong binding to gold immobilisation surfaces [168]. Others have exploited the carbohydrate moieties on the Fc portion of the antibody to orientate them on an Affi-gel matrix [167].

Many immobilisation strategies rely on non-covalent interactions between the protein to be immobilised and the array surface. These may be based on the affinity of the base layer for the PRM, as in the case of Protein A, or on hydrophobic, ionic, or van der Waal's interactions between the PRM and the absorptive surface [161, 169]. In order to covalently bind the PRMs, metal or silica surfaces are photolithographically etched and proteins are attached via a chemical linker such as aminosilane capable of forming covalent bonds with both array surface and PRM [161]. Alternatively, substances such as glutaraldehyde and N-succinimidyl-4-maleimidobutyrate may be used as cross-linkers [161, 170, 171].

Notwithstanding these considerations, attempts to array proteinaceous PRMs have been initiated successfully [161]. In one such, acrylamide gel has been photopolymerised into a grid pattern using photolithography, followed by transfer of the PRM (IgG or BSA) to the gel surface by a multi-pin device in much the same way as the cDNA microarrays were prepared [172]. In later systems, the acrylamide was replaced by an absorptive protein layer such as streptavidin. The streptavidin was micropatterned using ink-jet printing techniques then exposed to biotin followed by the protein to be adsorbed [173]. Alternatively, a streptavidin layer was overlaid with photobiotin and the grid generated by exposure to UV light [173]. Both of these techniques required multiple layers to be deposited sequentially and were prone to poor specificity because of cross-contamination of the squares of the grid [173]. However, the fact that commercially available ink-jet printers may be modified to generate arrays on cellulose paper makes them accessible to any interested party [174].

One of the most recent microfabrication methods to have been proposed for patterning biomolecules is termed 'soft-lithography' [104, 105, 161, 175]. This comprises an elastomeric material or 'hydrogel' that can be moulded to form stamps or channels for the transfer of proteins to appropriate surfaces. For instance, Gaber and colleagues have used drawn-out capillary tubes filled with freeze-dried disaccharide acrylate polymers that, when exposed to aqueous protein solution, re-swell and form a protein-saturated nib [173]. The method is said to be suitable for arraying small, well-defined areas of delicate biomolecules with relative ease [104, 105, 175]. Electrospray deposition may also be effective for arraying delicate biomolecules since it only requires a support that is slightly conductive, such as damp membrane, to achieve immobilisation [176]. Whether the arraying methods described above will be as versatile and practicable as those currently used to generate DNA arrays remains to be seen.

14.4.2.2 **Arraying of Nucleic Acid PRMs**

Methodologies already exist to array nucleic acids but these are constantly being upgraded in an attempt to increase the number of targets per unit area, thereby increasing the data generated per array or chip. The simplest, and therefore probably the most accessible arrays, comprise cDNA inserts or PCR products that have been 'dot-blotted' onto a nylon membrane of approximately the size of a 96-well plate. The DNA binds strongly to the membrane and in a manner that enables its hybridisation to single stranded probes [177]. Preparation of these arrays has been both automated and miniaturised so that thousands of cDNAs may be spotted onto a surface the size of a postage stamp [178, 179].

The arraying protocol has been modified through the use of glass microscope slides as supports for the DNA instead of membranes. The glass is silanated and the cDNA is spotted onto the surface in a grid pattern using a computer-controlled tri-directional robot to which capillary-tipped pens are attached [180]. The DNA is then immobilised by exposure to ultraviolet light. Glass microarrays such as these are able to represent between 5000 and 10,000 genes per square centimetre [6, 178, 179].

The most sophisticated method developed to date for arraying nucleic acids is the *in situ* synthesis of oligonucleotides such that up to 64,000 'features', each containing several million oligonucleotides, may be represented on a small glass chip. The oligonucleotides are synthesized from modified photo-labile deoxynucleosides that polymerise on exposure to ultraviolet light. The sequence of the DNA is directed by masking from irradiation areas of the chip where the available nucleoside is not required [6, 179]. A recent modification of this technology involves virtual masking in which a computer describes the areas of the chip to be irradiated [181].

Each of these technologies could be adapted as a platform for an array for protein expression profiling in which the PRMs were nucleic acid aptamers. The selected nucleic acid sequences could be spotted onto a membrane or synthesized *in situ*. Membranes are accessible and manageable, but suffer from potential background problems and relatively large surface areas. Glass chips are less susceptible to high backgrounds, would require considerably less target protein mixture, and are more amenable to automation. Thus, the future prospects for immobilising aptamers for protein arrays mirrors the current position for gene expression arrays.

14.4.2.3 **Arraying of Synthetic PRMs**

The development of MIPs is still at an early stage compared to organic PRMs, yet they would have many advantages over proteins and nucleic acids in an array format. There would be no difficulty with array longevity nor would the ability to capture target be compromised by assay conditions. Combinatorial libraries of MIPs are already under construction, and the process has been semi-automated by the use of liquid handling robots [159], and miniaturised to produce microcolumns of 1mm internal diameter [182].

14.5
Detection Methods

Detection of the interaction of proteins with recognition molecules on a 'protein chip' in a rapid, highly parallel and sensitive manner is likely to prove very challenging. Many of the methods currently in development are very effective for less complex devices but are probably not applicable to 'protein chips' at least as envisaged here. For example, methods that require a second fluorescently labelled recognition molecule to detect the captured protein are unlikely to prove applicable. Despite the potential drawbacks, development of a fluorescence-based fibre-optic biosensor in which the capture antibodies are coated directly onto a fibre-optic probe is interesting [183]. In this approach detection is achieved by application of a detection antibody labelled with cyanine dye and laser excitation of the fibre optic probe to enable quantification of the bound analyte by fluorimetry. The system developed by Tempelman and colleagues was able to analyse four samples simultaneously for the presence of a single analyte and had an optimum detection range of 5–200 ng/ml for Staphylococcal enterotoxins [183]. Another interesting fluorescence-based detection method has also been described in which an antibody-coated matrix is saturated with fluorescently labelled target antigen. The fluorescent target antigen is displaced by unlabelled target present in the test sample and the released fluorescent material measured. The method has been used by the US Federal Drugs Administration to quantify cocaine metabolites in urine [184] and by the Office of Naval Research to detect explosives [185]. Both of these studies reported detection limits in the femtomole-picomole range. Although, in present form this would not be translatable to a 'protein chip,' one could foresee a system in which the protein recognition molecules arrayed on the surface are designed such that they all bind a generic fluorescent moiety that would be released on binding of the individual proteins. The fluorescence remaining at an individual site would then be inversely related to the amount of relevant protein in the sample mixture.

Biosensors, including surface plasmon resonance (SPR)/resonant mirror biosensors, allow real time detection of protein interactions with other proteins or ligands (see [166, 186]). These devices comprise a glass surface onto which a thin layer of metal, usually gold, has been deposited and is subsequently covered by a self-assembled monolayer (SAM) of matrix such as carboxymethylated hydrogel [187], carboxymethylated dextran [188, 189] or a mixture of alkanethiolates [190, 191]. The chemistry of the matrix is such that proteins and ligands may be immobilised on the sensor and when the target protein binds, the refractive index of the matrix changes. The change in refractive index may be measured by a diode array detector [190]. Clearly, if such devices could be made more parallel, i.e. with a larger number of protein recognition sites, and miniaturised they could possess considerable potential.

Quartz crystal microbalance (QCM) devices are an interesting method for detecting antibody antigen interactions. In these, antibodies are bound to quartz crystals through amino groups incorporated into ethylenediamine plasma-polymerised films [171]. The QCM is then exposed to the fluid to be analysed, binding of target antigen results in small changes in mass on the surface of the crystal and these changes are measured by detecting changes in resonance frequency via two gold electrodes. The sensitivity of these devices is such that they have the potential to detect the interaction of a single protein molecule with a single antibody molecule.

Other devices that have the potential to detect single molecule interactions include the force amplified biological sensor being developed by Colton and colleagues [192]. This is a modification of the atomic force microscope in which micron-sized magnetic particles are used in combination with micromachined piezoresistive cantilevers to measure antibody-antigen interactions. Finally, the possibility of incorporating biosensors into microelectronic circuits has been explored using molecular ion channels as sensing devices [193]. In this approach, a gold electrode was attached to gramicidin molecules in the lower layer of a synthetic lipid membrane with the ion channel being formed by a second gramicidin molecule present in mobile form in the upper membrane layer. The mobile gramicidin molecules were linked to an antibody such that binding of antigen to antibody altered the conductance properties of the channel. There are many advantages offered by such a device, including the amplification that occurs as part of the detection event; thus, a single antigen-antibody interaction results in fluxes of millions of ions per second (see [194]).

14.6
Conclusions

Despite the technical limitations associated with 2-DE, there is little doubt that, at present, it is unrivalled as a method to resolve up to several thousand proteins simultaneously (see [14]). Furthermore, the automation of at least part of the work flow from 2DE gels to protein identification has increased throughput and rapid developments in the technology continue. It seems likely therefore that 2-DE and these supporting techniques will remain in widespread use for some time yet. The technical limitations are sufficient to have motivated many to attempt to develop alternative methods of protein expression mapping. We have reviewed some of these developments and in doing so, it has become evident that any future technology for high-throughput protein expression analysis will almost certainly require a multidisciplinary approach and the further development of novel methods. It remains to be seen whether the potential of 'protein chips' can be translated into an accessible, versatile and rigorous method for proteomic analyses.

14.7
References

1 E. S. LANDER *et al.* 'Initial sequencing and analysis of the human genome.' *Nature* **2001**, *409*, 860–921.

2 J. D. McPHERSON *et al.* 'A physical map of the human genome.' *Nature* **2001**, *409*, 934–941.

3 J. C. VENTER *et al.* 'The sequence of the human genome.' *Science* **2001**, *291*, 1304–1351.

4 M. CHEE *et al.* 'Accessing genetic information with high-density DNA arrays.' *Science* **1996**, *274*, 610–614.

5 M. B EISEN, P. T. SPELLMAN, P. O. BROWN, D. BOTSTEIN. 'Cluster analysis and display of genome-wide expression patterns.' *Proceedings of the National Academy of Sciences of the United States of America* **1998**, *95*, 14863–14868.

6 D. GERHOLD, T. RUSHMORE, C. T. CASKEY. 'DNA chips: promising toys have become powerful tools.' *Trends in Biochemical Science* **1999**, *24*, 168–173.

7 S. FIELDS, O. SONG. 'A novel genetic system to detect protein-protein interactions.' *Nature* **1989**, *340*, 245–246.

8 M. FROMONT-RACINE, J. C. RAIN, P. LE-GRAIN. 'Toward a functional analysis of the yeast genome through exhaustive two-hybrid screens [see comments].' *Nature Genetics* **1997**, *16*, 277–282.

9 N. LECRENIER, F. FOURY, A. GOFFEAU. 'Two-hybrid systematic screening of the yeast proteome.' *Bioessays* **1998**, *20*, 1–5.

10 U. SCHERF *et al.* 'A gene expression database for the molecular pharmacology of cancer.' *Nature Genetics* **2000**, *24*, 236–244.

11 T. GAASTERLAND, S. BEKIRANOV. 'Making the most of microarray data.' *Nature Genetics* **2000**, *24*, 204–206.

12 D. D. L. BOWTELL. 'Options available – from start to finish – for obtaining expression data by microarray.' *Nature Genetics* **1999**, *21*, 25–32.

13 K. M. KURIAN, C. J. WATSON, A. H. WYL-LIE. 'DNA chip technology.' *Journal of Pathology* **1999**, *187*, 267–271.

14 A. ABBOTT. 'A post-genomic challenge: learning to read patterns of protein synthesis.' *Nature* **1999**, *402*, 715–720.

15 L. ANDERSON, J. SEILHAMER. 'A comparison of selected mRNA and protein abundances in human liver.' *Electrophoresis* **1997**, *18*, 533–537.

16 N. L. ANDERSON, N. G. ANDERSON. 'Proteome and proteomics: new technologies, new concepts, and new words.' *Electrophoresis* **1998**, *19*, 1853–1861.

17 W. P. BLACKSTOCK, M. P. WEIR. 'Proteomics: quantitative and physical mapping of cellular proteins.' *Trends in Biotechnology* **1999**, *17*, 121–127.

18 A. DOVE. 'Proteomics: translating genomics into products?' *Nature Biotechnology* **1999**, *17*, 233–236.

19 R. PAREKH. 'Proteomics and molecular medicine.' *Nature Biotechnology* **1999**, *17 supplement*, BV19–BV20.

20 S. R. PENNINGTON, M. R. WILKINS, D. F. HOCHSTRASSER, M. J. DUNN. 'Proteome analysis: from protein characterisation to biological function.' *Trends in Cell Biology* **1997**, *7*, 168–173.

21 M. R. WILKINS *et al.* 'Progress with proteome projects: why all proteins expressed by a genome should be identified and how to do it.' *Biotechnology* **1995**, *13*, 19–50.

22 S. M. DHANASEKARAN *et al.* 'Delineation of prognostic biomarkers in prostate cancer.' *Nature* **2001**, *412*, 822–826.

23 A. T. BRÜNGER. 'X-ray crystallography and NMR reveal complementary views of structure and dynamics.' *Nature Structural Biology* **1997**, *4 Suppl*, 862–865.

23 S. CUSACK *et al.* 'Small is beautiful: protein micro-crystallography.' *Nature Structural Biology* **1998**, *5 Suppl*, 634–637.

25 M. GERSTEIN, H. HEGYI. 'Comparing genomes in terms of protein structure: surveys of a finite parts list.' *FEMS Microbiology Reviews* **1998**, *22*, 277–304.

26 B. ROBSON. 'Beyond proteins.' *Trends in Biotechnology* **1999**, *17*, 311–315.

27 A. SALI, J. KURIYAN. 'Challenges at the frontiers of structural biology (Reprinted from Trends in Biochemical Science, vol 12, Dec., 1999).' *Trends in Genetics* **1999**, *15*, M20–M24.

28 M. B SWINDELLS, C. A. ORENGO, D. T. JONES, E. G. HUTCHINSON, J. M. THORNTON. 'Contemporary approaches to protein structure classification.' *Bioessays* 1998, *20*, 884–891.

29 J. P. WERY, R. W. SCHEVITZ. 'New trends in macromolecular X-ray crystallography.' *Current Opinion in Chemical Biology* 1997, *1*, 365–369.

30 E. M. MARCOTTE *et al.* 'Detecting protein function and protein-protein interactions from genome sequences.' *Science* 1999a, *285*, 751–753.

31 E. M. MARCOTTE, M. PELLEGRINI, M. J. THOMPSON, T. O. YEATES, D. EISENBERG. 'A combined algorithm for genome-wide prediction of protein function.' *Nature* 1999b, *402*, 83–86.

32 P. BORK *et al.* 'Predicting function: from genes to genomes and back.' *Journal of Molecular Biology* 1998, *283*, 707–725.

33 M. PELLEGRINI, E. M. MARCOTTE, M. J. THOMPSON, D. EISENBERG, T. O. YEATES. 'Assigning protein functions by comparative genome analysis: protein phylogenetic profiles.' *Proceedings of the National Academy of Sciences of the United States of America* 1999, *96*, 4285–4288.

34 P. UETZ *et al.* 'A comprehensive analysis of protein-protein interactions in Saccharomyces cerevisiae.' *Nature* 2000, *403*, 623–627.

35 A. C. GAVIN *et al.* 'Functional organization of the yeast proteome by systematic analysis of protein complexes.' *Nature* 2002, *415*, 141–147.

36 O. PUIG *et al.* 'The tandem affinity purification (TAP) method: A general procedure of protein complex purification.' *Methods* 2001, *24*, 218–229.

37 J. J. TASTO, R. H. CARNAHAN, W. H. MCDONALD, K. L. GOULD. 'Vectors and gene targeting modules for tandem affinity purification in Schizosaccharomyces pombe.' *Yeast* 2001, *18*, 657–662.

38 A. J. ENRIGHT, I. ILIOPOULOS, N. C. KYRPIDES, C. A. OUZOUNIS. 'Protein interaction maps for complete genomes based on gene fusion events.' *Nature* 1999, *402*, 86–90.

39 P. ROSSMACDONALD *et al.* 'Large-scale analysis of the yeast genome by transposon tagging and gene disruption.' *Nature* 1999, *402*, 413–418.

40 A. H. Y. TONG *et al.* 'Systematic genetic analysis with ordered arrays of yeast deletion mutants.' *Science* 2001, *294*, 2364–2368.

41 P. JAMES. 'Protein identification in the post-genome era: the rapid rise of proteomics.' *Quarterly Reviews of Physics* 1997, *30*, 279–331.

42 V. SANTONI, M. MOLLOY, T. RABILLOUD. 'Membrane proteins and proteomics: Un amour impossible?' *Electrophoresis* 2000, *21*, 1054–1070.

43 G. L. CORTHALS, V. C. WASINGER, D. F. HOCHSTRASSER, J. C. SANCHEZ. 'The dynamic range of protein expression: A challenge for proteomic research.' *Electrophoresis* 2000, *21*, 1104–1115.

44 M. J. DUNN. 'Quantitative two-dimensional gel electrophoresis: from proteins to proteomes.' *Biochemical Society Transactions* 1997, *25*, 248–254.

45 A. GORG *et al.* 'The current state of two-dimensional electrophoresis with immobilized pH gradients.' *Electrophoresis* 2000, *21*, 1037–1053.

46 J. M. HILLE, A. L. FREED, H. WATZIG. 'Possibilities to improve automation, speed and precision of proteome analysis: A comparison of two-dimensional electrophoresis and alternatives.' *Electrophoresis* 2001, *22*, 4035–4052.

47 H. J. ISSAQ. 'The role of separation science in proteomics research.' *Electrophoresis* 2001, *22*, 3629–3638.

48 S. E. ONG, A. PANDEY. 'An evaluation of the use of two-dimensional gel electrophoresis in proteomics.' *Biomolecular Engineering* 2001, *18*, 195–205.

49 B. R. HERBERT *et al.* 'What place for polyacrylamide in proteomics?' *Trends in Biotechnology* 2001, *19*, S3–S9.

50 J. KLOSE. 'Large-Gel 2D Electrophoresis.' *Methods in Molecular Biology* 1999, *112: 2–D Proteome Analysis Protocols*, 147–172.

51 J. E. CELIS *et al.* 'Human and mouse proteomic databases: novel resources in the protein universe.' *FEBS Letters* 1998, *430*, 64–72.

52 T. K. CHATAWAY *et al.* 'Development of a two-dimensional gel electrophoresis data-

base of human lysosomal proteins.' *Electrophoresis* **1998**, *19*, 834–836.

53 G. Friso, L. Wikström. 'Analysis of proteins from membrane-enriched cerebellar preparations by two-dimensional gel electrophoresis and mass spectrometry.' *Electrophoresis* **1999**, *20*, 917–927.

54 R. E. Jenkins *et al.* 'Regulation of growth factor induced gene expression by calcium signalling: Integrated mRNA and protein expression analysis.' *Proteomics* **2001**, *1*, 1092–1104.

55 X. P. Li *et al.* 'A two-dimensional electrophoresis database of rat heart proteins.' *Electrophoresis* **1999**, *20*, 891–897.

56 A. C. Shaw *et al.* 'Mapping and identification of HeLa cell proteins separated by immobilized pH-gradient two-dimensional gel electrophoresis and construction of a two-dimensional polyacrylamide gel electrophoresis database.' *Electrophoresis* **1999**, *20*, 977–983.

57 H. Schagger, K. Pfeiffer. 'The ratio of oxidative phosphorylation complexes I–V in bovine heart mitochondria and the composition of respiratory chain supercomplexes.' *Journal of Biological Chemistry* **2001**, *276*, 37861–37867.

58 B. Thiede, C. Dimmler, F. Siejak, T. Rudel. 'Predominant identification of RNA-binding proteins in Fas- induced apoptosis by proteome analysis.' *Journal of Biological Chemistry* **2001**, *276*, 26044–26050.

59 A. S. Vercoutter-Edouart *et al.* 'Proteomic detection of changes in protein synthesis induced by fibroblast growth factor-2 in MCF-7 human breast cancer cells.' *Experimental Cell Research* **2001**, *262*, 59–68.

60 F. Kernec, M. Unlu, W. Labeikovsky, J. S. Minden, A. P. Koretsky. 'Changes in the mitochondrial proteome from mouse hearts deficient in creatine kinase.' *Physiological Genomics* **2001**, *6*, 117–128.

61 R. E. Jenkins, S. R. Pennington. 'Arrays for protein expression profiling: Towards a viable alternative to two-dimensional gel electrophoresis?' *Proteomics* **2001**, *1*, 13–29.

62 H. Gmuender *et al.* 'Gene expression changes triggered by exposure of Haemo-philus influenzae to novobiocin or ciprofloxacin: Combined transcription and translation analysis.' *Genome Research* **2001**, *11*, 28–42.

63 W. F. Patton. 'A thousand points of light: The application of fluorescence detection technologies to two-dimensional gel electrophoresis and proteomics.' *Electrophoresis* **2000**, *21*, 1123–1144.

64 W. F. Patton, in *Proteomics: From protein sequence to function* S. P. Pennington, M. J. Dunn, Eds. (Bios Scientific Publishers Ltd, Oxford, 2000).

65 F. Gharahdaghi, C. R. Weinberg, D. A. Meagher, B. S. Imai, S. M. Mische. 'Mass spectrometric identification of proteins from silver-stained polyacrylamide gel: a method for the removal of silver ions to enhance sensitivity.' *Electrophoresis* **1999**, *20*, 601–605.

66 W. M. Lauber *et al.* 'Mass spectrometry compatibility of two-dimensional gel protein stains.' *Electrophoresis* **2001**, *22*, 906–918.

67 K. Klarskov, S. Naylor. 'India Ink staining after sodium dodecyl sulfate polyacrylamide gel electrophoresis and in conjunction with Western blots for peptide mapping by matrix-assisted laser desorption/ionization time-of-flight mass Spectrometry.' *Rapid Communications in Mass Spectrometry* **2002**, *16*, 35–42.

68 M. Quadroni, P. James. 'Proteomics and automation.' *Electrophoresis* **1999**, *20*, 664–677.

69 R. Tonge *et al.* 'Validation and development of fluorescence two-dimensional differential gel electrophoresis proteomics technology.' *Proteomics* **2001**, *1*, 377–396.

70 M. Unlü, M. E. Morgan, J. S. Minden. 'Difference gel electrophoresis: a single gel method for detecting changes in protein extracts.' *Electrophoresis* **1997**, *18*, 2071–2077.

71 R. A. Craven, R. E. Banks. 'Laser capture microdissection and proteomics: Possibilities and limitation.' *Proteomics* **2001**, *1*, 1200–1204.

72 D. K. Ornstein *et al.* 'Proteomic analysis of laser capture microdissected human prostate cancer and in vitro prostate cell lines.' *Electrophoresis* **2000**, *21*, 2235–2242.

73 M.P. Molloy *et al.* 'Extraction of membrane proteins by differential solubilization for separation using two-dimensional gel electrophoresis.' *Electrophoresis* **1998**, *19*, 837–844.

74 M.F. Lopez, in *Proteomics: From protein sequence to function* S.R. Pennington, M.J. Dunn, Eds. (Bios Scientific Publishers Ltd, Oxford, 2000).

75 M.F. Lopez. 'Better approaches to finding the needle in a haystack: Optimizing proteome analysis through automation.' *Electrophoresis* **2000**, *21*, 1082–1093.

76 M. Traini *et al.* 'Towards an automated approach for protein identification in proteome projects.' *Electrophoresis* **1998**, *19*, 1941–1949.

77 A.J. Nicola, A. Gusev, A. Proctor, D.M. Hercules. 'Automation of data collection for matrix assisted laser desorption/ionization mass spectrometry using a correlative analysis algorithm.' *Analytical Chemistry* **1998**, *70*, 3213–3219.

78 O.N. Jensen, P. Mortensen, O. Vorm, M. Mann. 'Automation of matrix-assisted laser desorption/ionization mass spectrometry using fuzzy logic feedback control.' *Analytical Chemistry* **1997**, *69*, 1706–1714.

79 D.S. Hage. 'Affinity chromatography: a review of clinical applications.' *Clinical Chemistry* **1999**, *45*, 593–615.

80 T. Manabe. 'Combination of electrophoretic techniques for comprehensive analysis of complex protein systems.' *Electrophoresis* **2000**, *21*, 1116–1122.

81 G.J. Opiteck, S.M. Ramirez, J.W. Jorgenson, M.A. Moseley III. 'Comprehensive two-dimensional high-performance liquid chromatography for the isolation of overexpressed proteins and proteome mapping.' *Analytical Biochemistry* **1998**, *258*, 349–361.

82 J.P.C. Vissers. 'Recent developments in microcolumn liquid chromatography.' *Journal of Chromatography a* **1999**, *856*, 117–143.

83 L.A. Colon, Y. Guo, A. Fermier. 'Capillary electrochromatography.' *Analytical Chemistry* **1997**, *69*, A461–A467.

84 S.P. Gygi *et al.* 'Quantitative analysis of complex protein mixtures using isotope-coded affinity tags.' *Nature Biotechnology* **1999**, *17*, 994–999.

85 T.J. Griffin *et al.* 'Quantitative proteomic analysis using a MALDI quadrupole time-of-flight mass spectrometer.' *Analytical Chemistry* **2001**, *73*, 978–986.

86 D.K. Han, J. Eng, H. Zhou, R. Aebersold. 'Quantitative profiling of differentiation-induced microsomal proteins using isotope-coded affinity tags and mass spectrometry.' *Nature Biotechnology* **2001**, *19*, 946–951.

87 M.B Smolka, H. Zhou, S. Purkayastha, R. Aebersold. 'Optimization of the isotope-coded affinity tag-labeling procedure for quantitative proteome analysis.' *Analytical Biochemistry* **2001**, *297*, 25–31.

88 V. Dolnik, S.R. Liu, S. Jovanovich. 'Capillary electrophoresis on microchip.' *Electrophoresis* **2000**, *21*, 41–54.

89 A.R. Dongre, J.K. Eng, J.R. Yates. 'Emerging tandem-mass-spectrometry techniques for the rapid identification of proteins.' *Trends in Biotechnology* **1997**, *15*, 418–425.

90 B.B. Feng, M.S. McQueney, T.M. Mezzasaima, J.R. Slemmon. 'An integrated ten-pump, eight-channel parallel LC/MS system for automated high-throughput analysis of proteins.' *Analytical Chemistry* **2001**, *73*, 5691–5697.

91 B. He, F. Regnier. 'Microfabricated liquid chromatography columns based on collocated monolith support structures.' *Journal of Pharmaceutical and Biomedical Analysis* **1998**, *17*, 925–932.

92 L.J. Jin, J. Ferrance, J.P. Landers. 'Miniaturized electrophoresis: An evolving role in laboratory medicine.' *Biotechniques* **2001**, *31*, 1332–+.

93 S.D. Patterson, R. Aebersold, D.R. Goodlett, in *Proteomics: From protein sequence to function* S.P. Pennington, M.J. Dunn, Eds. (Bios Scientific Publishers Ltd, Oxford, 2000).

94 F.E. Regnier, B. He, S. Lin, J. Busse. 'Chromatography and electrophoresis on chips: critical elements of future integrated, microfluidic analytical systems for life science.' *Trends in Biotechnology* **1999**, *17*, 101–106.

95 D. Figeys, R. Aebersold. 'Microfabricated modules for sample handling, sample concentration and flow mixing: application to protein analysis by tandem mass spectrometry.' *Journal of Biomechanical Engineering* **1999**, *121*, 7–12.

96 A. I. LAMOND, M. MANN. 'Cell biology and the genome projects – A concerted strategy for characterizing multiprotein complexes by using mass spectrometry.' *Trends in Cell Biology* **1997**, *7*, 139–142.

97 P. J. MINTZ, S. D. PATTERSON, A. F. NEUWALD, C. S. SPAHR, D. L. SPECTOR. 'Purification and biochemical characterization of interchromatin granule clusters.' *EMBO Journal* **1999**, *18*, 4308–4320.

98 C. S. SPAHR *et al.* 'Towards defining the urinary proteome using liquid chromatography-tandem mass spectrometry I. Profiling an unfractionated tryptic digest.' *Proteomics* **2001**, *1*, 93–107.

99 G. SCHMITT-ULMS *et al.* 'Binding of neural cell adhesion molecules (N-CAMs) to the cellular prion protein.' *Journal of Molecular Biology* **2001**, *314*, 1209–1225.

100 J. M. PENG, S. P. GYGI. 'Proteomics: the move to mixtures.' *Journal of Mass Spectrometry* **2001**, *36*, 1083–1091.

101 E. C. KOC, W. BURKHART, K. BLACKBURN, A. MOSELEY, L. L. SPREMULLI. 'The small subunit of the mammalian mitochondrial ribosome – Identification of the full complement of ribosomal proteins present.' *Journal of Biological Chemistry* **2001**, *276*, 19363–19374.

102 A. J. GAWRON, R. S. MARTIN, S. M. LUNTE. 'Microchip electrophoretic separation systems for biomedical and pharmaceutical analysis.' *European Journal of Pharmaceutical Sciences* **2001**, *14*, 1–12.

103 D. FIGEYS, D. PINTO. 'Proteomics on a chip: Promising developments.' *Electrophoresis* **2001**, *22*, 208–216.

104 M. A. UNGER, H.-P. CHOU, T. THORSEN, A. SCHERER, S. R. QUAKE. 'Monolithic microfabricated valves and pumps by multilayer soft lithography.' *Science* **2000**, *288*, 113–116.

105 J. C. MCDONALD *et al.* 'Fabrication of microfluidic systems in poly(dimethylsiloxane).' *Electrophoresis* **2000**, *21*, 27–40.

106 H. BECKER, K. LOWACK, A. MANZ. 'Planar quartz chips with submicron channels for two-dimensional capillary electrophoresis applications.' *Journal of Micromechanics and Microengineering* **1998**, *8*, 24–28.

107 M. MERCHANT, S. R. WEINBERGER. 'Recent advancements in surface-enhanced laser desorption/ionization- time of flight-mass spectrometry.' *Electrophoresis* **2000**, *21*, 1164–1177.

108 M. CZADER *et al.* 'The application of protein chip surface-enhanced laser desorption/ionization (SELDI) mass spectrometry for the identification of proteins in chronic lymphocytic leukemia.' *Blood* **2000**, *96*, 2488.

109 C. P. PAWELETZ *et al.* 'Rapid protein display profiling of cancer progression directly from human tissue using a protein biochip.' *Drug Development Research* **2000**, *49*, 34–42.

110 S. Y. WANG, D. L. DIAMOND, G. M. HASS, R. SOKOLOFF, R. L. VESSELLA. 'Identification of prostate specific membrane antigen (PSMA) as the target of monoclonal antibody 107–1A4 by proteinchip (R); Array, surface-enhanced laser desorption/ionization (SELDI) technology.' *International Journal of Cancer* **2001**, *92*, 871–876.

111 J. D. WULFKUHLE *et al.* 'New approaches to proteomic analysis of breast cancer.' *Proteomics* **2001**, *1*, 1205–1215.

112 M. F. TEMPLIN *et al.* 'Protein microarray technology.' *Trends in Biotechnology* **2002**, *20*, 160–166.

113 A. A. ALIZADEH *et al.* 'Distinct types of diffuse large B-cell lymphoma identified by gene expression profiling.' *Nature* **2000**, *403*, 503–511.

114 U. ALON *et al.* 'Broad patterns of gene expression revealed by clustering analysis of tumor and normal colon tissues probed by oligonucleotide arrays.' *Proceedings of the National Academy of Sciences of the United States of America* **1999**, *96*, 6745–6750.

115 A. BHATTACHARJEE *et al.* 'Classification of human lung carcinomas by mRNA expression profiling reveals distinct adenocarcinoma subclasses.' *Proceedings of the National Academy of Sciences of the United States of America* **2001**, *98*, 13790–13795.

116 M. BITTNER *et al.* 'Molecular classification of cutaneous malignant melanoma by gene expression profiling.' *Nature* **2000**, *406*, 536–540.

117 E. A. CLARK, T. R. GOLUB, E. S. LANDER, R. O. HYNES. 'Genomic analysis of metastasis reveals an essential role for RhoC.' *Nature* 2000, *406*, 532–535.

118 S. RAMASWAMY *et al.* 'Multiclass cancer diagnosis using tumor gene expression signatures.' *Proceedings of the National Academy of Sciences of the United States of America* 2001, *98*, 15149–15154.

119 G. MACBEATH, S. L. SCHREIBER. 'Printing proteins as microarrays for high-throughput function determination.' *Science* 2000, *289*, 1760–1763.

120 D. CAHILL. 'Protein and antibody arrays and their medical applications.' *Journal of Immunological Methods* 2001, *250*, 81–91.

121 H. ZHU *et al.* 'Global analysis of protein activities using proteome chips.' *Science* 2001, *293*, 2101–2105.

122 J. MADOZ-GURPIDE, H. WANG, D. E. MISEK, F. BRICHORY, S. M. HANASH. 'Protein based microarrays: A tool for probing the proteome of cancer cells and tissues.' *Proteomics* 2001, *1*, 1279–1287.

123 T. R. HUGHES *et al.* 'Functional discovery via a compendium of expression profiles.' *Cell* 2000, *102*, 109–126.

124 T. O. JOOS *et al.* 'A microarray enzyme-linked immunosorbent assay for autoimmune diagnostics.' *Electrophoresis* 2000, *21*, 2641–2650.

125 B. B. HAAB, M. J. DUNHAM, P. O. BROWN. 'Protein microarrays for highly parallel detection and quantitation of specific proteins and antibodies in complex solutions.' *Genome Biology* 2001, *2*, 4.1–4.13.

126 K. BUSSOW *et al.* 'A method for global protein expression and antibody screening on high-density filters of an arrayed cDNA library.' *Nucleic Acids Research* 1998, *26*, 5007–5008.

127 R. M. DE WILDT, C. R. MUNDY, B. D. GORICK, I. M. TOMLINSON. 'Antibody arrays for high-throughput screening of antibody-antigen interactions.' *Nature Biotechnology* 2000, *18*, 989–994.

128 M. L. BULYK, E. GENTALEN, D. J. LOCKHART, G. M. CHURCH. 'Quantifying DNA-protein interactions by double-stranded DNA arrays.' *Nature Biotechnology* 1999, *17*, 573–577.

129 C. A. K. BORREBAECK *et al.* 'Protein chips based on recombinant antibody fragments: A highly sensitive approach as detected by mass spectrometry.' *Biotechniques* 2001, *30*, 1126–+.

130 B. S. DUNBAR, S. M. SKINNER. 'Preparation of Monoclonal-Antibodies.' *Methods in Enzymology* 1990, *182*, 670–679.

131 E. HARLOW, D. LANE, *Antibodies. A laboratory manual*, (Cold Spring Harbour Laboratory, New York, 1988.

132 J. DEKRUIF, L. TERSTAPPEN, E. BOEL, T. LOGTENBERG. 'Rapid Selection of Cell Subpopulation-Specific Human Monoclonal-Antibodies From a Synthetic Phage Antibody Library.' *Proceedings of the National Academy of Sciences of the United States of America* 1995, *92*, 3938–3942.

133 C. F. BARBAS *et al.* 'In-Vitro Evolution of a Neutralizing Human-Antibody to Human-Immunodeficiency-Virus Type-1 to Enhance Affinity and Broaden Strain Cross-Reactivity.' *Proceedings of the National Academy of Sciences of the United States of America* 1994, *91*, 3809–3813.

134 A. D. GRIFFITHS *et al.* 'Isolation of High-Affinity Human-Antibodies Directly From Large Synthetic Repertoires.' *EMBO Journal* 1994, *13*, 3245–3260.

135 J. GUZMAN, K. FREI, D. NADAL. 'In-Vitro Immunization – Generation of Neutralizing Monoclonal- Antibodies to Human Interleukin-10.' *Journal of Immunological Methods* 1995, *179*, 265–268.

136 G. A. HULS *et al.* 'A recombinant, fully human monoclonal antibody with antitumor activity constructed from phage-displayed antibody fragments.' *Nature Biotechnology* 1999, *17*, 276–281.

137 J. MCCAFFERTY, A. D. GRIFFITHS, G. WINTER, D. J. CHISWELL. 'Phage Antibodies – Filamentous Phage Displaying Antibody Variable Domains.' *Nature* 1990, *348*, 552–554.

138 J. THOMPSON *et al.* 'Affinity maturation of a high-affinity human monoclonal antibody against the third hypervariable loop of human immunodeficiency virus: Use of phage display to improve affinity and broaden strain reactivity.' *Journal of Molecular Biology* 1996, *256*, 77–88.

139 T. J. VAUGHAN *et al.* 'Human antibodies with sub-nanomolar affinities isolated from a large non-immunized phage display library.' *Nature Biotechnology* 1996, *14*, 309–314.

140 P. MARTINEAU, P. JONES, G. WINTER. 'Expression of an antibody fragment at high levels in the bacterial cytoplasm.' *Journal of Molecular Biology* **1998**, *280*, 117–127.

141 C. MIYAZAKI et al. 'Changes in the specificity of antibodies by site-specific mutagenesis followed by random mutagenesis.' *Protein Engineering* **1999**, *12*, 407–415.

142 P. M. O'BRIEN, R. AITKEN, EDS., *Antibody Phage Display*, vol. 178 (Humana Press Inc., Totowa, New Jersey, 2002).

143 M. FAMULOK, M. BLIND, G. MAYER. 'Intramers as promising new tools in functional proteomics.' *Chemistry & Biology* **2001**, *8*, 931–939.

144 F. HOPPE-SEYLER et al. 'Peptide aptamers: new tools to study protein interactions.' *Journal of Steroid Biochemistry and Molecular Biology* **2001**, *78*, 105–111.

145 U. REINEKE, R. VOLKMER-ENGERT, J. SCHNEIDER-MERGENER. 'Applications of peptide arrays prepared by the SPOT-technology.' *Current Opinion in Biotechnology* **2001**, *12*, 59–64.

146 J. SCHNEIDER-MERGENER. 'Synthetic peptide and protein domain arrays prepared by the SPOT technology.' *Comparative and Functional Genomics* **2001**, *2*, 307–309.

147 A. LAWTON. 'PROfusion: A broad-based proteomics platform.' *British Journal of Pharmacology* **2002**, *135*, 144P.

148 L. C. MATTHEAKIS, J. M. DIAS, W. J. DOWER. 'Cell-free synthesis of peptide libraries displayed on polysomes.' *Methods in Enzymology* **1996**, *267*, 195–207.

149 E. N. BRODY et al. 'The use of aptamers in large arrays for molecular diagnostics.' *Molecular Diagnosis* **1999**, *4*, 381–388.

150 J. C. COX, A. D. ELLINGTON. 'Automated selection of anti-protein aptamers.' *Bioorganic & Medicinal Chemistry* **2001**, *9*, 2525–2531.

151 T. HERMANN, D. J. PATEL. 'Biochemistry-adaptive recognition by nucleic acid aptamers.' *Science* **2000**, *287*, 820–825.

152 A. D. ELLINGTON, J. W. SZOSTAK. 'Invitro Selection of Rna Molecules That Bind Specific Ligands.' *Nature* **1990**, *346*, 818–822.

153 T. S. ROMIG, C. BELL, D. W. DROLET. 'Aptamer affinity chromatography: combinatorial chemistry applied to protein purification.' *Journal of Chromatography B* **1999**, *731*, 275–284.

154 K. HAUPT, K. MOSBACH. 'Plastic antibodies: developments and applications.' *Trends in Biotechnology* **1998**, *16*, 468–475.

155 R. J. ANSELL, K. MOSBACH. 'Magnetic molecularly imprinted polymer beads for drug radioligand binding assay.' *Analyst* **1998**, *123*, 1611–1616.

156 L. YE, O. RAMSTROM, M. O. MANSSON, K. MOSBACH. 'A new application of molecularly imprinted materials.' *Journal of Molecular Recognition* **1998**, *11*, 75–78.

157 S. A. PILETSKY et al. 'Substitution of antibodies and receptors with molecularly imprinted polymers in enzyme-linked and fluorescent assays.' *Biosensors & Bioelectronics* **2001**, *16*, 701–707.

158 P. K. OWENS, L. KARLSSON, E. S. M. LUTZ, L. I. ANDERSSON. 'Molecular imprinting for bio-and pharmaceutical analysis.' *Trac-Trends in Analytical Chemistry* **1999**, *18*, 146–154.

159 T. TAKEUCHI, D. FUKUMA, J. MATSUI. 'Combinatorial molecular imprinting: an approach to synthetic polymer receptors.' *Analytical Chemistry* **1999**, *71*, 285–290.

160 W. M. MULLETT, E. P. C. LAI. 'Molecularly imprinted solid phase extraction microcolumn with differential pulsed elution for theophylline determination.' *Microchemical Journal* **1999**, *61*, 143–155.

161 A. S. BLAWAS, W. M. REICHERT. 'Protein patterning.' *Biomaterials* **1998**, *19*, 595–609.

162 D. STOLL et al. 'Protein microarray technology.' *Frontiers in Bioscience* **2002**, *7*, C13–C32.

163 C. D. HODNELAND, Y. S. LEE, D. H. MIN, M. MRKSICH. 'Selective immobilization of proteins to self-assembled monolayers presenting active site-directed capture ligands.' *Proceedings of the National Academy of Sciences of the United States of America* **2002**, *99*, 5048–5052.

164 C. WILLIAMS, T. A. ADDONA. 'The integration of SPR biosensors with mass spectrometry: possible applications for proteome analysis.' *Trends in Biotechnology* **2000**, *18*, 45–48.

165 A. P. TURNER. 'Array of hope for biosensors in Europe.' *Nature Biotechnology* **1998**, *16*, 824.

166 C. ZIEGLER, W. GOPEL. 'Biosensor development.' *Curr Opin Chem Biol* **1998**, *2*, 585–91.

167 R. VANKOVA et al. 'Comparison of oriented and random antibody immobilization in immunoaffinity chromatography of cytokinins.' *Journal of Chromatography a* **1998**, *811*, 77–84.

168 S. KANNO, Y. YANAGIDA, T. HARUYAMA, E. KOBATAKE, M. AIZAWA. 'Assembling of engineered IgG-binding protein on gold surface for highly oriented antibody immobilization.' *Journal of Biotechnology* **2000**, *76*, 207–214.

169 J.F. MOONEY et al. 'Patterning of functional antibodies and other proteins by photolithography of silane monolayers.' *Proceedings of the National Academy of Sciences of the USA* **1996**, *93*, 12287–12291.

170 L.C. SHRIVERLAKE et al. 'Antibody immobilization using heterobifunctional cross-linkers.' *Biosensors & Bioelectronics* **1997**, *12*, 1101–1106.

171 K. NAKANISHI, H. MUGURUMA, I. KARUBE. 'A novel method of immobilising antibodies on a quartz crystal microbalance using plasma-polymerized films for immunosensors.' *Analytical Chemistry* **1996**, *68*, 1695–1700.

172 D. GUSCHIN et al. 'Manual manufacturing of oligonucleotide, DNA, and protein microchips.' *Analytical Biochemistry* **1997**, *250*, 203–211.

173 B.P. GABER, B.D. MARTIN, D.C. TURNER. 'Create a protein microarray using a hydrogel 'stamper'.' *Chemtech* **1999**, *29*, 29–24.

174 A. RODA, M. GUARDIGLI, C. RUSSO, P. PASINI, M. BARALDINI. 'Protein microdeposition using a conventional ink-jet printer.' *Biotechniques* **2000**, *28*, 492–496.

175 R.S. KANE, S. TAKAYAMA, E. OSTUNI, D.E. INGBER, G.M. WHITESIDES. 'Patterning proteins and cells using soft lithography.' *Biomaterials* **1999**, *20*, 2363–2376.

176 V.N. MOROZOV, T.Y. MOROZOVA. 'Electrospray deposition as a method for mass fabrication of mono- and multicomponent microarrays of biological and biologically active substances.' *Analytical Chemistry* **1999**, *71*, 3110–3117.

177 E. SOUTHERN, K. MIR, M. SHCHEPINOV. 'Molecular interactions on microarrays.' *Nature Genetics* **1999**, *21*, 5–9.

178 S. GRANJEAUD, F. BERTUCCI, B.R. JORDAN. 'Expression profiling: DNA arrays in many guises.' *Bioessays* **1999**, *21*, 781–790.

179 A. SCHULZE, J. DOWNWARD. 'Navigating gene expression using microarrays – a technology review.' *Nature Cell Biology* **2001**, *3*, E190–E195.

180 V.G. CHEUNG et al. 'Making and reading microarrays.' *Nature Genetics* **1999**, *21*, 15–19.

181 S. SINGHGASSON et al. 'Maskless fabrication of light-directed oligonucleotide microarrays using a digital micromirror array.' *Nature Biotechnology* **1999**, *17*, 974–978.

182 T. TAKEUCHI, J. MATSUI. 'Miniaturized molecularly imprinted continuous polymer rods.' *Hrc-Journal of High Resolution Chromatography* **2000**, *23*, 44–46.

183 L.A. TEMPELMAN, K.D. KING, G.P. ANDERSON, F.S. LIGLER. 'Quantitating Staphylococcal enterotoxin B in diverse media using a portable fibre-optic biosensor.' *Analytical Biochemistry* **1996**, *233*, 50–57.

184 H. YU, A.W. KUSTERBECK, M.J. HALE, F.S. LIGLER, J.P. WHELAN. 'Use of the USDT flow immunosensor for quantification of benzoylecgonine in urine.' *Biosensors and Bioelectronics* **1996**, *11*, 725–734.

185 U. NARANG, P.R. GAUGER, A.W. KUSTERBECK, F.S. LIGLER. 'Multianalyte detection using a capillary-based flow immunosensor.' *Analytical Biochemistry* **1998**, *255*, 13–19.

186 E. GIZELI, C.R. LOWE. 'Immunosensors.' *Current Opinion in Biotechnology* **1996**, *7*, 66–71.

187 L. LARICCHIA-ROBBIO et al. 'Mapping of monoclonal antibody- and receptor-binding domains on human granulocyte-macrophage colony-stimulating factor (rhGM-CSF) using a surface plasmon resonance based biosensor.' *Hybridoma* **1996**, *15*, 343–350.

188 T.P. VIKINGE, A. ASKENDAL, B. LIEDBERG, T. LINDAHL, P. TENGVALL. 'Immobilized chicken antibodies improve the

detection of serum antigens with surface plasmon resonance (SPR).' *Biosensors and Bioelectronics* **1998**, *13*, 1257–1262.

189 H. J. WATTS, C. R. LOWE, D. V. POLLARD-KNIGHT. 'Optical biosensor for monitoring microbial cells.' *Analytical Chemistry* **1994**, *66*, 2465–2470.

190 J. LAHIRI, L. ISAACS, J. TIEN, G. M. WHITESIDES. 'A strategy for the generation of surfaces presenting ligands for studies of binding based on an active ester as a common reactive intermediate: A surface plasmon resonance study.' *Analytical Chemistry* **1999**, *71*, 777–790.

191 T. B. DUBROVSKY, Z. HOU, P. STROEVE, N. L. ABBOTT. 'Self-assembled monolayers formed on electroless gold deposited on silica gel: a potential stationary phase for biological assays.' *Analytical Chemistry* **1999**, *71*, 327–332.

192 D. R. BASELT, G. U. LEE, K. M. HANSE, L. A. CHRISEY, R. J. COLTON. 'A high-sensitivity micromachined biosensor.' *Proc. IEEE* **1997**, *85*, 672–680.

193 B. A. CORNELL *et al.* 'A biosensor that uses ion-channel switches.' *Nature* **1997**, *387*, 580–583.

194 A. P. F. TURNER. 'Switching channels makes sense.' *Nature* **1997**, *387*, 555–557.

15

Recent Applications of Functional Proteomics: Investigations in Smooth Muscle Cell Physiology

Justin A. MacDonald and Timothy A. J. Haystead

15.1
Introduction to Functional Proteomics

Studies on the proteome, the complete set of proteins that are encoded and expressed by a genome [1, 2], were initiated in the early 1970s with the goal of mapping the entire cellular protein complement. The purpose of this endevour, termed the human protein index, was to use 2-dimensional gel electrophoresis (2DE) to map and catalog all human proteins. Unfortunately, technical limitations prevented this project from progressing beyond an early-development stage. It was not until the mid to late 1990s when advances in bioinformatics, protein sequencing technologies, and the annotation of entire genomes permitted the "rebirth" of proteomic studies. More powerful computers and software design has driven bioinformatics, more sensitive protein sequencing technologies were developed- first, Edman microsequencing from electroblotted proteins, and later, highly sensitive mass spectrometry methods. Finally, the enormous progress in characterizing the genomes of a wide variety of organisms has made proteomics the current buzz-word in biological studies.

Although projects such as the human genome project are defining the genes involved in human disease, it is increasingly clear that this genetic information provides only a rudimentary beginning to unraveling the function of cells at the molecular level [3–5]. Proteins are not invariant products of genes, but are subject to a high degree of processing at the protein level that is a critical component of cellular function and regulation. Synthesis of mRNA is several steps removed from the cell phenotype. Message is subject to post-transcriptional control in the form of mRNA editing and/or alternative splicing. Regulation also occurs at the level of protein translation; many studies now show a poor correlation between mRNA levels and protein expression levels. Quantification of mRNA (i.e. with DNA micro-arrays) is not a direct reflection of the protein content or biological activity in the cell [3, 6]. Finally, proteins are subject to numerous post-translational modifications (e.g. lipidation, proteolysis, phosphorylation, glycation, and oligomerization to name a few). Thus, although there may only be ∼ 30,000 human genes [7], there may well be 200,000 to 2 million human proteins once splice variants and essential modifications following transcription are included [2].

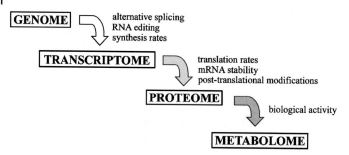

Fig. 15.1 The connectivity of biological studies. Biological information encoded in the genome is linked through the transcriptome and the proteome to the metabolome. Multiple protein isoforms (proteome) can be generated by RNA processing. The mRNA population (transcriptome) can, in addition, be regulated by message stability and efficacy of translation. Post-translational processing can alter the protein's biological activity and thereby alter the cellular concentration of metabolic substrates, products, and effectors (metabolome).

Because the number of proteins can strongly exceed the number of genes, the goal of assigning functional roles to all of these proteins is a daunting task. As a result, the proteomics field has split into a few key approaches.

15.1.1
Expression Proteomics (Catalogue)

A major undertaking to exhaustively map and identify all of the proteins in "the human proteome" in a fashion similar to genome sequencing projects continues. Post translational modifications and isoforms can potentially also be determined. This project will provide a functional annotation of the genome, integrating data with genomic information to confirm the existence of putative gene products.

15.1.2
Functional Proteomics (Applied)

Functional proteomics is a loosely appointed term that applies to the use of proteomic methods to monitor and analyze the spatial and temporal properties of the molecular networks and fluxes involved in living cells [8]. In practice, this approach allows for a select group of proteins to be studied and characterized with respect to a particular set of criteria (e.g. disease state, signaling pathway, or protein/drug interaction). In contrast to the static genome, where all information could in principle be obtained from the DNA of a single cell, the proteome is dynamic and highly dependent not only on the type of cell, but also on the *state* of the cell. Expression proteomics will not provide adequate information to define the molecular mechanisms of biological function. Integrated genomic and proteomic expression profiling may be able to define "the players", but will not be able to ascertain key elements which define their function in complex networks of protein activities. Proteomic analysis is complicated since it is not only the changing amounts of different proteins but also the widespread post-translational processes such as phosphorylation that alter dynamic function at the cellular level. As a result, an important motivation for functional proteomics is to find ways of concentrating on those proteins that are involved in a particular biological function of interest.

15.1.3
Structural Proteomics

Some proteomic approaches are mapping protein-protein interactions, determining the structure of protein complexes, and elucidating the 3-dimensional topology of proteins [9]. Currently, 2-hybrid analysis and protein array chips are being used for screening protein binding partners. While, automated and reproducible protein production, purification, crystallization and structure determination by X-ray beams from synchrotron radiation sources are allowing the high-throughput elucidation of protein structure. The deduction of protein structure is crucial for rational drug design. Co-crystallization and structural determination of hit compounds with target proteins may enable prioritization and optimization of lead drug molecules. The information obtained from structural proteomic studies will provide important information about the overall architecture of cells.

15.2
Proteomic Tools for Smooth Muscle Physiologists

15.2.1
Protein Separation: The 2DE Technology

Traditionally, the "workhorse" for protein expression profiling has been two-dimensional polyacrylamide gel electrophoresis (2DE). 2DE is currently the most generally applicable technique for separating and visualizing a large number of proteins. A protein mixture can be resolved into its individual components by first separating proteins on the basis of their isoelectric points using ampholyte gradients in one dimension. Proteins are further resolved on the basis of their mobility (dependent upon molecular weight) through a polyacrylamide gel matrix in the second dimension. The initial set of 2DE experiments was able to resolve ~1,000 proteins from *E. coli*, and it was predicted that up to 5,000 distinct protein spots would be discernable [10]. In addition, the application was able to detect post-translational modifications and amino acid mutations that altered a protein"s characteristic mobility in the 2D-gel. Protein abundance could also be quantified by using ^{35}S-methionine labeling; proteins differing in concentration by ratios of 10^{-4} to 10^{-5} could be detected. Technological advances in the electrophoresis field has greatly improved 2DE resolution. Over 10 000 protein spots can now be resolved on a single gel [11], although reports are normally below 2000 protein spots. The resolution continues to be improved through the development of new isoelectric focusing media with narrower pH ranges and new gel matrices [12]. By resolving the same protein mixture on multiple narrow range 2DEs, the total number of resolvable spots has increased. Despite the outstanding properties and ease of use of 2D-gels systems, they did not become an integral part of protein expression profiling for nearly two decades. 2DE technology did not reach its potential for the simple reason that methods for the rapid, routine identification of the protein spots in the gels was not available. Three crucial developments have emerged to address these challenges. First, whole genome sequencing from a myriad of different organisms is defining at the gene level all the proteins that exist in that organism. This has meant that, instead of requiring time consuming complete Edman N-terminal sequence to identify a protein, partial information on fragments or sequence is adequate to identify the protein at the gene level within a genomic sequence database. Second, protein sequencing by mass spectrometric methods can now supply the needed information to identify the protein at sensitivities well below 1 pmol. And third, the development of algorithms for the identification of proteins by mass spectrometric data matched to genomic databases. It is now clear that the sequencing of multiple whole genomes and the field of bioinformatics along with corresponding advances in mass spectrometric methods have drastically altered the landscape of biological research at the protein level.

Although 2DE is currently the best protein analog of cDNA arrays, it does not produce a comprehensive display of the entire cellular proteome. At least part of this problem arises from the well-documented difficulty in resolving certain types of proteins (i.e. membrane proteins, proteins with mass greater than 100 kDa, highly basic proteins, and highly acidic proteins) by 2DE [13, 14]. Many spots are likely to contain multiple overlapping proteins that are not resolved in gels that attempt to cover most of the pI range appropriate for all cellular proteins. One approach to improve resolution has been to use immobilized pH gradient (IPG) strips to cover narrower ranges of pI. Other disadvantages of 2DE include a lack of reproducibility and a lack of dynamic range. The detection of proteins that occur with low copy number remains a major limitation [11, 15]. Indeed one of the main impediments to easily observe all soluble proteins is the large variation in the concentrations of cellular proteins with estimates placing the dynamic range of protein expression at between 7 to 9 orders of magnitude [11]. With limitations on the amount of total protein able to be loaded on a gel that is consistent with good resolution of individual protein spots on analytical 2DEs, detection of low abundance proteins remains difficult if not impossible. Currently, the total protein load that can be applied to analytical 2DE and the chemical staining methods limit analysis to less than half of the yeast proteome [16]. This problem is even more serious in proteomes isolated from more complex organisms such as humans.

As the field of proteomics evolves rapidly away from a simple visual representation of cellular proteins on 2DE, the availability of new data and technologies has triggered many initiatives to start large-scale proteomic studies. 2DE involves a great deal of expertise and hands-on-time to execute, thus there is a strong need to work on replacement technologies that are easily automated and highly reproducible.

15.2.2
Mass Spectrometers

Mass spectrometry (MS) was originally developed in chemistry laboratories as a tool for small molecule analysis and structure determination. Mass spectrometers originally required small charged gaseous molecules as analytes. Proteins are large and polar molecules in comparison, and ionization methods were not available to introduce proteins into the MS instrumentation. Beginning in the early 1980s as the development of techniques capable of presenting proteins in an ionized gaseous form to the MS instrument were perfected, MS slowly emerged from chemistry laboratories to become an important tool in protein biochemistry and the key technology driving the proteomics field. Electrospray ionization (ESI) [17] and matrix-assisted laser desorption ionization (MALDI) [18] are the two ionization techniques, which are most responsible for the success of mass spectrometry in proteomics. MALDI creates ions by using a laser to excite a crystalline mixture of analyte molecules and energy desorbing matrix into the gas phase. ESI uses a potential difference between a capillary and the inlet of the mass spectrometer to cause charged droplets to be released from the tip of the capillary. As the droplets evaporate, gas phase charged ions are desorbed from the droplets. In addition to the ionization source, a mass analyzer and an ion detector are also present in all mass spectrometers.

A number of excellent and exhaustive reviews have addressed new technical developments in mass spectrometry [19, 20]. See also chapters 12 and 13. Rather than engage in a lengthy description of the current advances in mass spectrometry, we have chosen to highlight some significant features, which have directed our current experience with mass spectrometers. Three different methods are used to measure the mass-to-charge ratio of analytes such as proteins, peptides, or peptide fragments. The most common mass spectrometer in use in the proteomics field is the time of flight (TOF) mass analyzer, which measures the mass/charge ratio of ions by the time it takes them to travel through a flight tube to the detector. The other frequently used mass detection methods are quadrupole MS, the separation of analytes by quadrupole electric fields generated by metal rods, and ion-trap MS the separation of analytes by selective ejection of ions from a three-dimensional electrical-trapping field. Both ESI and MALDI ionizations can be used to create ions for analysis in TOF-MS. Typically, however, MALDI is coupled with TOF-MS and ESI with quadrupole and ion-trapping MS because MALDI produces short bursts of ions and ESI produces a continuous beam of ions. For peptide sequencing, two steps of mass spectrometry are completed in tandem (tandem mass spectrometry or MS/MS). These mass spectrometers are able to select a particular peptide ion, fragment it in a collision chamber, and then detect the resulting fragment ions. The actual amino acid sequence is reconstructed from the fragmentation pattern with computer software. The primary advantage of tandem mass spectrometry is the ability to select a particular peptide ion from a mixture, allowing for the identification of components of complex mixtures.

MS/MS is performed by using the same mass separation principle twice or by combining two different separation principles. To date, the quadrupole-TOF (qTOF) has been an effective MS/MS performer. Two recent advances are the MALDI-qTOF [21] and the two section TOF separated by a high energy collision chamber (MALDI-TOFTOF) [22] mass spectrometers. These instruments provide very fast, automated analysis of large numbers of samples with high sensitivity, and the ability to obtain amino acid sequence from any selected peptide.

15.2.3
Protein Identification by Database Searching

Both ESI-MS/MS and MALDI-MS are routinely used for protein identification from 2DE although different search strategies have been developed depending on the type of data obtained from the mass spectrometer.

Peptide Mass Fingerprinting. Mass fingerprinting uses the characteristic distribution of peptide masses obtained by digesting proteins with a sequence-specific protease such as trypsin. From the masses of the resulting peptides, usually obtained by a MALDI-TOF instrument, a mass map or mass fingerprint of the original protein is obtained [23]. A number of computer programs (e.g. ProteinProspector[TM] and MASCOT[TM]) are available for using the observed peptide masses to search gene sequence databases for proteins that fit the mass fingerprint. The proteins can be ranked according to the number of peptide matches; sophisticated scoring algorithms take the mass accuracy and the percentage of the protein sequence covered into account and calculate a level of confidence for the identification. Mass fingerprinting when coupled with automated 2DE spot pickers and MALDI-TOF instrumentation has become the primary technology for large-scale proteomics work.

While mass fingerprinting is a very powerful methodology of database searching, it has several important limitations. Peptide fingerprints are generally sufficient for identification of proteins from completely sequenced genomes; however, splice variants, edited transcripts, significant numbers of post-translational modifications, and other differences leading to unaccounted peptide masses can prevent correct identification. The most significant limitation of peptide fingerprint based database searching is the difficulty in assigning the correct identity of proteins in complex mixture. Mixtures of three or more proteins increase the level of ambiguity in the mass fingerprint to the extent that usually only the most abundant protein component is identified. More sophisticated search programs are being designed to address this limitation; for example, databases can now be searched iteratively by removing the peptides associated with an unambiguous match [24].

MS/MS Identification with Sequence Tags and Primary Amino Acid Sequence. Tandem mass (MS/MS) spectrometric data obtained on peptides from proteins of interest can also be used to search databases. Because MS/MS spectra contain structural information related to the primary amino acid sequence of the peptide, rather than just its mass, these searches are generally more specific and discriminating. However, manual evaluation of peptide MS/MS spectra is generally required which tends to be a tedious and time consuming endevour. While more laborious and more difficult to automate than MALDI-TOF analysis, ESI-MS/MS generally provides a more reliable method for the positive identification of proteins from complex silver-stained 2D-gels. This is especially true of when multiple proteins co-migrate within the same 2DE spot and when the database to be searched lacks adequate coverage of the organism''s genome.

When protein sequence information is needed to find the gene within the organism, several search approaches exist. The peptide sequence tag method used partial sequence interpretation to search databases [25]. Nearly every MS/MS spectrum contains a short "ladder" of fragment ions that are easily interpreted. When a few amino acids are combined with the start mass and the end mass of the series, which specify the exact location of the "ladder" sequence in the peptide and the known cleavage sites of the enzyme. This sequence tag is used to retrieve candidate proteins from the database which have theoretical fragmentation patterns that match the experimental one. Another method for searching MS/MS data attempts to match the experimental MS/MS spectrum against predicted spectra for all peptides in the database without extracting any sequence information. SEQUEST^{TM}, a database search tool, performs this function and provides scores in-

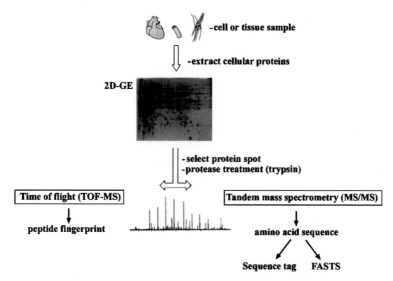

Fig. 15.2 Strategies for protein identification. A mixture of proteins is first resolved by 2DE and visualized by staining. Individual spots are excised and subjected to proteolytic digestion. The resulting peptide mixture can be analyzed either by TOF-MS to determine the mass fingerprint or by tandem mass spectrometry (MS/MS) to determine the amino acid sequence of individual peptides. Protein identity can be determined by matching the peptide mass fingerprint for TOF-MS data or by database searching with sequence "tags" or with FASTS (http://fasta.bioch.virginia.edu) for MS/MS data.

dicating to what degree the MS/MS spectrum matches the theoretical spectra derived from the database sequence and how differently the next most similar sequence in the database fits the spectrum.

If there is no corresponding genome sequence for the organism (or protein) under investigation, then it becomes necessary to identify the protein by comparison of *de novo* sequence data with proteins of other organisms in the sequence databases. A reliable search protocol uses the FASTS search algorithms [26]. FASTS searches with multiple peptide sequences of unknown order, as obtained by MS/MS-based sequencing, evaluating all possible arrangements of the peptides. This program is capable of searching peptide sequence information across both DNA (TFASTS) and protein databases (FASTS). The programs can be accessed at http://fasta.bioch.virginia.edu.

15.3
Defining Signal Transduction Modules in Smooth Muscle

One area of cardiovascular disease research that has received considerable attention through proteomics is the global characterization of proteins in vascular tissue [27]. Indeed, most proteomic studies of cardiovascular biology have focused on the construction of 2DE protein databases from cardiomyocytes [28–30]. Determining global changes in protein expression levels of cardiomyocytes in response to environmental and disease states such as ischemia, hypertension, hypertrophy, infarction, restenosis, atherosclerosis, obesity and diabetes may provide unique insight and understanding of the respective molecular mechanisms of disease. Equally important is the discovery of novel proteins that may play an integral function in normal and disease processes. Such identifications may provide basic information regarding molecular mechanisms as well as provide targets for novel drug discovery and biomarkers for a disease state. In comparison to other organs, the heart is a relatively homogenous organ, consisting mainly of myocardial muscle cells, and is therefore well suited for a proteomic investigation by 2DE. Perhaps for this reason, major initiatives have been highly successful in the development of master myocardial 2DE databases. These myocardial 2DE databases have been constructed and are available over the World Wide Web: HSC-2DPAGE [31] and HEART-2D-PAGE [32]. High performance 2DE procedures have resulted in the resolution of 3,300 myocardial protein species with ~300 proteins identified. Studies undertaken to compare healthy versus disease state myocardium have identified many non-reproducible spot intensity variations, probably caused by different forms of cardiovascular disease or by parameters including but not limited to disease stage, medication used, age, gender, and nutritional status. Despite the fact that these studies were undertaken with different sample preparations, different isoelectric focusing conditions, and different gel sizes, the inter-laboratory comparison has shown that spots identified by sequencing methods appeared at the same positions and identification by pattern comparison was successful in many cases. The continued refinement of 2DE technology, bioinformatics and MS sequencing sensitivity should continue to foster advancements in the cardiovascular expression proteomics.

Rather than use protein expression proteomics to simply characterize the large number of proteins present in SMC, we recognized the true power of this technology in its ability to enhance the output of existing approaches currently used by the modern physiologist. We prefer to let the biological question drive the application of proteomics. Thus, rather than define an entire cellular proteome (a laborious process that is perhaps beyond the scope of an academic lab [33]) we have defined experiments to select a number of relevant proteins in the mixture of interest. Examples of such experiments that we routinely perform are the definition of early phosphorylation events after agonist treatment of intact SMC, and the comparison of protein complex components in affinity pull down experiments from stimulated and non-stimulated SMC. Only proteins that are demonstrated to be specifically phosphorylated or associated with a protein complex in response to the stimulus are examined. Thus, we can eliminate a great deal of extraneous information by sequencing only those proteins that are demonstrated to be regulated; these select protein targets then direct all subsequent biological investigations.

Recent work in our laboratory has been using a proteomic approach to analyze calcium sensitization and desensitization in smooth muscle and vascular tissue. The major mechanism for adjusting the sensitivity of the contractile apparatus in smooth muscle is through the sensitization of the 20-kDa myosin light chain (MLC20) phosphorylation and contractile force to $[Ca^{2+}]_i$ [34–36]. However, stimulation of contraction can be induced with several agonists at submaximal, fixed Ca^{2+} levels [37]. This regulation of force and MLC20 phosphorylation that is independent of changes in $[Ca^{2+}]_i$ is referred to as calcium sensitivity [37, 38]. The extent of MLC20 phosphorylation depends on the relative activities of myosin light chain kinase (MLCK) and myosin light chain phosphatase (SMPP-1M). "Ca^{2+}-sensitization" is chiefly mediated via a G-protein linked pathway that acts to inhibit SMPP-1M activity [39, 40]. SMPP-1M is composed of three subunits: the 37 kDa catalytic subunit of PP-1 (PP-1C); a 110–130 kDa regulatory myosin phosphatase targeting subunit (MYPT1) and a 20 kDa subunit of undetermined function [41]. The myosin phosphatase activity of SMPP-1M is known to be regulated by phosphorylation of the MYPT1 subunit at an inhibitory site of phosphorylation by an unidentified endogenous kinase [42].

To identify the relevant endogenous kinase responsible for the inhibitory phosphorylation of SMPP-1M, we used the "purine binding cassette sub-proteome" isolated from bladder tissue. The most critical stage of any functional proteomics approach is the strategic design for the isolation of protein targets. Much of the focus of proteomics has been on the advancement of mass spectrometry sequencing technology (i.e., automation, speed, and sensitivity) which has resulted in a de-emphasis on the design of proteomic experiments. Our general strategy has been to devise techniques that enrich for low abundance proteins of significant relevance to the biological question at hand. The reduction in proteome complexity saves us from identifying hundreds of irrelevant proteins which could consume a majority of our time and effort. We used γ-linked ATP Sepharose affinity chroma-

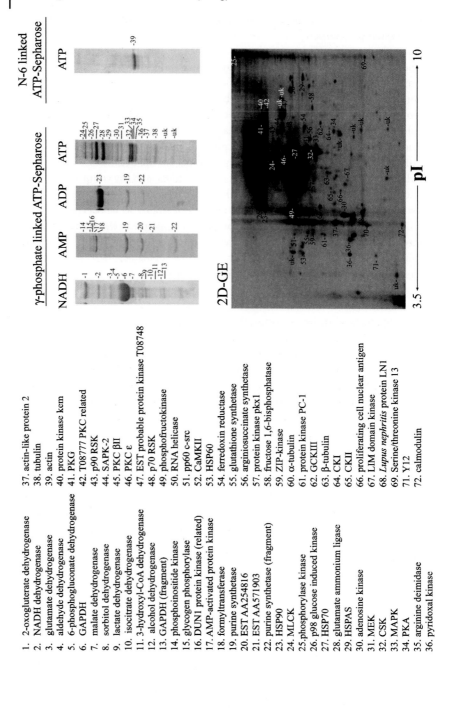

1. 2-oxogluterate dehydrogenase
2. NADH dehydrogenase
3. glutamate dehydrogenase
4. aldehyde dehydrogenase
5. 6-phosphogluconate dehydrogenase
6. GAPDH
7. malate dehydrogenase
8. sorbitol dehydrogenase
9. lactate dehydrogenase
10. isocitrate dehydrogenase
11. 3-hydroxyl-CoA dehydrogenase
12. alcohol dehydrogenase
13. GAPDH (fragment)
14. phosphoinositide kinase
15. glycogen phosphorylase
16. DUN1 protein kinase (related)
17. AMP-activated protein kinase
18. formyltransferase
19. purine synthetase
20. EST AA254816
21. EST AA571903
22. purine synthetase (fragment)
23. HSP90
24. MLCK
25. phosphorylase kinase
26. p98 glucose induced kinase
27. HSP70
28. glutamate ammonium ligase
29. HSPAS
30. adenosine kinase
31. MEK
32. CSK
33. MAPK
34. PKA
35. arginine deimidase
36. pyridoxal kinase
37. actin-like protein 2
38. tubulin
39. actin
40. protein kinase kem
41. PKG
42. T08777 PKC related
43. p90 RSK
44. SAPK-2
45. PKC βII
46. PKC ε
47. EST probable protein kinase T08748
48. p70 RSK
49. phosphofructokinase
50. RNA helicase
51. pp60 c-src
52. CaMKII
53. HSP60
54. ferredoxin reductase
55. glutathione synthetase
56. arginiosuccinate synthetase
57. protein kinase pkx1
58. fructose 1,6-bisphosphatase
59. ZIP-kinase
60. α-tubulin
61. protein kinase PC-1
62. GCKIII
63. β-tubulin
64. CKI
65. CKII
66. proliferating cell nuclear antigen
67. LIM domain kinase
68. *Lupus nephritis* protein LN1
69. Serine/threonine kinase 13
70. Y12
71.
72. calmodulin

tography to create a sub-proteome that was enriched with protein kinases. This affinity resin was developed with ATP immobilized in the "protein kinase orientation" (via its γ-phosphate); the entire SMC complement of ATP binding proteins can be isolated following high stringency washes to remove non-specific proteins [43]. Using this methodology, we were successful in sequencing the previously unidentified MYPT1-kinase following affinity purification. Additional proof from expression and structure function studies has confirmed a role for the MYPT1-kinase as a regulator of SMC contractile state [44, 45]. The kinase was shown to associate with MYPT1, to phosphorylate MYPT1 at the inhibitory site in intact smooth muscle, and to be regulated by a yet unidentified upstream-kinase.

Several hormones and clinically administered nitrovasodilators exert their actions to widen vessels and inhibit vasoconstriction through activation of smooth muscle guanylate cyclase to produce cGMP [46]. One of the primary effects of elevated intracellular cGMP is the reduction of $[Ca^{2+}]_i$. The mechanisms by which cGMP lowers intracellular Ca^{2+} are well studied [47, 48]; they include stimulation of Ca^{2+} extrusion by activation of the calmodulin-stimulated Ca^{2+}-ATPase, a decrease in Ca^{2+} influx due to decreased Ca^{2+}-channel activity, a decrease in Ca^{2+} influx through hyperpolerization as a result of increased K^+-channel activity, stimulation of calcium uptake into the sarcoplasmic reticulum by phosphorylation of phospholamban, and antagonism of hormone triggered generation of second messengers. A second relevant physiological effect of elevated intracellular [cGMP] (and cAMP) is the phenomenon of Ca^{2+}-desensitization (reviewed in [49]). Elevation of [cGMP] at fixed $[Ca^{2+}]$ causes muscle relaxation in a manner that does not disrupt the LC20 phosphorylation/force relationship. The molecular mechanisms that bring about this phenomenon, presumably through the activation of cyclic-GMP dependent protein kinase (PKG) are unknown. We are using a functional proteomic approach to identify novel kinase substrates that are phosphorylated in response to cGMP administration. The identification of these physiologically-relevant, phosphorylated targets is a critical step to our understanding of the regulation of SMC tone. Several previously characterized targets were identified during cyclic nucleotide-evoked relaxation including but not limited to HSP20, HSP27 [50, 51], and telokin [49, 52]. Several additional phosphorylation targets are detected after cGMP-dependent relaxation of ill longitudinal smooth muscle and vascular smooth muscle (MacDonald and Haystead, un-

Fig. 15.3 Catching the adenine nucleotide binding proteome on γ-phosphate linked ATP-Sepharose. Mouse extract was prepared and passed over 50 ml of γ-phosphate linked ATP-Sepharose containing 10 μmols/ml of linked ATP or N-6 linked ATP-Sepharose (Sigma Chemical Co., St. Louis MO). Following washing, the column was eluted sequentially with the indicated nucleotides and fractions collected (10 ml). Column fractions were separated by 1D SDS-PAGE and silver-stained. A portion of the ATP eluate was concentrated 100 fold and 10 μl analyzed by 2DE. Proteins were identified by ESI-MS/MS sequencing and were matched against the entire published protein or DNA data bases with the FASTS or TFASTS algorithms respectively. Expectation scores for the identified proteins ranged from 2.6 e^{-7} for PKA to 1.2 e^{-54} for GAPDH. Expectation scores after each search for the next highest scoring non-related protein were generally <2.3 e^{-4}. (Redrawn from P. R. Graves, et al., 2002; submitted for publication.)

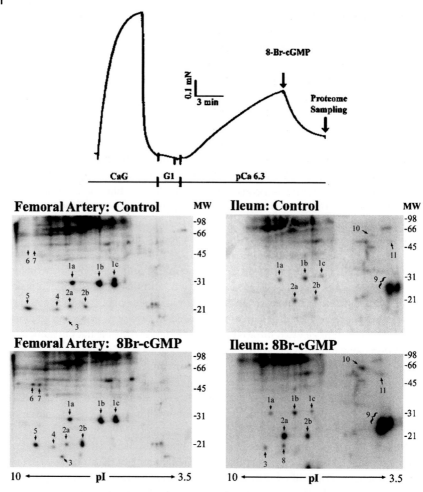

Fig. 15.4 2DE profile of phosphoprotein targets in ileum and femoral artery SMC following cGMP-treatment. Smooth muscle strips were permeabilized with α-toxin, contracted with exposure to pCa6.3 solution with 1 μM calmodulin, and treated with vehicle (control) or 100 μM 8-bromo-cGMP solution (8Br-cGMP) in the presence of [γ-³²P]ATP. Cellular extracts were applied to 2DE, and proteins of interest were identified by ESI-MS/MS sequencing with FASTS or TFASTS searching of protein and DNA databases. 1a,b,c) HSP27; 2a,b) p20; 3) ubiquitin-conjugating enzyme; 4) below detection limit; 5) actin depolymerizing factor; 6) dbEST putative gene product, 7) unknown, 8) unknown; 9) telokin; 10) RIKEN cDNA putative gene product; and 11) unknown. (Redrawn from J.A. MacDonald et al., 2002; submitted for publication.)

published observations). The proteins have been identified in the dbEST and RIKEN cDNA libraries as novel putative gene products. Our initial observations suggest that there are different molecular mechanisms that mediate cGMP-dependent relaxation in tonic versus phasic SMC.

Proteomic Approaches to Substrate Profiling. A significant hindrance to the signal transduction field has been the difficulty in identifying true *in vivo* substrates to individual protein kinases and phosphatases. This problem has been perpetuated by the *in vitro* promiscuity of kinases [53] and phosphatases [54] toward protein substrates. Two significant problems plague kinase substrate preference characterization. First, there are no methods available for the systematic, unbiased characterization of substrate preferences; existing techniques for substrate profiling (i.e. site prediction from phosphorylation consensus site mapping) must start from known determinants. Even peptide libraries cannot generate all possible amino acid combinations for testing due to the shear number of combinatorial possibilities and the technological limitations those numbers impose. Second, there is no simple way to compare the general pattern of kinase substrate preferences. Both of these issues can be partially addressed through proteomics by taking advantage of the fact that only a small fraction of all possible amino acid combinations are actually produced as proteins. By defining a substrate proteome as a set of proteins in a cell which could be *in vitro* substrates for a particular kinase, we can use a cellular protein extract to explore differences in that set between kinases. This limits substrate possibilities to sequences which are available for phosphorylation, at least in one cell type, and increases the complexity relative to a combinatorial library, since each sequence is independent of the next. Moreover, because an extract contains proteins not peptides, we further restrict substrate specificity to secondary and tertiary structures, and have added the complexity of post-translational modifications. These differences may actually reduce the number of possible sequences tested, but do so in a way that corresponds to biological possibilities. As a screen, it is unbiased in any assumption of substrate sequence, although choices of source material for the extract and of biochemical techniques to solubilize the proteins will affect the results.

Test conditions were designed to provide the best possible definition of the substrate proteome. This was primarily achieved by using high concentrations of exogenous kinase relative to the substrate proteome concentration, so that the kinase tested could overwhelm kinases present in the extract. By fractionating the extract prior to the kinase reaction, we are able to significantly increase the sensitivity toward less abundant substrates. Recently developed methodology for the selective enhancement of phosphoproteins from the cellular proteome [55–57] could also be used for the selective isolation of phosphoproteins from a highly complex mixture.

High kinase concentrations, short reactions times that approximate initial rate conditions, and high specific radioactivity all increase sensitivity by increasing radioactive incorporation into any give substrate. Importantly, these assays were not designed to identify solely physiological substrates. While the proteins phosphorylated should reflect the best targets for phosphorylation among the proteins presented to the kinase, this methodology, without further studies to present *in vivo* validation, cannot attribute physiological relevance to any particular phosphorylation in the assay. Local concentrations of kinase and substrate in a cell are regulated and vary widely from the total cellular concentrations [58], further em-

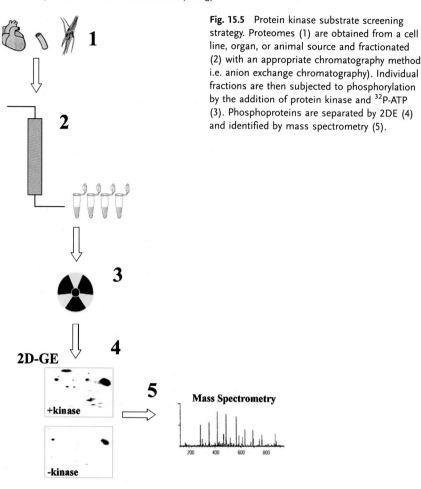

Fig. 15.5 Protein kinase substrate screening strategy. Proteomes (1) are obtained from a cell line, organ, or animal source and fractionated (2) with an appropriate chromatography method i.e. anion exchange chromatography). Individual fractions are then subjected to phosphorylation by the addition of protein kinase and ^{32}P-ATP (3). Phosphoproteins are separated by 2DE (4) and identified by mass spectrometry (5).

phasizing the need to validate the relevance of a phosphorylation event by further biochemical study.

Our prototypical application of the methodology has been in the delineation of distinct substrate pools for the individual Ca^{2+}/CaM-dependent protein kinase (CaMK) family members (E. E. Corcoran, J. Joseph, J. A. MacDonald, T. A. J. Haystead, & A. R. Means, data submitted for publication). Missing from many studies of CaMK signaling is the differentiation between CaMKI and CaMKIV isozymes; the substrate preferences of CaMKI and CaMKIV are similar and intersect with the so-called multi-functional CaMK, CaMKII [59]. Unfortunately, the kinase inhibitors currently available inhibit CaMKI, II, and IV similarly and so they cannot be used alone to define a role for a specific CaMK isotype. Our preliminary studies have identified several novel substrates of CaMKI and IV. Our analysis, by comparing the overall similarities and differences between kinases, can answer more general questions about the biological function of these enzymes. For example, if CaMKI and CaMKIV were essentially the same in their substrate preferences, differences in function could be wholly attributed to differences in localization and regulation by other pathways.

One powerful application of functional proteomics is the ability to juxtapose with genetics to delineate signaling events *in vivo*. Currently, genetic manipulation is most easily accomplished in yeast. We have combined proteomics and yeast genetics to identify physiological targets of protein phosphatases. Protein phosphatase-1 (Glc7p) and its binding protein Reg1p are essential for the regulation of glucose repression pathways *in Saccharomyces cerevisiae* [60]. In order to identify the physiological substrates for the Reg1p/Glc7p complex, the effects of deletion of the *REG1* gene on the yeast phosphoproteome were examined. Mapping of the phosphoproteome by 2DE revealed two distinct proteins that were greatly increased in phosphate content in *REG1Δ* mutants. Microsequencing identified one of these proteins as hexokinase II (Hxk2p). A comprehensive biochemical study indicated that Reg1p targeted PP-1C to dephosphorylate Hxk2p *in vivo* and that the peptide motif (K/R)(X)(I/V)F was necessary for its PP-1C targeting function. The types of proteomic approaches that were utilized in this study to identify Hxk2p as a physiological target of the Reg1p/Glc7p complex herald a new age in which the functional consequences of genetic manipulations can be assessed directly and fully on the proteome as a whole. Logical genetic and biochemical experiments can be carried out to probe the function and address the physiological relevance in the context of the intact cell. Clearly, a combined proteomics and genetics approach greatly enhances the ability to ascertain the molecular mechanisms underlying complex biological phenomena. We believe that a similar strategy could be adopted for SMC systems with transgenic or knock-out mouse work, particularly where there is no obvious phenotype.

15.4
Proteome Mining of the Smooth Muscle Cell

Genomic powerhouses have invested billions to identify relevant drug targets by conventional high-throughput drug screening protocols. These screening efforts have exploited the power of combinatorial chemistry (combichem) by generating libraries containing millions of molecules. Regrettably, the massive size of these libraries delayed the identification of new drugs due to the time required to screen through them. With reductional iterations the absolute size of combichem libraries has been minimized to a few hundred thousand diverse compounds. Drug screening initially used these libraries to generate lead molecules. The leads then serve as scaffolds for iterative substitution using rational combinatorial approaches to improve selectivity and bioavailability. The conventional high-throughput (HT) screen against a designated protein target is still the preferred path to novel drug discovery. A designated target, usually a cell or a protein, is used to complete a screen against a combichem library for a selective biological effect

A Wild type

B Phosphatase activity toward ^{32}P-Hxk2p

PP-1C Western Blot

Column fractions containing Hxk2p phosphatase activity

PP1bp KO (_REG1Δ_)

(i.e., modulation of enzyme activity, inhibition of cell growth, reduction in parasite/viral infection etc.). Many compounds are identified as preliminary candidates, but these often fail in cell and animal studies because they are non-selective or non-reproducible. Ultimately the failure of traditional drug screening methodology rests on the fact that there is no clue as to the mechanism of action [61]. Invariably, when a molecule finds its way to clinical trials, deleterious side-effects are revealed which eliminate the molecule as a drug candidate. Serendipity many play a role in determining a drug"s ultimate success by identifying the real biological target via some unexpected but desirable side effect [62]. The list of drugs in which serendipity played a role in the final clinical outcome includes the majority of our pharmaceutical repertoire (a highly popularized example being Viagra™ and penile erectile dysfunction).

The more information we have concerning the biology of different SMC diseases the more we are able to reduce the role of serendipity in the discovery of new pharmacotheraputics. Toward this end, the various genome projects will play a significant role by identifying a host of new genes in SMC. However, the benefit of a strictly genomic-based approach to the identification of new targets in SMC disease is not so obvious. Recent advances in DNA sequencing, genotyping, and DNA microarray chip technologies enable the identification of specific nucleotide point mutations (SNPs) in genes that can, in principle, be the molecular basis of disease. One scenario is to sequence the entire genome of each disease sub-population and compare it with the genomes of individuals without the disease. A monumental task that will most surely identify many genomic differences between the individuals, any one of which could be the molecular basis for the disease, but with equal probability could be harmless polymorphisms. In addition, the vastness of the human genome may obscure the defective gene. Genomic strategies make sense as long as one gene produces one messenger RNA that in turn codes for one protein, as the conventional dogma has dictated. But genes clearly do not tell the whole story. The etiology of diseases involve ill-defined environmental factors that can trigger inappropriate activation of rogue genes. These factors cannot be measured by examining the genome alone.

Fig. 15.6 A. The effects of REG1 deletion on the yeast phosphoproteome. Yeast were labeled to steady state with [^{32}P] orthophosphate and whole cell extracts characterized by 2DE and autoradiography. Spots of interest were identified by microsequencing. Molecular weight (kDa) and isoelectric points (pI) are theoretical values calculated by the Expasy compute pI/MW tool (*http://expasy.hcuge.ch/ch2d/pi_tool.html*) and are derived from the primary amino acid sequence of the identified protein. Spots C and D are exposure artifacts are were not reproduced in additional experiments. B. Identification of REG1 as a PP-1 binding protein by biochemical study. Assay of phosphatase activity toward Hxk2p in yeast extracts. Cell extracts were prepared from wild-type and *REG1Δ* yeast and fractionated by anion-exchange chromatography. Column fractions were assayed for Hxk2p phosphatase activity. Column fractions were also Western blotted with rabbit anti-PP-1C antibody. (Redrawn from [60].)

The foundation underlying the recent growth of the proteomic field was the mass of genomics data delivered by the Human Genome Project and the major advances in bioinformatics in the late 1990s. With the newfound ability to identify almost any protein isolated from a tissue sample, researchers are able to devise logical genetic and biochemical experiments to probe the function and address the physiological relevance in the context of the intact cell. Genomics and proteomics are now being used as synergistic partners rather than competitors in drug discovery. A mixture of genomic and proteomic technologies are increasingly being used to identify relevant drug targets for conventional HT screens, but they do nothing to provide key information needed to further advance the HT screen. The conventional HT screen (proteomic or genomic) lacks information on possible serendipitous interactions with other protein targets. Also, the true value of the combichem library is not fully exploited. Theoretically, a structurally diverse combichem library contains within it every molecular shape that could fit selectively into a binding site on every protein target in the cell. By screening with a single protein target rather than with the entire cellular complement of proteins (i.e., the proteome), important biological information is lost for future analysis. Technologies need to be developed that maximize the information obtained from HT combichem library screening when coupled with genomic and proteomic protocols.

Proteome mining technologies may provide a solution to this challenge [62]. The principles of proteome mining are based on the assumption that all drug-like molecules selectively compete with a natural cellular ligand for a binding site on a protein target. In a proteome mine, natural ligands are immobilize on beads at high density and in an orientation that sterically favors interaction with their protein targets. The immobilized ligand is exposed to whole animal or tissue extract, and bound proteins are evaluated for specificity by protein sequencing by mass spectrometry. In the prototypical example in our laboratory, the proteome mine is charged with ATP ligand. The ATP molecule is immobilized in the protein kinase orientation via its γ-phosphate [43]. Microsequencing of the proteins that are eluted from the γ-phosphate-linked ATP Sepharose column with free MgATP revealed that the nucleotide selectively recovered purine-binding proteins, including all protein kinases, dehydrogenases, purine dependent metabolic enzymes, DNA ligases, heat shock proteins and a variety of miscellaneous ATP-binding enzymes. This immobilized proteome is estimated to represent about 5% of the expressed eukaryotic genome.

The "ATP-binding cassette proteome" can be utilized to test the selectivity of purine analogs that have been shown to inhibit proteins kinases *in vitro*. Using proteome mining ATP affinity array apparatus constructed in our laboratory, sufficient biomass is applied to ensure recovery of 1 fmol/column of any protein expressed at 100 copies/cell (10^7 cells). After stringent washing, each column in the array is eluted in parallel with molecules from a purine-based iterative library and fractions collected. Eluates are screened for protein, and positive fractions generally contain: (1) a single protein, (2) a small number of structurally related proteins, or (3) a complex mixture of unrelated proteins. Only the first two categories are considered to be of use since the third category suggests elution with a non-selective molecule. Eluted

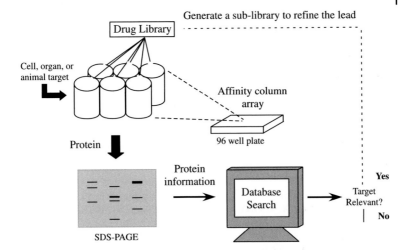

Fig. 15.7 Proteome Mining Strategy. Proteins from a cell line, organ, or animal source are isolated on affinity column arrays to remove non-specific adherents. Combichem libraries are passed over the array; any proteins that are eluted are analyzed by protein electrophoresis. Protein sequence obtained by mass spectrometry is then used to search DNA and protein databases. If a relevant target is identified, a sub-library of compounds can be evaluated to refine the lead. In this manner, a protein target, a drug lead, and any potential side-targets can be simultaneously identified. (Redrawn from [62].)

proteins from the first two situations are identified and their biological significance are considered. If a protein has no obvious use as a drug target then it is ignored, and if the protein is deemed relevant, one immediately has a lead molecule, a defined target protein, and any potential side-targets. In the cases where a single protein is eluted, the candidate drug molecule is likely to be selective since it had an equal opportunity to interact with the rest of the captured proteome. Selectivity can be investigated by changing the concentration of the drug molecule during elution (i.e., from nano- to micromolar concentrations). Information concerning potential toxicity is gained upon sequencing of other proteins that are simultaneously eluted. If these are undesirable then iterative substitutions can be made around the lead molecular scaffold to improve sensitivity and selectivity.

Screening combichem libraries with a proteome mining approach maximally exploits the serendipitous nature of drug discovery by accelerating the hit rate over a conventional screen by a factorial of the proteome that is bound. In the case of purine-binding proteins this may be a factor of 10^3. Rational interpretation of the outcome is enabled by protein microsequencing, use of genome project data and the ability to instantaneously search the literature for relevance. We are currently using proteome mining to refine new anti-hypertensive drugs that specifically target proteins identified by our functional proteomic screens.

15.5
Acknowledgements

The authors wish to thank the past and present members of the Haystead laboratory. We also thank Applied Biosystems for their generous support of our laboratory.

15.6
References

1 WASINGER, V. C., CORDWELL, S. J., CERPA-POLJAK, A., YAN, J. X., GOOLEY, A. A., WILKINS, M. R., DUNCAN, M. W., HARRIS, R., WILLIAMS, K. L., HUMPHERY-SMITH, I. Progress with gene-product mapping of the Mollicutes: Mycoplasma genitalium. *Electrophoresis.* **1995**, *16*, 1090–1094.

2 WILKINS, M. R., SANCHEZ, J. C., WILLIAMS, K. L., HOCHSTRASSER, D. F. Current challenges and future applications for protein maps and post-translational vector maps in proteome projects. *Electrophoresis.* **1996**, *17*, 830–838.

3 IDEKER, T., THORSSON, V., RANISH, J. A., CHRISTMAS, R. BUHLER, J., ENG, J. K., BUMGARNER, R., GOODLETT, D. R., AEBERSOLD, R., HOOD, L. Integrated genomic and proteomic analyses of a systematically perturbed metabolic network. *Science.* **2001**, *292*, 929–934.

4 FELL, D. A. Beyond genomics. *TRENDS in Genetics.* **2001**, *17*, 680–682.

5 TER KUILE, B. H. V., WESTERHOFF, H. V. Transcriptome meets metabolome: hierarchical and metabolic regulation of the glycolytic pathway. *F.E.B.S. Lett.* **2001**, *500*, 169–171.

6 GYGI, S. P., RIST, B., GERBER, S. A., TURECEK, F. GELB, M. H., AEBERSOLD, R. Quantitative analysis of complex protein mixtures using isotope-coded affinity tags. *Nat. Biotechnol.* **1999**, *17*, 994–999.

7 VENTER, J. C., ADAMS, M. D., MYERS, E. W., LI, P. W. et al. The sequence of the human genome. *Science.* **2001**, *291*, 1304–1351.

8 GODOVAC-ZIMMERMANN, J., BROWN, L. R. Perspectives for mass spectrometry and functional proteomics. *Mass Spectrometry Rev.* **2001**, *20*, 1–57.

9 LEUNG, Y. F., PANG, C. P. Trends in proteomics. *TRENDS in Biotechnology.* **2001**, *19*, 480–481.

10 O"FARRELL, P. H. High resolution two-dimensional electrophoresis of proteins. *J. Biol. Chem.* **1975**, *250*, 4007–4021.

11 CORTHALS, G. L., WASINGER, V. C., HOCHSTRASSER, D. F., SANCHEZ, J. C. The dynamic range of protein expression: a challenge for proteomic research. *Electrophoresis.* **2000**, *21*, 1104–1115.

12 GORG, A., OBERMAIER, C., BOGUTH, G., HARDER, A., SCHEIBE, B., WILDGRUBER, R., WEISS, W. The current state of two-dimensional electrophoresis with immobilized pH gradients. *Electrophoresis.* **2000**, *21*, 1037–1053.

13 SANTONI, V., MOLLOY, M., RABILLOUD T. Membrane proteins and proteomics: un amour impossible. *Electrophoresis.* **2000**, *21*, 1054–1070.

14 LANGEN, H., TAKACS, B., EVERS, S., BERNDT, P., LAHM, H. W., WIPF, B., GRAY, C., FOUNTOULAKIS, M. Two-dimensional map of the proteome of Haemophilus influenzae. *Electrophoresis.* **2000**, *21*, 411–429.

15 LOPEZ, M. F. Better approaches to finding the needle in the haystack: optimizing proteome analysis through automation. *Electrophoresis.* **2000**, *21*, 1082–1093.

16 GYGI, S. P., CORTHALS, G. L., ZHANG, Y., ROCHON, Y., AEBERSOLD, R. Evaluation of two-dimensional gel electrophoresis-based analysis technology. *Proc. Natl. Acad. Sci. USA.* **2000**, *97*, 9390–9395.

17 FENN, J. B., MANN, M., MENG, C. K., WONG, S. F., WHITEHOUSE C. M. Electrospray ionization for mass spectrometry of large biomolecules. *Science.* **1989**, *246*, 64–71.

18 KARAS, M., HILLENKAMP, F. Laser desorption ionization of proteins with molecular masses exceeding 10,000 daltons. *Anal. Chem.* **1988**, *60*, 2299–2301.

19 MANN, M., HENDRICKSON, R.C., PANDEY, A. Analysis of proteins and proteomes by mass spectrometry. *Annu. Rev. Biochem.* **2001**, *70*, 437–473.

20 McDONALD, W.H., YATES, J.R. III. Proteomic tools for cell biology. *Traffic.* **2000**, *1*, 747–754.

21 SHEVCHENKO, A., LOBODA, A., ENS, W., STANDING, K.G. MALDI quadrupole time-of-flight mass spectrometry: a powerful tool for proteomic research. *Anal. Chem.* **2000**, *72*, 2132–2141.

22 MEDZIHRADSZKY, K.F., CAMPBELL, J.M., BALDWIN, M.A., FALICK, A.M., JUHASZ, P., VESTAL, M.L., BURLINGAME, A.L. The characteristics of peptide collision-induced dissociation using a high-performance MALDI-TOF/TOF tandem mass spectrometer. *Anal. Chem.* **2000**, *72*, 552–558.

23 GEVAERT, K., VANDEKERCKHOVE, J. Protein identification methods in proteomics. *Electrophoresis.* **2000**, *21*, 1145–1154.

24 JENSEN, O.N., PODTELEJNIKOV, A.V., MANN, M. Identification of the components of simple protein mixtures by high-accuracy peptide mass mapping and database searching. *Anal. Chem.* **1998**, *66*, 4741–4750.

25 MANN, M., WILM, M.S. Error-tolerant identification of peptides in sequence databases by peptide sequence tags. *Anal. Chem.* **1994**, *66*, 4390–4399.

26 MACKEY, A.J., HAYSTEAD, T.A.J., PEARSON, W.R. Getting more from less: algorithms for rapid protein identification with multiple short peptide sequences. *Mol. Cell. Proteomics.* **2001**, *1*, 139–147.

27 JUNGBLUT, P.R., ZIMNY-ARNDT, U., ZEINDL-EBERHART, E., STULIK, J., KOUPILOVA, K., PLEISSNER, K.P., OTTO, A., MULLER, E.C., SOKOLOWSKA-KOHLER, W., GRABHER, G., STOFFLER, G. Proteomics in human disease: cancer, heart and infectious diseases. *Electrophoresis.* **1999**, *20*, 2100–2110.

28 BAKER, C.S., CORBETT, J.M., MAY, A.J. YACOUB, M.H., DUNN, M.J. A human myocardial two-dimensional electrophoresis database: protein characterisation by microsequencing and immunoblotting. *Electrophoresis.* **1992**, *13*, 723–726.

29 JUNGBLUT, P., OTTO, A., ZEINDL-EBERHART, E., PLEISSNER, K.P., KNECHT, M., REGITZ-ZAGROSEK, V., FLECK, E., WITTMANN-LIEBOLD, B. Protein composition of the human heart: the construction of a myocardial two-dimensional electrophoresis database. *Electrophoresis.* **1994**, *15*, 685–707.

30 MACRI, J., RAPUNDALO, S.T. *Trends Cardiovasc. Med.* Application of proteomics to the study of cardiovascular biology. **2001**, *11*, 66–75.

31 EVANS, G.W., WHEELER, C.H., CORBETT, J.M., DUNN, M.J. Construction of HSC-2DPAGE: a two-dimensional gel electrophoresis database of heart proteins. *Electrophoresis.* **1997**, *18*, 471–479.

32 PLEISSNER, K.P., SANDER, S., OSWALD, H., REGITZ-ZAGROSEK, V., FLECK, E. Towards design and comparison of World Wide Web-accessible myocardial two-dimensional gel electrophoresis protein databases. *Electrophoresis.* **1997**, *18*, 480–483.

33 SERVICE, R.F. High speed biologists search for gold in proteins. *Science.* **2001**, *294*, 2074–2078.

34 SOMLYO, A.P., SOMLYO A.V., in *The Heart and Cardiovascular System*, H.A. FOZZARD (Ed), Raven Press, New York, 1991, pp. 1–30.

35 HARTSHORNE, D.J., in *Physiology of Gastrointestinal tract*, L.R. JOHNSON (Ed.), Raven Press, New York, 1987, pp. 423–482.

36 SELLERS, J.R., ADELSTEIN, R.S. The mechanism of regulation of smooth muscle myosin by phosphorylation. *Curr. Top. Cell. Regul.* **1985**, *27*, 51–62.

37 SOMLYO, A.P., SOMLYO, A.V. Signal transduction and regulation in smooth muscle. *Nature.* **1994**, *372*, 231–236.

38 HARTSHORNE, D.J., ITO, M., ERDODI, F. Myosin light chain phosphatase: subunit composition, interactions and regulation. *J. Muscle Res. Cell Motil.* **1998**, *19*, 325–341.

39 KIMURA, K., ITO, M., AMANO, M., CHIHARA, K., FUKATA, Y., NAKAFUKU, M., YAMAMORI, B. FENG, J., NAKANO, T., OKAWA, K., IWAMATSU, A., KAIBUCHI, K. Regulation of myosin phosphatase by Rho and Rho-associated kinase (Rho-kinase) *Science.* **1996**, *273*, 245–248.

40 SOMLYO, A. P., KITAWAWA, T., HIMPENS, B., MATTHIJS, G., HORIUTI, K., KOWYASHI, S., GOLDMAN, Y. E., SOMLYO, A. V. Modulation of Ca^{2+}-sensitivity and of the time course of contraction in smooth muscle: a major role of protein phosphatase? *Adv. Protein Phosphatases.* **1989**, *5*, 181–195.

41 SHIRAZI, A., IZUKA, K., FADDEN, P., MOSSE, C., SOMLYO, A. P., SOMLYO, A. V., HAYSTEAD, T. A. Purification and characterization of the mammalian myosin light chain phosphatase holoenzyme. The differential effects of the holoenzyme and its subunits on smooth muscle. *J. Biol. Chem.* **1994**, *269*, 31598–31606.

42 ICHIKAWA, K., ITO, M., HARTSHORNE, D. J. Phosphorylation of the large subunit of myosin phosphatase and inhibition of phosphatase activity. *J. Biol. Chem.* **1996**, *271*, 4733–4470.

43 HAYSTEAD, C. M., GREGORY, P., STURGILL, T. W., HAYSTEAD, T. A. J. Gamma-phosphate-linked ATP-sepharose for the affinity purification of protein kinases. Rapid purification to homogeneity of skeletal muscle mitogen-activated protein kinase kinase. *Eur. J. Biochem.* **1993**, *214*, 459–467.

44 MACDONALD, J. A., BORMAN, M. A., MURANYI, A., SOMLYO, A. V., HARTSHORNE, D. J., HAYSTEAD, T. A. Identification of the endogenous smooth muscle myosin phosphatase-associated kinase. *Proc. Natl. Acad. Sci. USA.* **2001**, *98*, 2419–2424.

45 MACDONALD, J. A., ETO, M., BORMAN, M. A., BRAUTIGAN, D. L., HAYSTEAD, T. A. J. Dual Ser and Thr phosphorylation of CPI-17, an inhibitor of myosin phosphatase, by MYPT-associated kinase. *F.E.B.S Lett.* **2001**, *493*, 91–94.

46 MONCADA, S., PALMER, R. M. J., HIGGS, E. A. Nitric oxide: physiology, pathophysiology, and pharmacology. *Pharmacol. Rev.* **1991**, *43*, 109–142.

47 CORNWELL, T. L., PRYZWANSKY, K. B., WYATT, T. A., LINCOLN, T. M. Regulation of sarcoplasmic reticulum protein phosphorylation by localized cyclic GMP-de-

pendent protein kinase in vascular smooth muscle cells. *Mol. Pharmacol.* **1991**, *40*, 923–931.

48 LINCOLN, T. M., KOMALAVILAS, P., CORNWELL, T. L. *Hypertension.* Pleiotropic regulation of vascular smooth muscle tone by cyclic GMP-dependent protein kinase. **1994**, *23*, 1141–1147.

49 SOMLYO, A. P., WU, X., WALKER, L. A., SOMLYO, A. V. Pharmacomechanical coupling: the role of calcium, G-proteins, kinases and phosphatases. *Rev. Physiol. Biochem. Pharmacol.* **1999**, *440*, 201–234.

50 BEALL, A., BAGWELL, D., WOODRUM, D., STOMING, T. A., KATO, K., SUZUKI, A., RASMUSSEN, H., BROPHY, C. M. The small heat shock-related protein, HSP20, is phosphorylated on serine 16 during cyclic nucleotide-dependent relaxation. *J. Biol. Chem.* **1999**, *274*, 11344–11351.

51 WOODRUM, D. A., BROPHY, C. M., WINGARD, C. J., BEALL, A., RASMUSSEN, H. *Am. J. Physiol.* Phosphorylation events associated with cyclic nucleotide-dependent inhibition of smooth muscle contraction. **1999**, *277*, H931–939.

52 WU, X., HAYSTEAD, T. A. J., NAKAMOTO, R. K., SOMLYO, A. V., SOMLYO, A. P. Acceleration of myosin light chain dephosphorylation and relaxation of smooth muscle by telokin. Synergism with cyclic nucleotide-activated kinase. *J. Biol. Chem.* **1998**, *273*, 11362–11369.

53 PINNA, L. A., RUZZENE, M. How do protein kinases recognize their substrates? *Biochim. Biophys. Acta.* **1996**, *1314*, 191–225.

54 SHENOLIKAR, S., NAIRN, A. C. Protein phosphatases: recent progress. *Adv. Second Messenger Phosphoprotein Res.* **1991**, *23*, 1–121.

55 FADDEN, P., HAYSTEAD, T. A. Quantitative and selective fluorophore labeling of phosphoserine on peptides and proteins: characterization at the attomole level by capillary electrophoresis and laser-induced fluorescence. *Anal. Biochem.* **1995**, *10*, 81–88.

56 ODA, Y., NAGASU, T., CHAIT, B. T. Enrichment analysis of phosphorylated proteins as a tool for probing the phosphoproteome. *Nat. Biotech.* **2001**, *19*, 379–382.

57 ZHOU, H., WATTS, J. D., AEBERSOLD, R. A systematic approach to the analysis of

protein phosphorylation. *Nat. Biotech.* **2001**, *19*, 375–378.

58 PAWSON, T., SCOTT., J. D. Signaling through scaffold, anchoring, and adaptor proteins. *Science.* **1997**, *278*, 2075–2080.

59 CORCORAN, E. E., MEANS, A. R. Defining Ca^{2+}/calmodulin-dependent protein kinase cascades in transcriptional regulation. *J. Biol. Chem.* **2001**, *276*, 2975–2978.

60 ALMS, G. R., SANZ, P., CARLSON, M., HAYSTEAD, T. A. J. Reg1p targets protein phosphatase 1 to dephosphorylate hexoki-

nase II in Saccharomyces cerevisiae: characterizing the effects of a phosphatase subunit on the yeast proteome. *EMBO J.* **1999**, *18*, 4157–4168.

61 CUNNINGHAM, M. J. Genomics and proteomics: the new millennium of drug discovery and development. *J. Pharm. Tox. Meth.* **2000**, *44*, 291–300.

62 HAYSTEAD, T. A. J. Proteome mining: exploiting serendipity in drug discovery. *Curr. Drug Discovery.* **2001**, *1*, 22–24.

16

Identification of Targets of Phosphorylation in Heart Mitochondria

ROBERTA A. GOTTLIEB and HUAPING HE

In this chapter, we will describe the approach that led to the identification of EF-Tu as a mitochondrial target of phosphorylation in the ischemic myocardium, which we published in *Circulation Research* [1]. In addition to discussing the methods and difficulties, we will comment on the broader applications of this experimental approach.

16.1
Experimental Approach and Pitfalls

Abundant evidence has shown immediate activation of cytosolic MAP kinases, protein kinase C isoforms, and tyrosine kinases during ischemia or ischemic preconditioning; some of these signals converge upon the mitochondria [2]. We hypothesized that we would be able to detect mitochondrial phosphorylation due to cytosolic protein kinases in a reconstituted system, using isolated mitochondria and cytosol from treated hearts. We assessed differential phosphorylation by com-

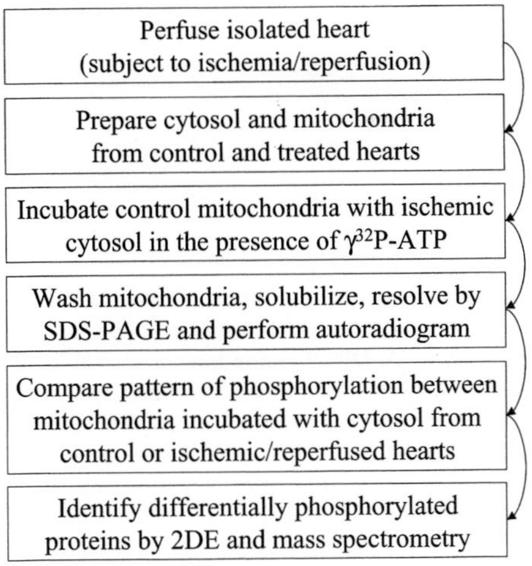

Fig. 16.1 Experimental approach utilized for the identification of differentially phosphorylated proteins in mitochondria.

paring continuously perfused control hearts with hearts that had been subjected to ischemia and reperfusion. A flow chart of the experimental approach is shown in Fig. 16.1.

We prepared a reconstituted system by incubating cytosol from control or ischemic/reperfused hearts with mitochondria from control hearts in labeling buffer with ^{32}P-γ-ATP. After the 10 min kinase reaction, the mitochondria were recovered by centrifugation and washed three times to eliminate cytosolic contamination. This gave us some differences in phosphorylation that could be detected on parallel lanes of a 1-D gel (Fig. 16.2 A). Most notably, there was increased phosphorylation of a 46 kDa protein in mitochondria that had been incubated with cytosol from ischemic/reperfused hearts.

One concern was that the heart cytosol was prepared from a mixed population of cells including cardiomyocytes, endothelial cells, and fibroblasts, while the mitochondria would be predominantly obtained from the cardiomyocytes. It was theoretically possible we were creating an artifact if the phosphorylation of cardiomyocyte mitochondria was in fact due to kinases that were activated in endothelial cells or fibroblasts. A second concern related to the fact that the mitochondria were prepared by Polytron homogenization; it was therefore possible that their

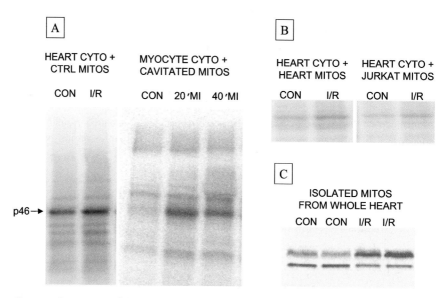

Fig. 16.2 Comparison of phosphoproteins in mitochondria subjected to various incubations. Panel A: Mitochondria from a normal heart (CTRL MITOS) or from cardiomyocytes disrupted by nitrogen cavitation (CAVITATED MITOS) were incubated with γ^{32}P-ATP and cytosol from a control heart (CON) or from a heart subjected to ischemia/reperfusion (I/R). Panel B: A similar reconstitution using heart mitochondria or mitochondria prepared from Jurkat cells disrupted by nitrogen cavitation. Panel C: Mitochondria were isolated from control hearts or from hearts subjected to ischemia/reperfusion, then incubated with γ^{32}P-ATP without added cytosol. *Reproduced with permission from Circulation Research 89:461–467, 2001.*

outer membranes were damaged, resulting in artifactual phosphorylation of target proteins in a compartment that would not ordinarily be accessible to a kinase. To address these concerns, we prepared cytosol from isolated cardiomyocytes that were subjected to simulated ischemia by metabolic inhibition with cyanide and 2-deoxyglucose, followed by recovery. We prepared mitochondria from isolated cardiomyocytes, using nitrogen cavitation. This method of preparing mitochondria preserves outer membrane integrity [3]. This preparation yielded the same pattern of differential phosphorylation of the 46 kDa protein, although some minor differences in other bands were noted (Fig. 16.2 A). We also confirmed that the same pattern of phosphorylation was obtained when heart cytosols were incubated with mitochondria prepared from Jurkat T-lymphoblast cells after disruption by nitrogen cavitation (Fig. 16.2 B).

A third concern was that the 46 kDa protein might represent a cytosolic contaminant, despite the fact that the mitochondria had been washed extensively after the kinase reaction. We reasoned that in ischemic hearts, some kinases would have already translocated to mitochondria (or become activated within the mitochondria), and could be demonstrated merely by incubating isolated mitochondria from ischemic (vs control) hearts in labeling buffer. This was indeed the case, as seen in Fig. 16.2 C.

Interestingly, the results seemed to suggest that the phosphorylation of p46 was not directly due to a cytosolic protein kinase, because even isolated mitochondria

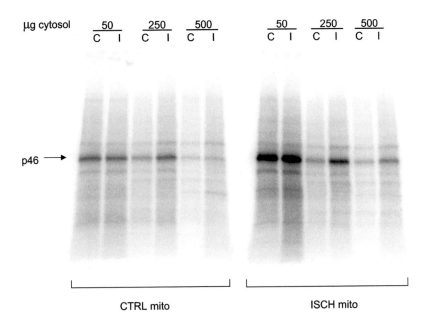

Fig. 16.3 Effect of varying the amount of cytosol on mitochondrial phosphorylation. Mitochondria from a control (CTRL) or ischemic (ISCH) heart (100 μg) were incubated with increasing amounts of cytosol from control (C) or ischemic hearts (I) in the presence of γ^{32}P-ATP.

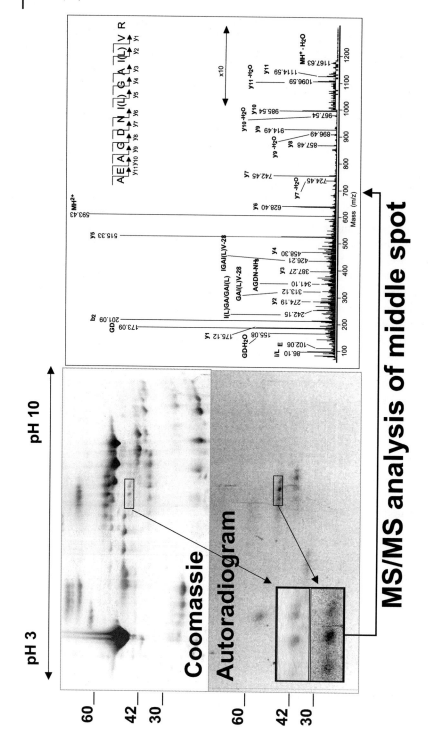

MS/MS analysis of middle spot

Coomassie

Autoradiogram

from a control heart exhibited a certain degree of phosphorylation of p46, which was suppressed by the addition of cytosol from control hearts and enhanced by cytosol from ischemic hearts. The addition of increasing amounts of cytosol tended to suppress the phosphorylation of p46, implying the presence of an inhibitory cytosolic factor that was diminished after ischemia (Fig. 16.3).

These studies indicated that the 46 kDa protein was a genuine mitochondrial phosphoprotein regulated by cytosolic signals. Specifically, the 46 kDa protein that we had demonstrated to be phosphorylated in the reconstituted system met the following criteria:

1. Differentially phosphorylated (greater in ischemia than control)
2. Resident in mitochondria (including mitochondria with intact membranes)
3. Present in cardiomyocytes.

These findings provided the impetus to pursue identification of the 46 kDa mitochondrial protein as a target of phosphorylation in the setting of ischemia/reperfusion.

Our initial approach to purify the 46 kDa protein was to use FPLC. The mitochondria were lysed in detergent-containing buffer and the soluble fraction was resolved on an anion exchange column. The 46 kDa protein eluted in multiple fractions, with excessive sample dilution. As we discovered later, this was probably due in part to the fact that p46 was multiply phosphorylated. We were unsuccessful in eluting the phosphoprotein from a hydrophobic interaction column. We turned instead to 2D gel electrophoresis. While many of the mitochondrial proteins are rather hydrophobic and require special treatment in order to successfully solubilize them, we were fortunate in that the 46 kDa phosphoprotein was readily solubilized and resolved nicely on a conventional IEF strip and second dimension SDS-PAGE.

2D spot mapping can be quite difficult, depending upon the complexity of the sample under analysis. We were optimistic because the autoradiogram of mitochondria after *in vitro* labeling was deceptively simple, revealing only ~ 20 spots. Encouragingly, the 46 kDa band resolved into ~ 3 closely spaced spots differing by isoelectric point (pI ~ 6.5), and which were nicely separated from other phosphoproteins. However, the complexity was greatly increased when we prepared a Coomassie-stained gel of the same material. Worse yet, the 46 kDa spots were not abundant enough to be detected by Coomassie staining of a gel loaded with up to 2 mg of mitochondrial protein. At this point, we were using total mitochondria.

Fig. 16.4 Preparation of sample for identification by mass spectrometry. Inner mitochondrial membrane from 7 mg of mitochondria were used for phosphorylation with "cold" or 'hot' ATP and resolved by 2DE for Coomassie staining and autoradiography, respectively. Inset shows the three spots of interest, which were removed and separately subjected to trypsin digestion and MALDI-MS. The middle spot was subsequently used for MS/MS analysis, shown on the right. *Reproduced with permission from Circulation Research 89:461–467, 2001.*

However, we reasoned that if we fractionated the mitochondria further, we could reduce the complexity and could also enrich for the 46 kDa protein. Submitochondrial fractionation revealed that the 46 kDa protein was present in the matrix and in the inner membrane.

Because it was easier to prepare and concentrate the inner membrane (IM), we used this as the material for subsequent analysis. This sub-fractionation was sufficient to enrich the 46 kDa protein so that it could be detected by Coomassie staining after loading IM derived from 7 mg of mitochondria on a large (18 cm) preparative gel. While having enough protein to detect by Coomassie staining is not essential, it is a reassuring indication that one will have enough material for successful identification by mass spectroscopy. Moreover, it will reveal the presence of additional contaminating proteins in the neighborhood that may complicate the analysis. In order to prepare the 46 kDa phosphoprotein for mass spectrometry, we prepared mitochondria, split into two aliquots, and performed two identical phosphorylation reactions, one with γ^{32}P-ATP and the other with non-radioactive ATP. The mitochondria were then washed and fractionated to obtain inner membranes. We then ran two identical 2-D gels and overlaid the autoradiogram of one on the Coomassie-stained non-radioactive gel, and cut out the three closely spaced spots of interest. We were fortunate that the region of interest did not have a lot of additional proteins (Fig. 16.4).

Karoline Scheffler and Brad Gibson at UCSF analyzed the three samples. Analysis of the tryptic peptides by MALDI-MS revealed that the three spots had identical mass signatures, confirming that they represented multiply phosphorylated forms of the same protein. A total of 12 mass fingerprints were used to query the database, but failed to match anything in the database. This disappointing outcome exemplifies one of the limitations of current analytical software. While the algorithm can tolerate one amino acid substitution in a fragment, it fails if there are two or more amino acid substitutions in a single tryptic fragment. It is further confounded if a tryptic cleavage site is missing. Since very little of the rabbit genome is in the database, one must rely on comparisons with other species such as mouse and human. Unfortunately, rabbit diverges just enough to confound such comparisons.

It was therefore necessary to sequence some of the more abundant peptide fragments in order to identify the 46 kDa phosphoprotein, using nanoelectrospray MS/MS (Fig. 16.4). By this approach, Drs. Scheffler and Gibson were able to obtain sequences from four peptides, consisting of 13, 12, 6, and 4 amino acids. The two longer peptides were subjected to a BLAST search and matched (100%) with the human and bovine mitochondrial precursor of the elongation factor Tu (EF-Tu$_{mt}$) (SwissProt accession numbers P49411 and P49410, respectively). We developed primers and used RT-PCR to obtain partial cDNA sequence for rabbit EF-Tu$_{mt}$. The overall sequence of mitochondrial EF-Tu is highly conserved, demonstrating 92% homology with human and 95% homology with bovine amino acid sequence. Despite this high degree of homology, detection by the mass fingerprint was unsuccessful. This points to the need to select starting material from a species that is well represented in the database, and also illustrates the value of sequencing the genomes of additional species that serve as important animal models, including rabbit, dog, and pig.

Having identified the three Coomassie-stained spots as EF-Tu$_{mt}$, it was necessary to verify that the phosphoprotein was indeed EF-Tu$_{mt}$, since it was possible that we had instead identified a more abundant protein that colocalized with the 46 kDa phosphoprotein. We were fortunate that antibody to EF-Tu$_{mt}$ had been generated by Dr. Linda Spremulli, at the University of North Carolina, Chapel Hill [4]. We collaborated with Dr. Spremulli in the next phase of the work. The most straightforward approach would have been to use anti-EF-Tu antibody to immunoprecipitate the phosphoprotein. Unfortunately, EF-Tu is a rather sticky protein and adhered to protein G-Sepharose even in the absence of primary antibody. We therefore resorted to ion exchange chromatography, and found that EF-Tu$_{mt}$ and the phosphoprotein eluted in the same fractions. Immunodetection of EF-Tu in those fractions could be superimposed on the autoradiogram of the same blot (Fig. 16.5A). Additionally, the submitochondrial distribution of the 46 kDa phos-

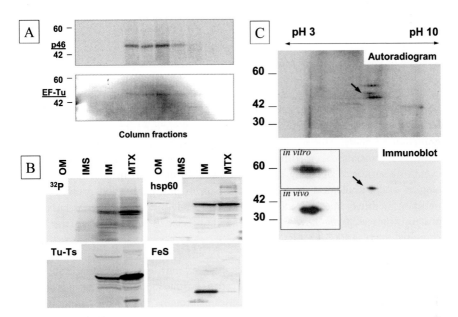

Fig. 16.5 Verification that p46 is EF-Tu$_{mt}$. Panel A: Column fractions from anion exchange were concentrated, resolved by SDS-PAGE, and subjected to autoradiography (upper panel) and immunoblotting for EF-Tu (lower panel). Panel B: Submitochondrial fractionation was performed to obtain outer membrane (OM), intermembrane space (IMS), inner membrane (IM), and matrix (MTX). Shown are the autoradiogram of p46 (32P), immunoblot of mitochondrial EF-Tu/Ts (Tu-Ts), and markers for IM (Rieske iron-sulfur protein, FeS) and matrix (hsp60). Panel C: 2D gel showing autoradiogram of p46 and immunoblot for EF-Tu. Inset shows an enlargement of the immunoblot, showing 2–4 isoelectric variants of EF-Tu from mitochondria incubated in vitro with cytosol (*in vitro*) and from mitochondria rapidly isolated from a fresh heart (*in vivo*). *Reproduced with permission from Circulation Research 89:461-467, 2001.*

phoprotein in inner membrane and matrix was identical with the distribution of EF-Tu$_{mt}$ (Fig. 16.5B). Additional evidence came from an immunoblot of a 2-D gel, in which EF-Tu immunoreactivity of three closely spaced spots could be superimposed on the autoradiogram of the phosphoprotein, thus confirming that EF-Tu$_{mt}$ exists as multiple isoforms which differ by isoelectric point, presumably representing multiply phosphorylated species (Fig. 16.5C). Comparison of the isoforms of EF-Tu$_{mt}$ observed in the reconstituted preparations of cytosol and mitochondria with those observed in a freshly isolated heart is shown in the inset (Fig. 16.5C, *inset*). The presence of multiply phosphorylated EF-Tu$_{mt}$ *in vivo* confirms that this is not merely an artifact of *in vitro* labeling.

16.2
Perspective and Comments

The recounting of our experience with EF-Tu$_{mt}$ allows us to offer several observations and speculations that may be applied to other systems.

16.2.1
Analysis of Post-translational Modifications

While most strategies in proteomics deal with changes in protein levels, post-translational modifications often represent pathological consequences at the molecular level without changes in abundance and, therefore represent a critical phenomenon to study. The present report elucidates an approach to studying one such post-translational modification, phosphorylation, which is an important event in a variety of signaling cascades initiated in the heart in response to physiological and pathological stimuli. Comparison of phosphorylation profiles in genetically modified mice may aid in identifying relevant mitochondrial targets of protein kinases, particularly those already suspected to translocate to the mitochondria, such as PKC$_\varepsilon$ [5, 6].

Comparison of phosphorylation profiles in different conditions, including drug treatments, receptor stimulation or blockade, etc., may lead to identification of relevant targets of signal transduction cascades. Furthermore, it may be possible to devise similar schemes to detect the subset of proteins modified by ubiquitinylation, lipid acylation, and so forth, and to compare differential patterns.

16.2.2
Detection of Phosphoproteins by Comparison of Two Conditions Can Be a Tool to Select Proteins of Potential Interest for Proteomic Analysis

By comparing mitochondria phosphorylated by cytosol from control and ischemic hearts, we were able to select only those proteins whose phosphorylation state changed in response to ischemia. A similar approach could be applied to other settings, such as drug treatment or receptor stimulation, in order to examine short-term posttranslational changes. Another example is illustrated in the comparison between the CalTG and control mice heart mitochondria.

16.2.3
Selection of an Organelle Simplifies the Analysis

We would not have been able to distinguish changes in the whole cell lysate, and inspection of cytosol after *in vitro* phosphorylation revealed too many bands to detect changes. However, relatively few proteins in the mitochondria underwent phosphorylation. Therefore the approach of identifying mitochondrial phosphoproteins that were targets of cytosolic signaling greatly simplified the analysis. In addition to organelles and cytoskeleton, it is possible to recover signaling scaffolds by immunoprecipitating one component under conditions that preserve the integrity of the entire complex. This provides some exciting functional relevance, as discussed in this book in the chapter by Peipei Ping.

The organellar approach could also be used to identify changes in the protein population of the outer mitochondrial membrane. Mitochondria from hearts subjected to two different treatments can be isolated and subjected to biotinylation under gentle conditions in which outer membrane proteins are selectively labeled [7]. Comparison of the 2-D spot patterns (after detection by streptavidin-peroxidase) may reveal the appearance of new spots in the experimental condition. The biotin provides a convenient handle for partial purification of the labeled proteins, and the two protein mixtures can next be analyzed with the use of isotope coded affinity tags [8].

16.2.4
Further Simplification of the Mixture of Proteins to be Analyzed is Desirable

The tools of cell biology (*e.g.*, sub-organellar fractionation) and biochemistry (chromatography) should be employed whenever possible. Mitochondria contain >1,000 different polypeptides of widely varying abundance. While detection of the most abundant proteins on 2D gels is straightforward, one is likely to miss many of the less abundant proteins. In addition membrane proteins are problematic due to the limiting factor of solubilization. Therefore it is useful to enrich for the subset of proteins one is most interested in.

16.2.5
Selection of a Protein Separation Strategy is Essential

While 2-D gels are one widely used strategy to separate complex mixtures of proteins for mass spectrometry, other approaches exist and are discussed elsewhere in this textbook. One approach that shows great promise is that developed by Yates and colleagues, utilizing multidimensional liquid chromatography coupled to tandem mass spectrometry [9]. In this approach, a complex mixture of proteins is subjected to tryptic digestion or cyanogen bromide hydrolysis, then loaded onto a microcapillary column packed with a C18 reverse-phase matrix and with a cation exchange material. Peptides are eluted through the column with an automated program using an acetonitrile gradient and a step gradient of ammonium acetate. The eluted peptides are injected directly into the mass spectrometer an analyzed by the SEQUEST algorithm [10].

16.3
Specific Methods Employed

16.3.1
Isolated Perfused Heart

Male New Zealand White rabbits (2.0–2.5 kg) were anesthetized and a midsternal thoracotomy was performed. The heart was rapidly excised and mounted onto a Langendorff heart perfusion apparatus using a protocol adapted from Tsuchida et al. [11]. The heart was perfused at a constant pressure of 60 mmHg with Krebs-Ringer buffer consisting of (in mmol/L) NaCl 118, KCl 4.75, KH_2PO_4 1.18, $MgSO_4 \cdot 7 H_2O$ 1.18, $CaCl_2 \cdot 2 H_2O$ 2.5, $NaHCO_3$ 25, and glucose 11. The perfusate was bubbled with a mixture of 95% O_2 and 5% CO_2 at 37 °C. Perfused hearts were stabilized for 15 min, then subjected to global ischemia for 30 min by turning off the perfusion system. After 30 min of ischemia the perfusion system was restarted, and the hearts were reperfused for up to 90 min. Ischemic preconditioning was induced by three 5 min cycles of no-flow ischemia and reperfusion immediately preceding the regular ischemia and reperfusion. The efficacy of these interventions was verified by measurement of creatine kinase release (Sigma Chemicals, St. Louis, MO) and infarct size measurement using triphenyl tetrazolium staining [12]. All animal procedures were approved by the Animal Care and Use Committee of The Scripps Research Institute.

16.3.2
Isolation of Cytosol and Mitochondria

Upon completion of global ischemia, the heart was removed from the cannula and the ventricles were minced in 20 ml per heart of ice-cold MSE buffer (225 mM mannitol, 75 mM sucrose, 1 mM EGTA, 1 mM Na_3VO_4, 20 mM HEPES-KOH, pH 7.4). The heart was further polytron homogenized for 5 s at maximal power output by PowerGen 125 (Fisher Scientific) equipped with a 10 mm diameter rotor knife. The homogenate was centrifuged for 10 min at 600 g, 4 °C. The pellet was discarded and the supernatant was centrifuged for 10 min at 10,000 g to pellet mitochondria and lysosomes. The postmitochondrial supernatant generally contains about 15 mg protein/ml. This supernatant was further centrifuged for 30 min at 100,000 g to obtain particulate-free cytosol (S100). The 10,000 g mitochondrial pellet from the previous centrifugation was resuspended in 10 ml of MSE buffer and centrifuged for 10 min at 8,000 g. This wash step was repeated once. The final pellet was resuspended in 3 ml of MSE buffer and was further purified by hybrid Percoll/metrizamide discontinuous gradient purification consisting of 5 ml of 6% Percoll, 2 ml of 17% metrizamide and 2 ml of 35% metrizamide, all prepared in 0.25 M sucrose and set up in 13 ml tubes [13]. Three ml of the sample were overlaid on top of the gradient and centrifuged for 20 min at 50,000 g, 4°C, using a Beckman SW41 rotor. The mitochondrial fraction at the interface between 17% and 35% metrizamide was collected and diluted at least 10-fold with MSE buffer, followed by centrifugation for 10 min at 10,000 g to remove metrizamide. The pellet was resuspended in 20 ml of MSE and centrifuged again. The final pellet was resuspended in 3 ml of MSE and aliquots were stored at − 80°C. Protein concentration was determined using the Bradford assay, and for all experiments, equal amounts of mitochondrial protein were loaded on the gels. Cytosol concentrations were adjusted to be equal in all conditions before incubating with mitochondria.

16.3.3
In Vitro Phosphorylation Reactions

For phosphorylation reactions, 100 µg of purified mitochondria were incubated in MSE supplemented with 25 mM HEPES-KOH, pH 7.5, 10 mM magnesium acetate, 10 µM ATP (cold) and 10 µCi [γ-^{32}P]ATP for 30 min at 30°C with or without 250 µg of cytosol. The reaction mixture was subsequently centrifuged for 5 min at 10,000 g. The mitochondrial pellet was resuspended in 500 µl MC buffer and was washed twice. The mitochondrial proteins were resolved on a 12% polyacrylamide gel, transferred to nitrocellulose, and detected by autoradiography.

16.3.4
Submitochondrial Fractionation

We used a modification of the method of Comte and Gautheron to fractionate mitochondria [14]. Freshly isolated purified mitochondria were pelleted by centrifugation for 5 min at 10,000 g. The mitochondrial pellet was resuspended in 10 mM KH$_2$PO$_4$, pH 7.4 and incubated on ice for 20 min for hypotonic swelling. The mitochondria were centrifuged for 15 min at 10,000 g, 4°C to pellet mitoplasts (inner membrane and matrix). The supernatant, containing outer membrane (OM) and intermembrane space (IMS), was centrifuged for 30 min at 100,000 g to separate OM (pellet) and IMS (supernatant). The mitoplast pellet was resuspended in 500 µl of MC buffer (300 mM sucrose, 1 mM EGTA, 1 mM Na$_3$VO$_4$, 20 mM MOPS, pH 7.4) and sonicated on ice in 5 cycles of 20 s bursts and 40 s rest intervals with output setting at 8–10 Watts. The sonicated mitoplast preparation was centrifuged for 10 min at 10,000 g to remove any remaining intact mitoplasts or mitochondria, followed by centrifugation at 100,000 g for 30 min to separate inner membrane (IM) and matrix (MTX).

16.3.5
Preparation of Samples for 2D PAGE and MS

To prepare p46 for identification by mass spectrometry, 14 mg of metrizamide-purified mitochondria were obtained from two untreated hearts. Two aliquots of mitochondria (7 mg each) were subjected to phosphorylation as above, except one of the aliquots was phosphorylated with non-radioactive "cold" ATP. Following the reaction, the mitochondria of both reactions were subjected to suborganellar fractionation as described above to obtain the IM. IM was chosen as the source for identification of p46 because of its small volume for subsequent 2D gel analysis and the reasonable abundance of p46 (Fig. 16.5). "Hot" and "cold" IM were resuspended in buffer containing 8 M urea and 20 mM Tris-HCl (pH 7.4) and resolved by 2D gel electrophoresis under identical conditions using the Pharmacia IPG-phor IEF system. Both gels were Coomassie-blue stained and dried. The gel containing the radiolabeled sample was exposed to X-ray film to localize p46. Using position markers, the autoradiogram was superimposed on the gel. Three closely spaced radioactive spots were found to precisely overlie three Coomassie blue-stained spots. These spots were located in a relatively clean portion of the gel with few other spots nearby, making their recognition straightforward. The same three spots on the non-radioactive gel were visually identified, confirmed by overlay of the autoradiograph, and excised for mass spectrometry analysis.

16.3.6
In-gel Digest of p46

Tryptic in-gel digest of all three spots was performed as described previously [15]. Unseparated tryptic peptide mixtures were diluted with 50% acetonitrile-5% trifluoroacetic acid to a final volume of 15 μl.

16.3.7
MALDI Analysis

MALDI analysis was performed as described previously [15] using a Voyager DE-Str MALDI-TOF instrument (Applied Biosystems, Framingham MA) equipped with a nitrogen laser (337 nm), operated in delayed-extraction [16] and reflectron mode [17]. Mass spectra were calibrated internally on the trypsin autolysis peptides.

16.3.8
Peptide Sequencing by MS/MS Analysis

For MS/MS analysis, the crude peptide mixture obtained after in-gel digest was purified over a C_{18} reversed-phase Zip-Tip® (Millipore; Bedford, MA). The purified sample was supplied into a nanospray needle (Protana, Odense; Denmark) and analyzed on a Q-Star quadrupole time-of-flight instrument (Sciex, Toronto; Canada) in nanospray mode. The ion spray voltage was set to 1,100 V. For MS/MS experiments the collision energy Q_0 was set to 50.

16.4
References

1 He, H., M. Chen, N. K. Scheffler, B. W. Gibson, L. L. Spremulli, R. A. Gottlieb. Phosphorylation of mitochondrial elongation factor Tu in ischemic myocardium: basis for chloramphenicol-mediated cardioprotection. Circ. Res. 2001, 89:461–467.

2 Cohen, M. V., C. P. Baines, J. M. Downey. Ischemic preconditioning: from adenosine receptor of K_{ATP} channel. Annu. Rev. Physiol. 2000, 62:79–109.

3 Gottlieb, R. A. S. Adachi. Nitrogen cavitation for cell disruption to obtain mitochondria from cultured cells. Meth. Enzymol. 2000, 322:213–221.

4 Woriax, V. L., W. Burkhart, L. L. Spremulli. Cloning, sequence analysis and expression of mammalian mitochondrial protein synthesis elongation factor Tu. Biochim. Biophys. Acta. 1995, 1264:347–356.

5 Vondriska, T. M., J. B. Klein, P. Ping. Use of functional proteomics to investigate PKC epsilon-mediated cardioprotection: the signaling module hypothesis. Am. J. Physiol. Heart. Circ. Physiol. 2001, 280:H1434–H1441.

6 Vondriska, T. M., J. Zhang, C. Song, X. L. Tang, X. Cao, C. P. Baines, J. M. Pass, S. Wang, R. Bolli, P. Ping. Protein kinase C epsilon-Src modules direct signal transduction in nitric oxide-induced cardioprotection: complex formation as a means for cardioprotective signaling. Circ. Res. 2001, 88:1306–1313.

7 Song, J., C. Midson, E. Blachly-Dyson, M. Forte, M. Colombini. The topology of VDAC as probed by biotin modification. J. Biol. Chem. 1998, 273:24406–24413.

8 Smolka, M. B., H. Zhou, S. Purkayastha, R. Aebersold. Optimization of the isotope-coded affinity tag-labeling procedure for quantitative proteome analysis. Anal. Biochem. 2001, 297:25–31.

9 Washburn, M. P., D. Wolters, J. R. 3. Yates. Large-scale analysis of the yeast proteome by multidimensional protein identification technology. Nat. Biotechnol. 2001, 19:242–247.

10 Eng, J. K., A. L. McCormack, J. R. 3. Yates. An approach to correlate tandem mass spectral data of peptides with amino acid sequences in a protein database. J. Am. Soc. Mass. Spectrom. 1994, 5:976–989.

11 Tsuchida, A., Y. Liu, G. S. Liu, M. V. Cohen, J. M. Downey. Alpha 1-adrenergic agonists precondition rabbit ischemic myocardium independent of adenosine by direct activation of protein kinase C. Circ. Res. 1994, 75:576–585.

12 Downey, J. M. Measuring infarct size by the tetrazolium method. http://www.usouthal.edu/ishr/help/ttc/. 2000

13 Storrie, B. E. A. Madden. Isolation of subcellular organelles. Meth. Enzymol. 1990, 182:203–225.

14 Comte, J. D. C. Gautheron. Preparation of outer membrane from pig heart mitochondria. Meth. Enzymol. 1979, LV:98–104.

15 Wong, D. K., B. Y. Lee, M. A. Horwitz, B. W. Gibson. Identification of fur, aconitase, and other proteins expressed by Mycobacterium tuberculosis under conditions of low and high concentrations of iron by combined two-dimensional gel electrophoresis and mass spectrometry. Infect. Immun. 1999, 67:327–336.

16 Vestal, M. L., P. Juhasz, S. A. Martin. Delayed extraction matrix-assisted laser desorption time-of-flight mass spectrometry. Rapid Commun. Mass Spec. 2001, 9:1044-1050.

17 Karas, M., U. Bahr, U. Giessmann. Matrix-assisted laser desorption ionization mass spectrometry. Mass Spectrom. Rev. 1991, 10:335–357.

17

Proteomic Characterization of Protein Kinase C Signaling Tasks in the Heart

Thomas M. Vondriska, Ph.D., Jun Zhang, M.S., and Peipei Ping, Ph.D.

Changes in an organism's phenotype are reflective of underlying alterations in protein function. Moreover, it has become increasingly clear that the molecular infrastructure supporting cellular function is composed of discrete multi-protein complexes that assemble and/or dissociate at given subcellular locations in order to accomplish specific tasks [1–13]. Thus, to understand organ phenotype in health and disease, one must understand the nature of these subcellular complexes. As a result, the study of disease is more than ever becoming the study of protein interactions.

The field of proteomics is poised to accept this challenge. In addition to biomarker discovery and expression profiling, a powerful application of proteomics is to facilitate the understanding of how changes in protein interactions engender disease. This is typified by proteomic investigations aimed to characterize protein interactions in specific disease states [4, 8, 14] and by more recent studies designed to map entire cellular networks in lower eukaryotes [2, 3]. Incumbent with these studies is the need for functional validation of proteomic data. In other words, if one wants to understand the relevance of observed protein interactions, one must understand whether these interactions occur in vivo, and if so, specifically *how* they contribute to the organism's phenotype.

17.1
Protein Kinase C Signaling Complexes

Cardiac tissue possesses an extremely limited (if any) ability to regenerate, and as a result, cell death in the heart is a particularly perilous occurrence [15]. Considerable research in the past has accordingly been aimed at preventing myocardial cell death, and more recently, with understanding the minute molecular changes responsible for myocyte death and survival. This is chiefly relevant with regard to cell death due to ischemia, or myocardial infarction, the most serious consequence of heart disease. Significant evidence supports a central role for the serine/threonine kinase protein kinase C epsilon (PKCε) in the protective signal transduction mechanism that prevents cell death due to ischemia [8, 16–21]. Importantly, recent evidence from proteomic studies suggests that PKCε forms multi-protein complexes with various different proteins to regulate key subcellular processes [8, 10, 11]. Interestingly, these processes appear to include metabolism and protein synthesis (both at the level of transcription and translation) [10]. Accordingly, this chapter aims to 1) discuss examples of task-targeted proteomics; 2) develop the concept of functional validation of proteomic data in the context of current research in the field; and 3) specifically discuss the use of proteomics to characterize PKCε complexes in the heart.

17.1.1
Metabolism-targeted Proteomics

There is a scarcity of studies reporting large scale proteomic characterizations of metabolic systems, but a few seminal examples are briefly discussed. It should be noted that these studies were chosen, as were those in the transcription and translation section below, both for the quality of the study and for the uniqueness of the proteomic approach used by the authors to conduct their investigation.

Seow and colleagues used a mammalian cell line in culture to study the effects of resource-induced metabolic shift [22]. Cells were incubated in a low nutrient medium (cells in normal media served as controls) – a protocol known to induce the aforementioned metabolic shift by decreasing the rate of lactate produced relative to glucose consumed. The cells were then lysed and analyzed using two-dimensional electrophoresis and matrix-assisted laser desorption ionization time-of-flight mass spectrometry (MALDI-TOF MS). Protein spots that were not readily identified with peptide mass fingerprinting were then subjected to electrospray ionization and MS/MS [22]. The investigators found that multiple proteins, some identifiable directly from the gel patterns and others identified by mass spectrometry, that had altered their expression and/or post-translational modification to allow the cell to metabolically accommodate the decreased nutrient environment [22]. This is a good example of a uniquely customized study that uses classical proteomic techniques to ask broad spectrum questions about cellular metabolism.

In a truly innovative study that samples both genomic and proteomic approaches, Ideker and colleagues investigated genetically altered yeast strains to understand global changes responsible for the metabolism of galactose [23]. This approach utilized extensive nucleotide microarray technology to understand the changes in mRNA levels that occurred in response to genetically induced galactose metabolism pathway perturbations. Importantly, the investigators employed state-of-the-art isotope coded affinity tag technology and tandem mass spectrometry to characterize proteomically which of these mRNA changes was matched by a protein change [23]. This step was extremely important, as the correlation between mRNA and protein levels continues to be a contentious issue regarding the interfacing of genomic and proteomic data. They found that there was a considerable correlation between mRNA and protein levels in this model, but that a significant number of proteins appeared to change independent of their mRNA counterparts, suggesting posttranscriptional modification. Lastly, the investigators also developed a model to integrate the observed protein changes with previously published changes associated with galactose metabolism, and to develop a visual representation of the galactose utilization network as it relates to other cellular processes [23]. By combining advanced genomic and proteomic tools in a novel manner, this study provides detailed information about the protein infrastructure responsible for a catabolic process, galactose metabolism.

17.1.2
Transcription-targeted Proteomics

Protein abundance is reliant upon dynamic interplay between transcription, translation, protein degradation, and other processes, and is therefore the study of no one organelle or specific family of molecules. Nonetheless, various reports in the past few years have examined sub-organelle proteomes [14, 24, 25] (e.g. the nucleolus) and protein complexes [1, 6] (e.g. nucleoporin) within the nucleus that relate to the control of this organelle's function. In addition, various studies have targeted other organelles such as the golgi apparatus [26], the ribosome [27], and the endoplasmic reticulum [28] which are known to be key regulators, along with the nucleus, of protein synthesis. This section will review two distinct proteomic approaches that share the common theme of characterizing transcriptional control.

RNA polymerase II is a critical enzyme for transcription and is well known to be regulated by a battery of transcription factors. Despite this knowledge, the nature of the interactions between the different members of a key regulator of RNA polymerase II, the Mediator complex, was unclear. Liu and colleagues used epitope-tagged-containing yeast strains to isolate Mediator complexes for analysis [6]. Affinity chromatography and gel filtration were used to isolate and separate the complexes, and mass spectrometry and immunoblotting were then employed to identify proteins within the isolated complexes. The investigators found that the complex existed primarily in two forms of markedly different molecular weight, and subsequently went on to characterize the protein components of each form of the complex. This type of approach is particularly useful if some of the components of a protein complex are known or suspected, such that they can be identified by western blotting, whilst a simultaneous non-biased analysis of all protein components (via mass spectrometry) allows for the detection of previously unknown members of a complex.

In contrast to focusing on a specific protein of interest that regulates transcription, Predic and colleagues investigated gene expression changes in response to agonist stimulation through a proteomic approach [29]. Fibroblasts were treated with endothelin-1 and protein expression was monitored by pulsing cells with radioactively labeled amino acids. This labeling technique facilitated detection of changes in extremely low abundance proteins on the gel, which is generally not possible with conventional stain-based labeling methods. This technique allowed for the visualization of proteins with abundance as low as 10 copies per cell. Furthermore, in that only newly synthesized proteins were labeled, the approach allowed for "gene expression" to be monitored by protein level, eliminating quantitative discrepancies between mRNA and protein. The investigators used this method to detect expression changes on the two dimensional gel and then identified the altered proteins by standard MALDI-TOF MS. This method also allowed for time-dependent characterization of protein expression changes following the treatment with endothelin-1, thus adding additional detail to the fibroblast regulatory profile provided by the study [29].

Having highlighted a number of distinct approaches that have been used by others to examine metabolic changes and the control of protein synthesis, the chapter will henceforth focus on the comprehensive proteomic platform developed by our laboratory to characterize molecular signaling tasks in the heart.

17.2
Subproteomic Analysis of PKCε Signaling

17.2.1
Introduction

Mechanisms to reduce the deleterious effects of ischemia and to prevent myocardial cell death are of particular clinical importance to prevent heart attack-induced mortality [15, 17, 30]. Intense research over the past decade has suggested that the serine/threonine kinase protein kinase C epsilon (PKCε) is a critical signaling kinase that, once activated, coordinates a protective response within the heart to prevent cell death [8, 16, 17]. Indeed, activation of PKCε in the heart is sufficient to significantly reduce myocardial infarct size due to coronary artery occlusion [31]. Importantly, multiple investigations have divulged a collection of molecules that are activated in the heart in a PKCε-dependent fashion, some of which also appear to be necessary for cardiac protection [8, 13, 16, 17, 20]. In other words, while it is clear that PKCε plays a necessary role in protection of the ischemic myocardium, it also seems apparent that there are other molecular players involved in the response orchestrated by PKCε. Thus, in the subproteome constituted by all molecules necessary to protect the myocyte against ischemic cell death, PKCε appears to be a central regulator. Accordingly, the following questions were precipitant from these findings: Does such a subproteome defined by PKCε and its associated proteins exist? And if so, how does one define such a subproteome in vivo?

To answer these questions, our laboratory has devised a comprehensive proteomic platform that allows for the functional characterization of the PKCε signaling subproteome in the normal and in the protected heart. For these studies, it was of paramount importance that the platform be versatile enough to directly link the observed protein interactions to the genesis of a cardiac protective phenotype in vivo. This approach, as diagrammed in Fig. 17.1, allows for an unbiased, yet focused, investigation of all molecules participating in a given biological process, and thereby provides a detailed blueprint of the entire signaling network [8, 10, 32]. This method can be easily modified to examine signaling complexes in other cell types and disease states.

FUNCTIONAL PROTEOMIC PLATFORM

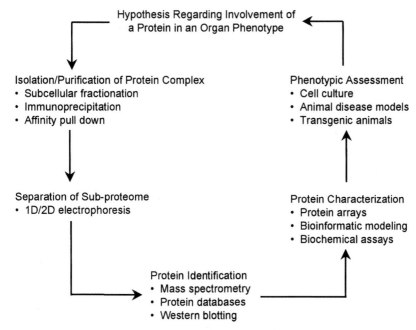

Fig. 17.1 Functional proteomic platform for the characterization of protein complexes. Please see text for explanation.

17.2.2
Functional Proteomic Platform

Isolation, Separation, and Identification of the Subproteome. In order to effectively characterize the proteins that associate with a protein of interest, one must first isolate this subproteome from the cell lysate. This process can be preceded by subcellular fractionation [11, 12], a method that has been described elsewhere [12] will not be discussed in detail in this chapter. Complex isolation is commonly done by one of two methods: immunoprecipitation [4, 8, 10–13] or recombinant protein-based affinity pull down [2, 3, 8, 12, 13]. With both of these techniques, the protein of interest is the bait with which one identifies the associated proteins (for a technical review of immunoprecipitation, please see reference 33). Briefly, immunoprecipitation requires that antibodies to the protein of interest exist, and must be controlled with IgG-based parallel experiments. Accordingly, affinity pull down experiments (e.g. GST-based pull down experiments) require the development of recombinant proteins and must be controlled with null proteins (i.e. by conducting parallel pull down experiments with the selection epitope, GST). Confirmation of results with both of these methods provides strong support for observed protein interactions, as will be discussed in greater detail below.

Regardless of which method is employed, proteins are then separated by one- or two-dimensional electrophoresis. Proteomic studies have classically relied heavily on the marriage of two-dimensional electrophoresis with mass spectrometry to identify proteins and post-translational modification. However, recent studies have illustrated the utility of standard large format one-dimensional SDS-PAGE for the identification of proteins of high molecular weight and of high and low isoelectric points [10], which have been traditionally problematic with a two-dimensional electrophoretic approach [34]. Furthermore, the advent of powerful and somewhat more affordable tandem mass spectrometric instruments and quantitative mass spectrometry approaches empowers the characterization of post-translational modification and protein expression at the level of the mass spectrometer [35–37]. An increasingly popular approach circumvents electrophoresis and directly subjects isolated protein complexes to mass spectrometric analysis [37, 38]. This method is particularly appealing in that it obviates concerns of protein resolution on a gel. We have found that all three of these methods (one- and two-dimensional electrophoresis followed by mass spectrometry, and direct subjection of isolated complexes to mass spectrometry following solution digest) yield novel data regarding the components of protein complexes [10]. As a result, signaling complex analysis is most thoroughly accomplished with an approach that combines these three methods.

Following separation, protein identification is then made by mass spectrometry. In the case of the two-dimensional gel, discernable gel features are excised and trypsinized in the traditional manner prior to mass spectrometric analysis. If a one-dimensional gel is used, protein spots from the entire continuum of the gel, as opposed to solely from densely stained regions of the gel, are excised, digested and analyzed. This is done because previous studies have indicated that the detection limit of the mass spectrometer exceeds that of the gel stain (e.g. coomassie, silver, or SYPRO) [10, 39], indicating that unstained regions of the gel may also contain members of the isolated subproteome, and therefore require analysis. This highlights another advantage of one- verses two-dimensional electrophoresis, in that many unstained regions of the two-dimensional gel presumably also contain proteins, but for reasons of practicality, are much more difficult to analyze then are unstained regions of a one-dimensional gel. Peptide mass fingerprinting is still a widely used technique to make protein identification [8, 36, 40], but one that is being increasingly replaced by more powerful tandem mass spectrometry approaches [2, 3, 10, 38]. These instruments allow for high resolution and mass accuracy and can be operated in data-dependent LC/MS/MS mode (such as a quadrupole TOF hybrid instrument, like the Q-Tof2), consequently increasing the confidence of database search results. In addition, tandem mass spectrometry instruments are far superior for the identification of post-translational modification.

Functional Characterization of the Subproteome. Numerous studies employing variations on the above described theme have been published in the past few years [2–4, 8, 10]. These include studies focused on a single molecule and its interacting proteins [4, 8, 10] and others that take a more large scale approach to examine multiple proteins and their associated complexes [2, 3]. Regardless of the

Different Organelles   **Distinct PKCε Modules**

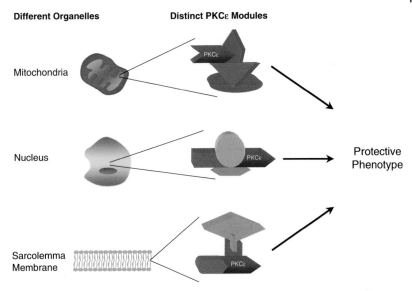

Fig. 17.2 Task specific modules as a means for signal transduction. In this model, distinct PKCε modules would exist at various subcellular compartments, containing different protein components, and all orchestrating different signaling tasks essential to the myocyte's resistance to ischemia.

approach, it is becoming increasingly clear that ancillary validation methods are required to not only confirm the members of a subproteome, but moreover, to understand the role that these proteins have in the genesis of a phenotype. In this key final step of the proteomic platform described herein, the role of the protein complex to engender a phenotype is discerned.

First, all proteins that are detected via mass spectrometry are confirmed as members of the subproteome of interest by co-immunoprecipitation followed by electrophoresis and western blotting. If antibodies to any of the identified proteins are not available, other methods of confirmation include recombinant protein-based interaction assays [8, 12, 13, 20]. The advent of antibody- and protein-based microarrays presents a powerful high-throughput method to screen for changes in protein expression and protein-protein interactions [41, 42]. This approach will likely allow for automation of these confirmatory experiments in the near future.

Next, one must discriminate proteins that form direct interactions with the protein of interest (i.e. binding partners) from those that associate indirectly with the target of complex isolation (Fig. 17.3). It is often overlooked in studies that employ immunoprecipitation and/or affinity pull down techniques that these methods can in fact isolate proteins that do not directly interact with the target of isolation. Nonetheless, these proteins may still be important members of the subproteome in question. Thus, after confirmation with immunoprecipitation and immunoblotting from cell lysates, the determination of whether direct protein-protein interactions are formed between each molecule in the complex and the pro-

Detection Via Mass Spectrometry
or Immunoblotting

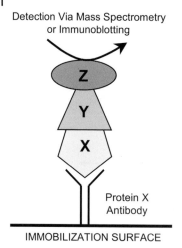

IMMOBILIZATION SURFACE

Protein X
Antibody

Fig. 17.3 Detection of direct and indirect protein-protein interactions. Whether the protein X complex is isolated via immunoprecipitation or affinity pull down, this complex may contain proteins that directly (protein Y) and indirectly (protein Z) associate with protein X. Thus, ancillary interaction studies are necessary to validate the nature of protein interactions within a multi-protein complex, as discussed in the text.

tein of interest is done in vitro either with recombinant proteins or by methods such as in vitro transcription/translation [8, 12, 13]. Commercial kits are available (e.g. Promega TnT) to expedite this process, and the interactions can be detected either by western blotting or through the use of radioactively-labeled amino acids/proteins. These processes are of paramount importance as recapitulation of protein interactions and detection with additional methods (i.e. other than mass spectrometry) greatly reduces the likelihood that the observed findings are artifactual.

The differences in complex formation that occur at distinct subcellular locations are studied using confocal microscopy and subcellular fractionation. As was addressed in the foregoing examples, the need to examine sub-cellular specific signal transduction events underlying a given phenotype cannot be over-emphasized. Often times, minute alterations in protein expression and/or activity at the level of a specific organelle are attenuated when the entire cellular component is analyzed [12, 13]. Thus, to insure a lucid characterization of protein function, organellar differences must be taken into consideration through subcellular fractionation (Fig. 17.2).

Lastly, the functional roles of the proteins within the complex are determined. This is the final step that links the proteomic data to a physiological phenotype. Accordingly, this step involves established biochemical and physiological assays to characterize the individual components of the subproteome [9]. This step includes the development of cell culture and/or transgenic mouse models to test the function of the protein within the context of the signaling system under examination. This is inherently the most time consuming step of the platform and is also the step that will vary the most depending on the system and the proteins being studied. Nonetheless, some basic tenets guide the determination of functional roles for the identified proteins: 1) assays for the activity of the protein are performed (e.g. kinase activity, phosphor-specific antibodies, and DNA binding) to determine the effects that association of the given molecule with the subproteome of interest has on its known activity; 2) inhibitors (e.g. pharmacological or dominant negative proteins) of the protein, if available, are used to perturb the function of the protein in vivo and to characterize how the subproteome in focus and the phenotype are affected; and 3) cell culture or transgenic animal models (e.g. knockout animals or other transgenic mutants) are created to understand the role of the protein in the given phenotype. This final step of the functional proteomic platform is essential in order to understand the relevance of the observed protein interactions within the context not only of the subproteome, but also with regard to how they affect the phenotype.

17.3
Metabolic and Transcription/Translation Functions of the PKCε Complex

Over the past four years, our laboratory has developed and implemented the above-described platform to characterize PKCε signaling complexes in the myocardium [8, 10]. Previous studies had shown that PKCε was a critical mediator of cardiac protection [16, 17], but the specific mechanism by which PKCε directed this protection remained unknown. We found that PKCε forms complexes with at least 93 proteins in the mouse heart, and that its association with these proteins is dynamically regulated by multiple factors [8, 10–13, 20]. We have so far identified at least six distinct classes of molecules in the PKCε subproteome, which are: signaling proteins, stress-activated proteins, structural proteins, metabolism-related proteins, transcription- and translation-related proteins, and PKCε binding domain containing proteins [8, 10]. These findings represent a conceptual framework within which many of the salubrious actions of PKCε and cardiac protection can be further investigated and understood. These findings are especially noteworthy in that the proteins identified in PKCε complexes are known to accomplish subcellular tasks necessary to prevent cell death during ischemia, thereby suggesting a role for PKCε in modulating these activities. We hypothesize that by forming distinct modules at different subcellular locations (Fig. 17.2), PKCε directs signal transduction to prevent cell death due to an ischemic insult [9–13]. In particular, the rest of the chapter will focus on the possible metabolic and protein synthesis regulatory actions of the PKCε complex.

Various stimuli that protect the heart from ischemia are known to act in part by improving cellular metabolism [43–46]. One recent example is a report by Jennings and colleagues, in which they found that the protected myocardium functions with a reduced energy demand [45]. This was measured as a decreased rate of lactate production and a slower depletion of the adenine nucleotide pool in the cardiac muscle. In another example, PKC was shown to be essential to restore high energy phosphate production following ischemia and to prevent glycolytic intermediate accumulation and lactate production [46]. However, the mechanism by which PKC contributes to this enhanced metabolic state during ischemia was unknown.

We found that PKCε associates with at least 15 proteins that are involved in the regulation of cellular metabolism [10]. These molecules include glycolytic enzymes such as glyceraldehyde phosphate dehydrogenase and enolase, and citric acid cycle enzymes such as succinate dehydrogenase and isocitrate dehydrogenase. These findings implicate PKCε in processes relating to the metabolism of fats, carbohydrates and proteins. In addition, we discovered a group of enzymes that are known to be critically involved in mitochondrial energy balance, including the adenine nucleotide transporter and the voltage-dependent anion channel. These proteins have subsequently been confirmed to be members of the PKCε subproteome by western blotting, and functional assays are underway to understand how association of these molecules with PKCε affects their activity. The ultimate goal of these functional studies is to fully detail the signaling blueprint through which PKCε mediates improved metabolism of the cardiomyocyte, and to understand how this contributes to enhanced cell survival during ischemia.

Cardiac protection regulated by PKCε has also been previously shown to require up-regulation of protective proteins [17, 47]. Some of these proteins include iNOS and Src tyrosine kinase, both of which are also members of the PKCε complex [8]. In addition, previous studies suggest that cardiac protection requires increased protein synthesis [47] and the activation of transcription factors such as AP-1, NFκB, and STAT1 [20, 48, 49]. Despite this, the complete visage of PKCε-dependent signaling to control protein synthesis remained undefined. We found that histones H1.3 and H4, proteins that participate in chromatin structure, and numerous ribosomal proteins (S3, S18, and S19), reside in PKCε complexes. The regulation of chromatin structure is a well characterized mechanism to control gene expression, and the association of PKCε with histones implies that one nuclear role for PKCε may be to influence the DNA-protein interactions in nucleosomes. The multiple ribosomal proteins that appeared in PKCε complexes support the concept that PKCε may also regulate protein expression through specific tasks at the ribosome. Notably, we also identified 10 heterogeneous nuclear ribonucleoproteins (hnRNPs), which have been shown to be involved in transcription, mRNA transport and regulation, and in the control of translation [50]. One study characterized the ability of PKC isoforms to interact with hnRNP K [51], however nothing is known regarding the functional consequences of an interaction between any of the hnRNPs and epsilon isoform of PKC. The findings of our studies suggest that PKCε may interact with sundry mRNA processing enzymes in order to specifically control protein expression and possibly splicing and thereby regulate the synthesis of protective proteins to resist cell death due to ischemia. Accordingly, studies are currently underway to understand the specific manner in which PKCε regulates the activity of these proteins to ultimately modulate the expression/post-translational modification of protective proteins in the heart. The long term goal of all of these studies is the functional validation of all PKCε complex members. In this manner, the proposed platform will allow for a truly annotated picture of how the PKCε subproteome is involved in protecting the myocardium during ischemic insult.

17.4
Conclusion

The concept of multi-protein complexes as a mechanism for signal transduction is gaining increased support from multiple scientific disciplines. Functional proteomic strategies, like the one described in this chapter, will enable researchers to further advance the characterization of these complexes in a thorough and non-biased manner. The platform described herein empowers the researcher to move the investigation "within the complex" and to understand the hierarchy of protein interactions that results in a phenotype. At the same time, this approach directly links signaling network information with organ phenotype. In this regard, the minute alterations that occur in protein complexes at different subcellular locations and in response to different stimuli can be determined directly correlated to the cellular response. Importantly, the platform is malleable enough to accommodate investigation of virtually any protein of interest. As a result, the approach described herein facilitates the investigation of how protein complexes regulate physiological phenotype.

17.5
References

1 N.P. ALLEN, L. HUANG, A. BURLINGAME, M. REXACH, Proteomic analysis of nucleoporin interacting proteins. *J. Biol. Chem.* **2001**, *276*, 29268–29274.

2 A.C. GAVIN, et al., Functional organization of the yeast proteome by systematic analysis of protein complexes. *Nature.* **2002**, *415*, 141–147.

3 Y. HO, et al., Systematic identification of protein complexes in *Saccharomyces cerevisiae* by mass spectrometry. *Nature.* **2002**, *415*, 180–183.

4 H. HUSI, M.A. WARD, J.S. CHOUDHARY, W.P. BLACKSTOCK, S.G. GRANT, Proteomic analysis of NMDA receptor-adhesion protein signaling complexes. *Nat. Neurosci.* **2000**, *3*, 661–669.

5 J.D. JORDAN, E.M. LANDAU, R. IYENGAR, Signaling networks: the origins of cellular multitasking. *Cell.* **2000**, *103*, 193–200.

6 Y. LIU, J.A. RANISH, R. AEBERSOLD, S. HAHN, Yeast nuclear extract contains two major forms of RNA polymerase II mediator complexes. *J. Biol. Chem.* **2001**, *276*, 7169–7175.

7 G. NEUBAUER, J. RAPPSILBER, C. CALVIO, M. WATSON, P. AJUH, J. SLEEMAN, A. LA-MOND, M. MANN, Mass spectrometry and EST-database searching allows characterization of the multi-protein spliceosome complex. *Nat. Genet.* **1998**, *20*, 46–50.

8 P. PING, J. ZHANG, W.M. PIERCE, JR., R. BOLLI, Functional proteomic analysis of protein kinas C ε signaling complexes in the normal heart and during cardioprotection. *Circ. Res.* **2001**, *88*, 59–62.

9 P. PING, C. SONG, J. ZHANG, Y. GUO, X. CAO, R.C. LI, W. WU, T.M. VONDRISKA, J.M. PASS, X.L. TANG, W.M. PIERCE, JR., R. BOLLI, Formation of protein kinase C ε-Lck modules confers cardioprotection. *J. Clin. Invest.* **2002**, *109*, 499–507.

10 R.D. EDMONDSON, T.M. VONDRISKA, K.J. BIEDERMAN, J. ZHANG, R.C. JONES, Y.T. ZHENG, D.L. ALLEN, J.X. XIU, E.M. CARDWELL, M.R. PISANO, P. PING, Protein kinase C epsilon signaling complexes include metabolism- and transcription/translation-related proteins: complimentary electrophoretic separation techniques with LC/MS/MS. *Mol. Cell. Proteomics.* **2002**, *1*, 421–433.

11 J. ZHANG, R. BOLLI, Y.T. ZHENG, J.X. XIU, G.W. WANG, W.M. PIERCE, JR., P.

Ping, Subcellular location is a major determinant of PKCε complex assembly: a view gained via functional proteomic analyses in cardioprotection. *Circulation.* **2001**, *II-209*, Abstract:1005.

12 C.P. Baines, J. Zhang, G.W. Wang, Y.T. Zheng, J.X. Xiu, E.M. Cardwell, R. Bolli, P. Ping, Mitochondrial PKCε and MAPK for signaling modules in the murine heart: enhanced mitochondrial PKCε-MAPK interactions and differential MAPK activation in PKCε-induced cardioprotection. *Circ. Res.* **2002**, *90*, 390–397.

13 T.M. Vondriska, C. Song, J. Zhang, X.L. Tang, X. Cao, C.P. Baines, J.M. Pass, S. Wang, R. Bolli, P. Ping, Protein kinase C ε-Src modules direct signal transduction in nitric oxide-induced cardioprotection: complex formation as a means for cardioprotective signaling. *Circ. Res.* **2001**, *88*, 1306–1313.

14 D.A. Berry, A. Keogh, C.G. dos Remedios, Nuclear membrane proteins in failing human dilated cardiomyopathy. *Proteomics.* **2001**, *1*, 1507–1512.

15 E. Braunwald, D.P. Zipes, P. Libby, Heart disease: a textbook of cardiovascular medicine. 6th Edition. W.B. Saunders Company, Philadelphia, PA, USA, **2001**.

16 C.P. Baines, J.M. Pass, P. Ping, Protein kinases and kinase-modulated effectors in the late phase of ischemic preconditioning. *Basic Res. Cardiol.* **2001**, *96*, 207–218

17 R. Bolli, The late phase of preconditioning. *Circ. Res.* **2000**, *87*, 972–983.

18 E.C. Dempsey, A.C. Newton, D. Mochly-Rosen, A.P. Fields, M.E. Reyland, P.A. Insel, R.O. Messing, Protein kinase C isozymes and the regulation of diverse cell responses. *Am. J. Physiol.* **2000**, *279*, L429–L438.

19 D.W. Dorn, 2nd, M.C. Souroujon, T. Liron, C.H. Chen, M.O. Gray, H.Z. Zhou, M. Csukai, G. Wu, J.N. Lorenz, D. Mochly-Rosen, Sustained in vivo cardiac protection by a rationally designed peptide that causes epsilon protein kinase C translocation. *Proc. Natl. Acad. Sci. USA.* **1999**, *96*, 12798–12803.

20 R.C. Li, P. Ping, J. Zhang, W.B. Wead, X. Cao, J. Gao, Y. Zheng, S. Huang, J. Han, R. Bolli, PKCε modulates NFκB and AP-1 via mitogen activated kinases in adult rabbit cardiomyocytes. *Am. J. Physiol.* **1999**, *279*, H1679–1689.

21 K. Ytrehus, Y. Liu, J.M. Downey, Preconditioning protects ischemic rabbit heart by protein kinase C activation. *Am. J. Physiol.* **1994**, *266*, H1145–H1152.

22 T.K. Seow, R. Korke, R.C. Liang, S.E. Ong, K. Ou, K. Wong, W.S. Hu, M.C. Chung, Proteomic investigation of metabolic shift in mammalian cell culture. *Biotechnol. Prog.* **2001**, *17*, 1137–1144.

23 T. Ideker, V. Thorsson, J.A. Ranish, R. Christmas, J. Buhler, J.K. Eng, R. Bumgarner, D.R. Goodlett, R. Aebersold, L. Hood, Integrated genomic and proteomic analyses of a systematically perturbed metabolic network. *Science.* **2001**, *292*, 929–934.

24 J.S. Andersen, C.E. Lyon, A.H. Fox, A.K. Leung, Y.W. Lam, H. Steen, M. Mann, A.I. Lamond, Directed proteomic analysis of the human nucleolus. *Curr. Biol.* **2002**, *12*, 1–11.

25 M. Dreger, L. Bengtsson, T. Schoneberg, H. Otto, F. Hucho, Nuclear envelope proteomics: novel integral membrane proteins of the inner nuclear membrane. *Proc. Natl. Acad. Sci. USA.* **2001**, *98*, 11943–11948.

26 A.W. Bell, M.A. Ward, W.P. Blackstock, H.N. Freeman, J.S. Choudhary, A.P. Lewis, D. Chotai, A. Fazel, J.N. Gushue, J. Paiement, S. Palcy, E. Chevet, M. Lafreniere-Roula, R. Solari, D.Y. Thomas, A. Rowley, J.J. Bergeron, Proteomics characterization of abundant Golgi membrane proteins. *J. Biol. Chem.* **2001**, *276*, 5152–5165.

27 A. Greco, W. Bienvenut, J.C. Sanchez, K. Kindbeiter, D. Hochstrasser, J.J. Madjar, J.J. Diaz, Identification of ribosome-associated viral and cellular basic proteins during the course of infection with herpes simplex virus type 1. *Proteomics.* **2001**, *1*, 545–549.

28 D.J. Maltman, W.J. Simon, C.H. Wheeler, M.J. Dunn, R. Wait, A.R. Slabas. Proteomic analysis of the endoplasmic reticulum from developing and germinating seed of castor (*Ricinus communis*). *Electrophoresis* **2002**, *23*, 626–639.

29 J. Predic, V. Soskic, D. Bradley, J. Godovac-Zimmermann, Monitoring of gene

expression by functional proteomics: response of human lung fibroblast cells to stimulation by endothelin-1. *Biochemistry.* **2002**, *41*, 1070–1078.

30 C. E. MURRY, R. B. JENNINGS, K. A. REIMER, Preconditioning with ischemia: a delay of lethal cell injury in ischemic myocardium. *Circulation.* **1986**, *74*, 1124–1136.

31 J. M. PASS, Y. ZHENG, W. B. WEAD, J. ZHANG, R. C. LI, R. BOLLI, P. PING, PKCε activation induces dichotomous cardiac phenotypes and modulates PKCε-RACK interactions and RACK expression. *Am. J. Physiol.* **2001**, *280*, H946–H955.

32 T. M. VONDRISKA, J. B. KLEIN, P. PING, Use of functional proteomics to investigate PKCε-mediated cardioprotection: the signaling module hypothesis. *Am. J. Physiol.* **2001**, *280*, H1434–H1441.

33 J. M. PASS, J. ZHANG, T. M. VONDRISKA, P. PING, Functional proteomic analysis of the PKC signaling system. In *Protein Kinase C Protocols* (Newton, A. C., ed.), Humana, Totowa, NJ, USA, **2002**.

34 S. M. HANASH, Biomedical applications of two-dimensional electrophoresis using immobilized pH gradients: current status. *Electrophoresis.* **2000**, *21*, 1202–1209.

35 S. P. GYGI, R. AEBERSOLD, Mass spectrometry and proteomics. *Curr. Opin. Chem. Biol.* **2000**, *4*, 489–494.

36 A. PANDEY, M. MANN, Proteomics to study genes and genomes. *Nature.* **2000**, *405*, 837–846.

37 J. PENG, S. P. GYGI, Proteomics: the move to mixtures. *J. Mass Spectrom.* **2001**, *36*, 1083–1091.

38 A. J. LINK, J. ENG, D. M. SCHIELTZ, E. CARMACK, G. J. MIZE, D. R. MORRIS, B. M. GARVIK, J. R. YATES, 3rd, Direct analysis of protein complexes using mass spectrometry. *Nat. Biotechnol.* **1999**, *17*, 676–682.

39 M. F. LOPEZ, K. BERGGREN, E. CHERNOKALSKAYA, A. LAZAREV, M. ROBINSON, W. F. PATTON, A comparison of silver stain and SYPRO Ruby Protein Gel Stain with respect to protein detection in two-dimensional gels and identification by peptide mass profiling. *Electrophoresis.* **2000**, *21*, 3673–3683.

40 T. S. LEWIS, J. B. HUNT, L. D. AVELINE, K. R. JONSCHER, D. F. LOUIE, J. M. YEH, T. S. NAHREINI, K. A. RESING, N. G. AHN, Identification of novel MAP kinase pathway signaling targets by functional proteomics and mass spectrometry. *Mol. Cell.* **2000**, *6*, 1343–1354.

41 J. MADOZ-GURPIDE, H. WANG, D. E. MISEK, F. BRICHORY, S. M. HANASH, Protein based microarrays: a tool for probing the proteome of cancer cells and tissues. *Proteomics.* **2001**, *1*, 1279–1287.

42 C. P. PAWELETZ, L. CHARBONEAU, V. E. BICHSEL, N. L. SIMONE, T. CHEN, J. W. GILLESPIE, M. R. EMMERT-BUCK, M. J. ROTH, E. F. PETRICOIN, III., L. A. LIOTTA, Reverse phase protein microarrays which capture disease progression show activation of pro-survival pathways at the cancer invasion front. *Oncogene.* **2001**, *20*, 1981–1989.

43 S. C. ARMSTRONG, C. E. GANOTE, Preconditioning of isolated rabbits cardiomyoctes: effects of glycolytic blockade, phorbol esters, and ischemia. *Cardiovasc. Res.* **1994**, *28*, 1700–1706.

44 H. R. CROSS, E. MURPHY, R. BOLLI, P. PING, C. STEENBERGEN, Expression of activated PKCε protects the ischemic heart, without attenuating ischemic H$^+$ production. *J. Mol. Cell. Cardiol.* **2002**, *34*, 361–367.

45 R. B. JENNINGS, L. SEBBAG, L. M. SCHWARTZ, M. S. CRAGO, K. A. REIMER, Metabolism of preconditioned myocardium: effect of loss and reinstatement of cardioprotection. *J. Mol. Cell. Cardiol.* **2001**, *33*, 1571–1588.

46 K. YABE, K. TANONAKA, M. KOSHIMIZU, T. KATSUNO, S. TAKEO, A role for PKC in the improvement of energy metabolism in preconditioned heart. *Basic Res. Cardiol.* **2000**, *95*, 215–227.

47 A. RIZVI, X. L. TANG, Y. QIU, Y. T. XUAN, H. TAKANO, A. K. JADOON, R. BOLLI, Increased protein synthesis is necessary for the development of late preconditioning against myocardial stunning. *Am. J. Physiol.* **1999**, *277*, H874–H884.

48 H. SASAKI, N. GALANG, N. MAULIK, Redox regulations of NF-κB and AP-1 in ischemic reperfused heart. *Antioxid. Redox. Signal.* **1999**, *1*, 317–324.

49 Y. T. XUAN, X. L. TANG, S. BANERJEE, H. TAKANO, R. C. LI, H. HAN, Y. QIU,

R. Bolli, Nuclear factor-κB plays an essential role in the late phase of ischemic preconditioning in conscious rabbits. *Circ. Res.* **1999**, *84*, 1095–1109.

50 A. M. Krecic, M. S. Swanson, hnRNP complexes: composition, structure, and function. *Curr. Opin. Cell Biol.* **1999**, *11*, 363–371.

51 D. S. Schullery, J. Ostrowski, O. N. Denisenko, L. Stempka, M. Shnyreva, H. Suzuki, M. Gschwendt, K. Bomsztyk, Regulated interaction of protein kinase C δ with the heterogenous nuclear ribonucleoprotein K protein. *J. Biol. Chem.* **1999**, *274*, 15101–15109.

18
Identification of Secreted Oxidative Stress-induced Factors (SOXF) and Associated Proteins: Proteomics in Vascular Biology

Zheng-Gen Jin, Duan-Fang Liao, Chen Yan, and Bradford C. Berk

18.1
Introduction

While reactive oxygen species (ROS), such as H_2O_2, O_2^-, and OH^-, are potentially toxic byproducts of mitochondrial respiration and other oxidases, ROS also have important physiological functions in the vasculature including regulating cell proliferation and vascular tone [1–4]. Within the vessel wall, ROS are generated by several mechanisms, including a vascular NAD(P)H oxidase [5, 6]. ROS formation can be stimulated by mechanical stress, environmental factors, platelet-derived growth factor (PDGF), angiotensin II (Ang II), and low-density lipoproteins [7-9]. Because many risk factors for coronary artery disease such as hyperlipidemia, hypertension, diabetes, and smoking increase production of ROS, it has been suggested that changes in vessel redox state are a common pathway involved in the pathogenesis of atherosclerosis [1, 6, 10]. We have previously reported that ROS stimulate vascular smooth muscle cell (VSMC) growth and DNA synthesis [11]. This proliferation was associated with stimulation of protein kinases, especially the extracellular signal-regulated kinases (ERK1/2) [4].

Activation of ERK1/2 by O_2^- generators such as the napthoquinolinedione LY83583, menadione, and xanthine/xanthine oxidase was biphasic: an early peak of ERK1/2 activity was present at 5 to 10 min, whereas a delayed ERK1/2 activation appeared at 2 hour [12]. A similar biphasic activation of ERK1/2 has been reported for mitogens such as fibroblast growth factor [13]. Recently, the delayed ERK1/2 activation has been reported to be mediated by different mechanisms than the early ERK1/2 activation and to be critical for cell cycle progression and cell proliferation [13, 14].

One logical mechanism for endothelial cells and VSMC to respond to ROS would be to produce autocrine/paracrine signals that enhance cell survival or stimulate pathways that protect cells from the damaging effects of ROS. VSMC are particularly likely to secrete protective factors that also promote cell survival based on previous studies that demonstrate secretion of a number of growth factors from VSMC in response to various stimuli. These growth factors include adrenomedullin, endothelin, epiregulin, FGF, Gas6, PDGF, and TGF-β [15–20]. For example, Gas6 is secreted from rat VSMC after stimulation by Ang II and thrombin, and exhibits growth factor activity [21, 22]. Epiregulin, an epidermal growth factor (EGF)-related growth factor, is a potent VSMC-secreted mitogen whose expression is regulated by Ang II, endothelin-1 and thrombin [23]. However, no factors have been identified as mediators of VSMC proliferation in response to ROS. In this paper we describe purification of factors secreted from VSMC exposed to oxidative stress that stimulate ERK1/2. These activities, termed SOXF for secreted oxidative stress induced factors are potentially important physiologic mediators of the vessel wall response to ROS.

18.2
Identification of SOXF by Tandem MS

18.2.1
Secreted Factors Are Involved in Regulation of ERK1/2 Activation by Oxidative Stress

To generate oxidative stress VSMC were exposed to LY83583, which generates O_2^- [4]. Production of O_2^- in VSMC to exposed to 1 μmol/L LY83583 was measured by lucigenin chemiluminescence as described previously [4]. LY83583-induced generation of O_2^- peaked at 15 min and returned to baseline by 120 min. Tiron (10 μmol/L), a membrane-permeant nonenzymatic O_2^- scavenger, completely abolished LY83583-induced O_2^-. Exposure of VSMC to LY83583 stimulated ERK1/2 activity with an initial peak at 10 min that paralleled O_2^- production and a second peak at 120 min, a time when O_2^- production was minimal. ERK1/2 activity at 120 min was greater than activity at 10 min, and was nearly equivalent to that observed with PDGF or with xanthine plus xanthine oxidase.

18.2.2
A Trypsin-sensitive Secreted Oxidative Stress-induced Factor (SOXF) Is Released in Response to LY83583

To identify the presence of SOXF from VSMC stimulated by LY83583, the ability of conditioned medium to stimulate ERK1/2 was assayed. Conditioned medium was prepared and then transferred to growth-arrested VSMC for 10 min. ERK1/2 activity, measured with a phospho-ERK1/2 specific antibody, demonstrated a significant increase with conditioned medium from cells treated with LY83583, but not control (Fig. 18.1). The activity was sensitive to trypsin (5 μg/ml for 30 min), and was inhibited by heating to 100 °C. To determine whether a few proteins comprised the majority of SOXF, we analyzed the proteins released into the medium in response to LY83583. Cells were labeled with [35S]methionine for 4 h and then exposed to 1 μmol/L LY83583 for 120 min. Conditioned medium was harvested, concentrated and proteins analyzed by 5–15% SDS-PAGE. Approximately 35 protein bands were detected by autoradiography suggesting that multiple proteins were released in response to LY83583.

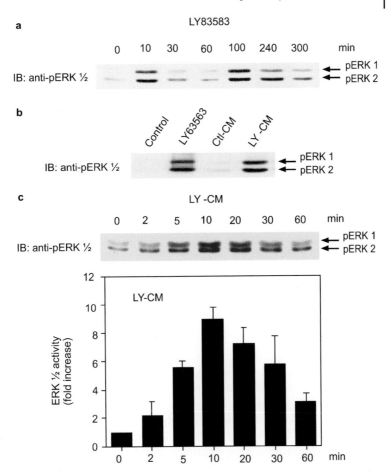

Fig. 18.1 Secreted factors are involved in regulation of ERK1/2 activation by oxidative stress. Growth-arrested vascular smooth muscle cells (VSMC) were exposed to 1 μM LY83583 for the indicated times (**a**), or treated with conditioned medium from LY83583-stimulated VSMC (LY-CM), control medium from HBSS-incubated cells (Ctl-CM) for 10 min, or 1 μM LY83583 for 2 h (**b**), or with concentrated LY-CM for the indicated times (**c**). Cell lysates were prepared and analyzed for ERK1/2 activity or total ERK1/2 proteins by Western blot using phospho-specific ERK1/2 (pERK1/2) antibody, or ERK1 plus ERK2 antibody. The results were quantified by densitometry of autoradiograms using NIH Image 1.49. Results were normalized to control (time=0 min) which was arbitrarily set to 1.0 (**c**). Results are representative or mean ±SD of 3 experiments.

18.2.3
Purification of SOXF and Identification of SOXF Candidate Proteins by Mass Spectrometry

To purify SOXF we used a sequential chromatographic approach that involved SP-Sepharose, heparin-Sepharose, phenyl Sepharose, and S-200 gel filtration chroma-

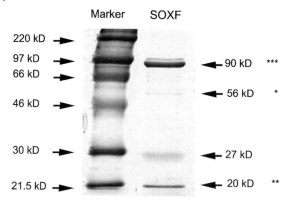

Marker   SOXF

220 kD →
97 kD →   ←— 90 kD   ***
66 kD →
                 ←— 56 kD   *
46 kD →

30 kD →   ←— 27 kD

21.5 kD →   ←— 20 kD   **

* Relative Stimulating Activity of ERK ½ on Gel Filtration

Fig. 18.2 Coomassie stain of the 0.15 M ammonium sulfate fraction from phenyl-Sepharose (SOXF). Fraction III from phenyl Sepharose chromatography containing ~25 µg protein was subjected to 12% SDS-PAGE. Proteins were stained with 0.1% Coomassie Brilliant Blue 250 in 40% methanol and 1% acetic acid, destained in 50% methanol. Approximate molecular weights were determined by logarithmic plot of migration of molecular weight markers. SOXF* indicates putative SOXF containing fraction from phenyl-Sepharose column.

tography [24]. For each step, activity was assayed for ERK1/2 stimulation by phospho-ERK1/2 Western blot using growth arrested VSMC. On S-200 chromatography, SOXF activity eluted in three peaks at approximate molecular weights (in order of relative ERK1/2 stimulating activity) of 80–100 kD > 20–30 kD > 45–65 kD. Proteins were identified by SDS-PAGE and staining with Coomassie Blue (Fig. 18.2). The proteins were digested by trypsin and the resulting peptides were analyzed by an electrospray triple quadrupole mass spectrometer (Finnigan-MAT TSQ 7000). The collision induced dissociation spectra generated were used to identify the proteins from which the peptide originated by database searching using the Sequest software program [25, 26]. Unambiguous identification of proteins in the 90 kD band revealed heat shock protein 90-α (HSP90-α). Unambiguous identification of proteins in the 20 kD band revealed cyclophilin B and cyclophilin A. No proteins were identified in the 56 kD and 27 kD bands. Other proteins identified included ezrin and moesin.

18.2.4
LY83583 Stimulates Release of HSP90-α Specifically from VSMC

To prove that proteins identified as putative SOXFs were released specifically from LY83583 treated VSMC, a conditioned medium experiment was analyzed. To prove specificity we compared the relative protein abundance of candidate SOXF proteins in the total cell lysate and in conditioned medium, before and after LY83583 stimulation. Because HSP90-α was the largest SOXF identified, and therefore least likely to be released non-specifically, we studied its abundance. Conditioned media from control and LY83583 treated cells were concentrated. SDS-PAGE followed by Western blot analysis for HSP90 (with an antibody that recognizes both HSP90-α and HSP90-β). The abundance of HSP90 was compared to two intracellular proteins of similar molecular weights; PKC-ξ (with an antibody which also recognizes PKC-α), and c-Raf-1. After treatment with LY83583 for 2 hr, there was a 10-fold increase in HSP90-α present in the conditioned medium, but no detectable HSP90-β, PKC or c-Raf-1. These results suggest that regulated secretion of HSP90-α occurred in response to LY83583.

18.2.5
Human Recombinant HSP90-α and Cyclophilin A (CyPA) Activate ERK1/2

To provide further evidence that HSP90-α and CyPA are SOXF, human recombinant HSP90-α (hrHSP90-α) and hrCyPA were studied. The preparations used for these studies were highly purified as shown by silver stain analysis, which revealed that >95% of total protein migrated at appropriate molecular weights of 90-kDa, and 18 kDa, respectively. Both CyPA and hrHSP90-α stimulated ERK1/2 activity in VSMC in a concentration-dependent manner [24, 27].

18.2.6
Human Recombinant Cyclophilin A Stimulates VSMC Growth
and Protects VSMC against Apoptosis

To determine the physiological significance of SOXF, we studied the effects of hrCyPA on VSMC DNA synthesis. 10 nmol/L hrCyPA significantly stimulated DNA synthesis in VSMC (2-fold increase versus 0.1% serum) assayed by [^{3}H]thymidine incorporation [27]. Thus, CyPA has growth promoting effects on VSMC which may contributes significantly to the growth promoting activity of ROS in VSMC. To further determine whether CyPA prevents VSMC apoptosis, we used sodium nitroprusside (SNP), which was shown to induce VSMC apoptosis [28, 29]. Incubating VSMC with 1 mmol/L SNP for 24 hours decreased cell viability to 19.4% of control, measured with a modified MTT assay. Addition of 10 nM hrCyPA in the presence of 1 mmol/L SNP blocked apoptosis, with cell viability returning to 47% of control [27]. In response to 0.5 mmol/L SNP for 24 hours, 10% of VSMCs were apoptotic as measured by nuclear morphology after DAPI staining, consistent with previous reports [28, 29]. Addition of hrCyPA significantly inhibited apoptosis induced by 0.5 mmol/L SNP with a decrease of 55%. These results indicated that SOXF have significant physiological effects on VSMC.

18.3
Identification of SOXF-associated Proteins by MALDI-TOF MS

18.3.1
Identification of SOXF-associated Proteins by MALDI-TOF MS

To understand the mechanisms responsible for secretion of CyPA and activation of VSMC by SOXF, we studied SOXF-associated proteins by MALDI-TOF MS (Matrix-assisted Laser Desorption/Ionization Time-Of-Flight Mass Spectrometry). In brief, VSMC cell lysates were incubated with recombinant GST-CyPA wild type (WT), GST-CyPA R55A (isomerase inactive mutant), or GST (negative control) in binding buffer [30]. The putative CyPA-binding proteins were pulled down by glutathione agarose beads, and separated on a SDS-PAGE gel. The proteins on the gel were visualized by silver staining. The stained protein band that corresponded to the position of putative CyPA-binding proteins was excised and subjected to tryptic hydrolysis. Tryptic peptides were spotted with α-cyano-4-hydroxycinnamic acid as matrix and analyzed using a MALDI-TOF mass spectrometer (PE Biosystems VOYAGER System 4187) [31]. Search of MS Fit with the resulting peptide masses in the NCB database, yielded several protein candidates. One group of proteins identified were cytoskeleton proteins, including β-spectrin and α-actin. Another group included proteins related to redox state such as AOP-2 (antioxidant protein 2, also termed 1–cysteine peroxiredoxin). AOP-2 belongs to a large family of bacterial and eukaryotic enzymes thought to be part of the cellular defense against oxidative stress [32, 33]. Several membrane proteins were also identified. The analysis appeared to be technically sound since proteins homologous to AOP-2 such as antioxidant protein 1 (AOP-1) and peroxiredoxin VI (Prx VI) were previously identified as CyPA binding proteins [34, 35]. Using the yeast two hybrid system Jaschke et al. found that human CyPA binds to AOP-1 and stimulates its enzymatic activity [34]. Lee at al. identified a 20-kD binding protein for Prx VI as CyPA, and demonstrated that CyPA enhances Prx VI peroxidase activity as an immediate electron donor [35].

18.3.2
Confirmation of 43-kD CyPA-binding Protein as a α-Actin

The SOXF-associated proteins identified by SDS-PAGE and MALDI-TOF MS were verified by GST-CyPA affinity pull-down assay and by immunoprecipitation with anti-CyPA antibodies followed by immunoblotting with antibodies against the co-precipitating proteins. As shown in Fig. 18.3, α-actin was detected in the pellets of GST-CyPA but not GST (negative control) precipitated from VSMC lysates. Of interest, α-actin did not bind to the GST-CyPA R55A mutant, which lacks isomerase activity (Fig. 18.3, lane 3), suggesting that isomerase activity is required for the interaction of CyPA with α-actin. CyPA binding to α-actin was further confirmed by immunoprecipitation from VSMC lysates. These results demonstrate that CyPA interacts with actin in VSMC. Since cytoskeletal proteins such as α-actin and β-

Fig. 18.3 Specific interactions of CyPA with actin. GST-CyPA proteins were prepared and used to co-precipitate CyPA binding proteins from VSMC lysates. Shown is an experiment in which VSMC lysates were incubated for 16 hours with GST-CyPA constructs, the beads were subjected to SDS-PAGE, and actin was detected by western blot. Wild type CyPA (W) was mutated to (R55A = R) to inhibit isomerase activity. G = GST.

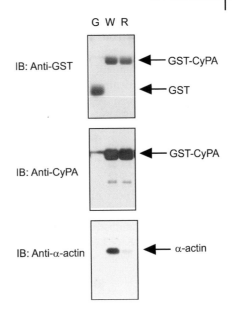

spectrin are thought to be involved in protein trafficking in the early secretory pathway [36], we speculate that these interactions might be a component of the mechanisms by which CyPA is secreted in response to oxidative stress. Further studies are needed to clarify the functional importance of these interactions for CyPA-mediated cellular effects.

18.4
Conclusions

Several general concepts arise from our study of SOXF in vascular smooth muscle cells. First, our approach was facilitated by analyzing a sub-proteome. We have studied two subproteomes to date. In our initial studies we analyzed conditioned medium from cells exposed to oxidative stress. Subsequently we studied proteins that co-precipitated with cyclophilin A from intracellular sources. By dramatically reducing the number of potential proteins that could interact with these proteins we simplified the proteomics analysis. Second, we designed our experiments to contain several controls that facilitated identification of important proteins. For example, we compared medium from cells exposed to oxidative stress to medium from untreated cells. For cyclophilin A we had access to proteins that lacked critical domains required for protein–protein interactions and enzyme activity. These biological controls significantly improved our ability to identify biologically important interacting proteins. Third, it is possible to use cross-linking reagents to increase the stability of protein-protein interactions which can increase the likelihood of finding transient interactions that are biologically important. Fourth we found that using early passage cultured cells yielded information that was relevant to the in vivo situation. These results indicate the utility of cultured cells and the importance of correlating in vitro results with in vivo experiments. Finally, and perhaps most important, a proteomics approach was excellent for discovery based research as we identified biologically important molecules that could not have been predicted based on current literature and knowledge.

18.5
Acknowledgements

We thank Drs. Ruedi Abersold and Steve Gygi for their invaluable expertise in mass spectrometry. We also thank Drs. Andreas Pahl and Holger Bang for their generous gift of antibody CyPA mAb7F1. We thank Drs. J. Abe and A. Baas, and members of the Berk laboratory for their valuable assistance. This work was supported by grants from National Institutes of Health (HL44721 and HL49192 to B.C. Berk).

18.6
References

1 ALEXANDER RW. Theodore Cooper Memorial Lecture. Hypertension and the pathogenesis of atherosclerosis. Oxidative stress and the mediation of arterial inflammatory response: a new perspective. *Hypertension.* 1995; *25*:155–161.

2 ABE J, BERK BC. Reactive Oxygen species as mediators of signal transduction in cardiovascular disease. *Trends Cardiovasc. Med.* 1998; 8:59–64.

3 OMAR HA, CHERRY PD, MORTELLITI MP, BURKE-WOLIN T, WOLIN MS. Inhibition of coronary artery superoxide dismutase attenuates endothelium-dependent and independent nitrovasodilator relaxation. *Circ. Res.* 1991; 69:601–608.

4 BAAS AS, BERK BC. Differential activation of mitogen-activated protein kinases by H_2O_2 and O_2^- in vascular smooth muscle cells. *Circ Res.* 1995; 77:29–36.

5 MOHAZZAB KM, KAMINSKI PM, WOLIN MS. NADH oxidoreductase is a major source of superoxide anion in bovine coronary artery endothelium. *Am. J. Physiol.* 1994;266:H2568–2572.

6 RAJAGOPALAN S, KURZ S, MUNZEL T, TARPEY M, FREEMAN BA, GRIENDLING KK, HARRISON DG. Angiotensin II-mediated hypertension in the rat increases vascular superoxide production via membrane NADH/NADPH oxidase activation. Contribution to alterations of vasomotor tone. *J. Clin. Invest.* 1996; *97*:1916–1923.

7 WUNG BS, CHENG JJ, HSIEH HJ, SHYY YJ, WANG DL. Cyclic strain-induced monocyte chemotactic protein-1 gene expression in endothelial cells involves reactive oxygen species activation of activator protein 1. *Circ. Res.* 1997; *81*:1–7.

8 SUNDARESAN M, YU ZX, FERRANS VJ, IRANI K, FINKEL T. Requirement for generation of H_2O_2 for platelet-derived growth factor signal transduction. *Science.* 1995; *270*:296–299.

9 GRIENDLING KK, MINIERI CA, OLLERENSHAW JD, ALEXANDER RW. Angiotensin II stimulates NADH and NADPH oxidase activation in cultured vascular smooth muscle cells. *Circ. Res.* 1994;7 4:1141–1148.

10 HALLIWELL B. Free radicals, reactive oxygen species and human disease: a critical evaluation with special reference to atherosclerosis. *Br. J. Exp. Pathol.* 1989; *70*:737–757.

11 RAO GN, BERK BC. Active oxygen species stimulate vascular smooth muscle cell growth and proto-oncogene expression. *Circ. Res.* 1992; *70*:593–599.

12 LIAO D-F, BAAS AS, DAUM G, BERK BC. Purification of a secreted protein factor induced by reactive oxygen species (ROS) in vascular smooth muscle cells (VSMC). *Circulation.* 1997; *96*:I-901.

13 MELOCHE S, SEUWEN K, PAGES G, POUYS-SÉGUR J. Biphasic and synergistic activation of p44^{mapk} (ERK1) by growth factors: correlation between late phase activation and mitogenicity. *Mol. Endocrinol.* 1992; *6*:845–854.

14 YORK RD, YAO H, DILLON T, ELLIG CL, ECKERT SP, McCLESKEY EW, STORK PJ. Rap1 mediates sustained MAP kinase activation induced by nerve growth factor. *Nature.* 1998; *392*:622–626.

15 LIU G, ESPINOSA E, OEMAR BS, LUSCHER TF. Bimodal effects of angiotensin II on migration of human and rat smooth muscle cells. Direct stimulation and indirect inhibition via transforming growth factor-beta 1. *Arterioscler. Thromb. Vasc. Biol.* 1997; *17*:1251–1257.

16 SUGO S, MINAMINO N, SHOJI H, KANGA-WA K, KITAMURA K, ETO T, MATSUO H. Production and secretion of adrenomedullin from vascular smooth muscle cells: augmented production by tumor necrosis factor-alpha. *Biochem. Biophys. Res. Commun.* 1994; *203*:719–726.

17 ITOH H, MUKOYAMA M, PRATT RE, GIBBONS GH, DZAU VJ. Multiple autocrine growth factors modulate vascular smooth muscle cell growth response to angiotensin II. *J. Clin. Invest.* 1993; *91*:2268–2274.

18 BATTEGAY EJ, RAINES EW, SEIFERT RA, BOWEN-POPE DF, ROSS R. TGF-beta induces bimodal proliferation of connective tissue cells via complex control of an autocrine PDGF loop. *Cell.* 1990; *63*:515–524.

19 NAKAMURA Y, MORISHITA R, HIGAKI J, KIDA I, AOKI M, MORIGUCHI A, YAMADA K, HAYASHI S, YO Y, MATSUMOTO K, et al. Expression of local hepatocyte growth factor system in vascular tissues. *Biochem. Biophys. Res. Commun.* 1995; *215*:483–488.

20 MOLLOY CJ, PAWLOWSKI JE, TAYLOR DS, TURNER CE, WEBER H, PELUSO M. Thrombin receptor activation elicits rapid protein tyrosine phosphorylation and stimulation of the raf-1/MAP kinase pathway preceding delayed mitogenesis in cultured rat aortic smooth muscle cells: evidence for an obligate autocrine mechanism promoting cell proliferation induced by G-protein-coupled receptor agonist. *J. Clin. Invest.* 1996; *97*:1173–1183.

21 NAKANO T, HIGASHINO K, KIKUCHI N, KISHINO J, NOMURA K, FUJITA H, OHARA O, ARITA H. Vascular smooth muscle cell-derived, Gla-containing growth-potentiating factor for Ca^{2+}-mobilizing growth factors. *J. Biol. Chem.* 1995; *270*:5702–5705.

22 MELARAGNO MG, WUTHRICH DA, POPPA V, GILL D, LINDNER V, BERK BC, CORSON MA. Increased expression of Axl tyrosine kinase after vascular injury and regulation by G protein-coupled receptor agonists in rats. *Circ. Res.* 1998; *83*:697–704.

23 TAYLOR DS, CHENG X, PAWLOWSKI JE, WALLACE AR, FERRER P, MOLLOY CJ. Epiregulin is a potent vascular smooth muscle cell-derived mitogen induced by angiotensin II, endothelin-1, and thrombin. *Proc. Natl. Acad. Sci. USA* 1999; *96*:1633–1638.

24 LIAO D-F, JIN Z-G, BAAS AS, DAUM G, GYGI SP, AEBERSOLD R, BERK BC. Purification and identification of secreted oxidative stress-induced factors from vascular smooth muscle cells. *J. Biol. Chem.* 2000; *275*:189–196.

25 YATES III JR, ENG JK, McCORMACK AL, SCHIELTZ D. Method to correlate tandem mass spectra of modified peptides to amino acid sequences in the protein database. *Anal. Chem.* 1995; *67*:1426–1436.

26 FIGEYS D, DUCRET A, YATES JR, AEBERSOLD R. Protein identification by solid phase microextraction-capillary zone electrophoresis-microspray-tandem mass spectrometry. *Nature. Biotechnol.* 1996; *14*:1579–1583.

27 JIN ZG, MELARAGNO MG, LIAO DF, YAN C, HAENDELER J, SUH YA, LAMBETH JD, BERK BC. Cyclophilin A is a secreted

growth factor induced by oxidative stress. *Circ. Res.* **2000**; *87*:789–796.

28 POLLMAN MJ, YAMADA T, HORIUCHI M, GIBBONS GH. Vasoactive substances regulate vascular smooth muscle cell apoptosis. Countervailing influences of nitric oxide and angiotensin II. *Circ. Res.* **1996**; *79*:748–756.

29 ZHAO Z, FRANCIS CE, WELCH G, LOSCALZO J, RAVID K. Reduced glutathione prevents nitric oxide-induced apoptosis in vascular smooth muscle cells. *Biochim. Biophys. Acta.* **1997**; *1359*:143–152.

30 LEHOUX S, ABE J, FLORIAN JA, BERK BC. 14-3-3 binding to Na$^+$/H+ exchanger isoform-1 is associated with serum-dependent activation of Na$^+$/H^{++} exchange. *J. Biol. Chem.* **2001**; *276*:15794–15800.

31 TEN HAGEN KG, BEDI GS, TETAERT D, KINGSLEY PD, HAGEN FK, BALYS MM, BERES TM, DEGAND P, TABAK LA. Cloning and characterization of a ninth member of the UDP-GalNAc:polypeptide N-acetylgalactosaminyltransferase family, ppGaNTase-T9. *J. Biol. Chem.* **2001**; *276*:17395–17404.

32 RHEE SG, KIM KH, CHAE HZ, YIM MB, UCHIDA K, NETTO LE, STADTMAN ER. Antioxidant defense mechanisms: a new thiol-specific antioxidant enzyme. *Ann. NY Acad. Sci.* **1994**; *738*:86–92.

33 CHAE HZ, UHM TB, RHEE SG. Dimerization of thiol-specific antioxidant and the essential role of cysteine 47. *Proc. Natl. Acad. Sci. USA.* **1994**; *91*:7022–7026.

34 JASCHKE A, MI H, TROPSCHUG M. Human T cell cyclophilin18 binds to thiol-specific antioxidant protein Aop1 and stimulates its activity. *J. Mol. Biol.* **1998**; *277*:763–769.

35 LEE SP, HWANG YS, KIM YJ, KWON KS, KIM HJ, KIM K, CHAE HZ. Cyclophilin a binds to peroxiredoxins and activates its peroxidase activity. *J. Biol. Chem.* **2001**; *276*:29826–29832.

36 PREKERIS R, MAYHEW MW, COOPER JB, TERRIAN DM. Identification and localization of an actin-binding motif that is unique to the epsilon isoform of protein kinase C and participates in the regulation of synaptic function. *J. Cell. Biol.* **1996**; *132*:77–90.

19

Myofilament Proteomics: Understanding Contractile Dysfunction in Cardiorespiratory Disease

J. A. Simpson, S. Iscoe, and J. E. Van Eyk

The function of an organ, including muscle, is impaired when its supply of energy (oxygen) is diminished. The best-known and best-studied examples include cerebrovascular accidents (stroke), myocardial infarction (heart attack), hemorrhage, sepsis, acute respiratory distress syndrome, and multi-organ system failure. While there is an understandable tendency to study these and other conditions in isolation in order to minimize confounding variables, diseases and/or injury in one organ system can and do affect others. Aliverti and Macklem [1], for example, have convincingly argued that limited oxygen delivery to limb (skeletal) muscles is *the* common element underlying their impaired contraction in health, chronic obstructive pulmonary disease (COPD), and heart failure (HF).

The finely tuned interplay of the cardiovascular and respiratory systems is critical for ensuring adequate oxygen delivery to individual organs. Inadequate oxygen delivery can result from low perfusion (ischemia), low arterial oxygen levels (hypoxemia), or some combination. The former is typical of reduced cardiac output, the latter of reduced oxygen uptake by the lungs. A reduction in cardiac output can be caused by impaired contraction due to myocardial stunning infarction or a genetic disorder. A reduced oxygen uptake is, in turn, the result of either inadequate alveolar ventilation due to reduced convection (impaired contractile performance of the respiratory muscles or high resistance and/or low compliance of the lung) or impaired diffusion. Whatever the cause, low perfusion or hypoxemia will result in further reductions in oxygen delivery, leading to a vicious circle in which the contractile performance of muscles, whether cardiac or skeletal (including respiratory), worsens, accelerating the patient's deterioration.

HF, the inability of the heart to maintain an adequate cardiac output, is a chronic progressive disease. It is often characterized by the remodeling of the left ventricle (hypertrophy) during the initial compensatory stage which can become detrimental and eventually end in failure and death. Breathlessness is a common symptom experienced by patients with HF due, in part, to changes in the position and function of the diaphragm as well as changes in the respiratory muscles [2–4]. In fact, the extent of reduction in inspiratory muscle strength with heart failure (HF) is an independent predictor of prognosis [5]. Skeletal muscle is affected in HF, leading to exercise intolerance, although the proximate cause remains controversial. Alterations to skeletal muscle (i.e., its proteome or the protein compliment of a biological sample at a given time) may be a cause of this exercise intolerance (for review, see [6]), but extensive analysis is required to determine the underlying mechanism for skeletal muscle dysfunction.

COPD, a term commonly applied to patients with chronic bronchitis, emphysema or some mixture of the two, is an insidiously progressive airway disease characterized by a gradual loss of pulmonary function that is not fully reversible [7]. In COPD, there can be a large strain on the right ventricle as a result of the increase in pulmonary vascular resistance. As the disease progresses, problems begin to develop because of right ventricular hypertrophy; the heart's rhythm is disrupted by abnormal beats and the interventricular septum is "pushed" leftward, thereby decreasing left ventricular volume and compliance. As a result, patients tend to tire more easily and additional strain (on either the lungs or heart) can result in an inadequate left ventricular output, leading to liver and kidney problems and edema [7]. Many COPD patients succumb to right ventricular failure as a result of pulmonary hypertension. All these problems are compounded by a decrease in skeletal, including respiratory, muscle function, especially in advanced COPD (for reviews, see [1, 8]).

The analysis of skeletal muscle injury is confounded, compared to that of cardiac muscle, by the presence of multiple fiber types. Rather than the injury being distributed among all fibers, it can be specific to a particular fiber type (slow oxidative, type I fibers or fast glycolytic, type II fibers). For example, one would not expect equal damage to fast and slow fibers of the quadriceps during a marathon. In addition, injury could occur within a specific group of muscles (limb muscles for movement, postural muscles for balance, and respiratory muscles for breathing). Moreover, injury can be distributed unequally within a synergistic group of muscles (e.g., diaphragm versus other respiratory muscles; [9]). Thus, unlike the cardiac proteome, the skeletal proteome involves different fiber types and/or groups of muscles that can be differentially affected.

Identification and characterization of both the skeletal and cardiac proteomes are important steps in understanding contractile function in health and disease. In disease, skeletal (including respiratory) and/or cardiac muscle function (contraction or contractility) is impaired. The role(s) of each of the various protein alterations associated with disease, either transcriptional (expression levels and/or isoform changes) or post-translational modification(s) (PTM), may contribute to or compensate for contractile dysfunction. Knowing if a modification is compensatory or detrimental (pathological) is crucial for the development of therapeutic strategies which may either prevent or delay the onset of disease.

19.1
Regulation of Muscle Contraction

Muscle contraction is the result of the complex interactions of many myofilament proteins. It is, ultimately, the interplay between the thick and thin filaments, in a Ca-dependant manner, which produces force at the expense of ATP hydrolysis (for reviews, see [10–13]). Even though many agents can affect contraction (e.g., catecholamines, angiotensin II), the interaction between the thick and thin filaments is tightly regulated. The thick filament is composed primarily of myosin, including the heavy chain and two associated light chains, which together form the mechanoeynzyme. The thin filament is comprised of filamentous actin (a polymerized filament composed of monomeric actin), tropomyosin (Tm, a dimer of the α isoform or a heterodimer composed of the α and β isoforms), and the troponin complex (Tn). Tn consists of troponin I (TnI), termed the inhibitory protein due to its ability to block actin-myosin interaction, troponin T (TnT), named for its extensive binding to Tm, and troponin C (TnC), which binds calcium and triggers contraction (for reviews, see [14–19]). In addition, such structural proteins as titin, desmin and α-actinin contribute to the spatial orientation of the thick and thin filaments and, possibly, the transmission of force. Changes in quantity, isoform expression, or the status of PTM of one or more of the myofilament proteins can dramatically influence protein-protein interactions, thus affecting force production and/or contractile economy.

The myofilament proteins are also called the contractile proteins; they make up the myofibrils. They produce force as a result of the formation of actin-myosin crossbridges between the thick and thin filaments. In striated (cardiac and skeletal) muscle, contraction occurs in response to increasing intracellular concentrations of Ca^{2+}. Tn works in concert with Tm to regulate Ca-dependent muscle contraction which is initiated by the binding of Ca^{2+} to TnC. TnC is an EF-hand Ca^{2+} binding protein with two domains (N- and C-terminals). When intracellular Ca^{2+} increases, Ca^{2+} binds to the regulatory site(s) within the N-terminus of TnC (cardiac TnC has one regulatory Ca^{2+} binding site; skeletal TnC has two). This alters the conformation of TnC (and hence the whole Tn complex), which induces the inhibitory region of TnI (the twelve amino acid sequence comprising residues 136–147 in human cTnI and 104–115 in human sTnI) to "switch" from its specific binding site on actin-Tm to a site on TnC. There are extensive interactions between the various troponin subunits, some with extremely high affinities, as well as between Tn and the other members of the thin filament, contributing to the complexity of this system.

The Ca-dependent change in the Tn-Tm complex results in its movement across the actin filament, increasing the probability of its interaction with myosin. Tm is a long coiled-coil dimer extending along the actin filament covering every seven actin monomers. There is one Tn complex per Tm (dimer) and their control of actin-myosin interactions involves the intricate interplay of both steric and allosteric mechanisms. The exact structural changes involved in signaling, and changes in the interaction between the myofilament proteins involved in regulation and muscle contraction, are complicated and still being elucidated. Clearly, the striated contractile system constitutes a finely-tuned array of proteins that regulates force production in response to changes in intracellular Ca^{2+} concentration.

The differences in functional and mechanical properties of cardiac and skeletal muscle are due, in part, to the different myofilament protein isoforms expressed. The isoforms have different amino acid sequences which affect their interactions with other proteins in the myofilament. For instance, cardiac muscle contains only *αα*-Tm while skeletal muscle contains varying ratios of *αα*-Tm and *αβ*-Tm depending on the fiber type. Lehman and colleagues [20] observed that the different Tm isoforms adopt different positions on the actin filament, indicating that the interaction between the various Tm isoforms and actin is distinctive despite the extensive amino acid sequence homology between these two isoforms. TnT and TnI have three isoforms (slow skeletal, fast skeletal, and cardiac) which are expressed in their corresponding fiber types, the single exception being the expression of the slow and cardiac isoforms of TnI in the embryonic heart until just after birth when the cardiac isoform predominates [21, 22]. The situation associated with TnT is more complex; it has many RNA splice sites with the potential to generate a remarkable diversity of isoforms. For example, there are at least six mammalian and up to thirteen avian exons that can be alternatively spliced for fast skeletal TnT [23–26]. Furthermore, these TnT isoforms change during development and disease (for reviews, see [27–30]). TnC has two isoforms arising from two separate genes [31]. The cardiac and slow skeletal isoforms share one gene while fast skeletal has the other. Thus, various combinations of isoforms of these different Tn subunits will influence the molecular regulation of contraction in the different striated muscle types.

Several PTMs are specific to cardiac myofilament proteins. Cardiac muscle is modulated by catecholamines through the activation of protein kinase A (PKA) (for review, see [32]). PKA phosphorylates several proteins, including cTnI at amino acid residues serine 22 and serine 23. These two amino acid residues are located in the unique 30 amino acid N-terminal extension of cTnI and are, therefore, not present in either the fast or slow skeletal isoforms. There is an interplay between PKA-mediated phosphorylation of cTnI and other functional regions of the molecule, as demonstrated by the changes in the extent of cTnI phosphorylation in the mutation cTnIR145G (a cardiomyopathy mutant) as compared to wild type cTnI [33]. Also, we recently identified two novel phosphorylation sites of cardiac myosin light chain 1 (MLC-1, a protein widely considered to be unphosphorylatable) while undertaking proteomic analysis of the myofilament proteins [34]. Not only was a basal level of MLC-1 phosphorylation observed (sham), the extent of phosphorylation was modulated during pharmacological preconditioning of rabbit cardiomyocytes. We have also observed novel PTM of skeletal myofilament proteins, including the *in vivo* phosphorylation of fast skeletal TnI (unpublished data). This illustrates the potential of proteomic analysis to uncover novel aspects in even a well-studied system like the myofilament proteins. One can only speculate on the number of changes that will be uncovered in various disease states.

19.2
Disease States and the Myofilament Proteins

A disease state (encompassing all types of injury), whether acute or chronic, is characterized by stress-induced (e.g., hypoxia, infection, toxins, drugs, hypertension, ischemia) dysfunction of the whole muscle system. This functional deficit can be global (e.g., type II skeletal muscle fibers in different locomotor muscles) or restricted (e.g., all muscle fibers but in a single muscle). Common disease states include HF, fatigue, myocardial infarction, myocarditis, COPD, sepsis, and inherited myopathies (e.g., skeletal muscle nemaline myopathy (NM) and cardiac muscle hypertrophic myopathy). For each disease state, there is a specific underlying molecular cause related to a proteomic change responsible for the dysfunction. As stated earlier, many studies indicate that myofilament protein alterations (e.g., mutations, isoform switching, re-expression of fetal isoforms, and PTMs) are associated with various disease states and most of these alterations are associated with contractile dysfunction (for reviews, see [35–37]). The next section deals with alterations specific to cardiac and skeletal muscle.

19.2.1
Myofilament Protein Mutations Cause Disease

Inherited muscle disease represents a spectrum of pathological conditions arising from many different etiologies with unique as well as overlapping characteristics. Familial hypertrophic cardiomyopathy (FHC) and NM are inherited disorders resulting from mutations in genes encoding specific cardiac and skeletal myofilament proteins, respectively. To date, all identified FHC and NM mutations are within the myofilament proteins (both structural and contractile proteins; e.g., α-actinin, nebulin, Tm, TnT, actin). They vary in severity (mild to lethal) with the time of onset (neonatal, childhood, or adult) reflecting, in part, the isoform(s) affected (adult vs. fetal), site of mutation, and the nature of the amino acid change. Mutations in cardiac myofilament proteins cause cardiac hypertrophy and are thought to be involved in some cases of sudden death syndrome. For example, some cases of FHC are attributed to different single amino acid mutations to cTnT (for review, see [38]) and/or cTnI [39–41] while five different known skeletal actin mutations can cause NM [42]. In skeletal muscle (for review, see [43]), rare (and perhaps under-represented) cases of NM occur as the result of mutations in skeletal isoforms that can cause facial, limb/locomotor, postural and respiratory muscle weakness. Although many of these problems affect quality of life (e.g., limb muscle weakness can limit activity and affect balance), they are generally not life threatening. In contrast, respiratory muscle dysfunction can, by impairing ventilation, predispose the individual to respiratory infections and, in severe cases, respiratory failure. Remarkably, FHC and NM are both caused by a change in just one amino acid of almost any one of the myofilament proteins. That changing one amino acid is sufficient to cause profound changes in contractile performance illustrates the importance of myofilament protein-protein interactions.

19.2.2
Myofilament Protein Isoform Alterations in Disease

Disease states are often associated with changes in the expression of various myofilament protein isoforms. In COPD and HF patients, depending on the severity of the disease, changes in both limb and respiratory muscles occur. In limb muscles, there is a shift to type II (fast glycolytic) fibers while the diaphragm shifts towards type I (slow oxidative) fibers (see [44] for review). However, a detailed analysis of alterations in expression of fiber type-specific proteins (e.g., co-expression of myosin isoforms in some cases of NM; [42]) or in myofilament PTM (e.g., phosphorylation, glycosylation) is still unavailable.

In the failing myocardium, there is selective re-expression of the fetal isoform of cardiac TnT (cTnT) but not cTnI. Furthermore, the extent to which the fetal isoforms are re-expressed varies between species and types of HF. In addition, *de novo* isoform expression can occur. For example, during HF, the atrial isoform of MLC-1 is abnormally expressed in the ventricles [45]. Remarkably, as little as 3% expression of the atrial isoform of MLC-1 in ventricle is sufficient to double *in vitro* force generation (when myosin light chain 2 (MLC-2) is completely dephosphorylated) (for review, see [46]).

19.2.3
Myofilament Protein Post-translational Modifications in Disease

There is an accumulating number of disease-induced PTMs of myofilament proteins that correlate with contractile dysfunction. For example, specific and progressive cTnI proteolysis occurs with ischemia/reperfusion injury in the isolated rat heart (see [47] for review). Proteolysis of cTnI has been reported in the myocardium of patients undergoing bypass surgery [48] and in the serum from patients with acute myocardial infarction (AMI) [49]. Interestingly, expression in mouse myocardium of the first proteolytic fragment observed with ischemia/reperfusion injury of the isolated rat heart (cTnI1-192) is sufficient to recapitulate the contractile dysfunction observed during HF in animal models and humans [50]. Moreover, this dysfunction occurs with only 9–17% expression of cTnI1-192; greater levels may not have been achieved due to lethality. In skeletal muscle, proteolysis of sTnI and sTnT occurs only in diaphragm, not in accessory respiratory muscles or limb muscles, in severely hypoxemic dogs [9]. In addition, proteolytic fragments of both fast and slow sTnI have also been reported in the serum of a patient with rhabdomyolysis [51]. Although the skeletal muscle pool (i.e., axial, limb, respiratory) from which the sTnI originated could not be identified, the proteolytic fragments observed in the serum likely originated from the diseased skeletal muscle proteome. Whether the observed protein alterations are the cause or the result of the disease remains to be investigated. Complete analysis of the myofilament proteome is critical for a full understanding of the contractile dysfunction associated with each disease state.

19.3
Proteomic Analysis of the Myofilament Proteins

Proteomics is the study of the proteome (the protein complement of a cell) at a given time. Proteomics most often involves separating all the proteins (and their alterations) for subsequent analysis (which "spots" are unique to disease and which unique to health), identification (e.g., myosin, Tm), and characterization (e.g., native-Tm and phosphorylated-Tm at serine 122). Two-dimensional gel electrophoresis (2-DE) is widely used for protein separation. In 2-DE, proteins are separated in the first (horizontal) dimension by isoelectric focusing (IEF) which separates proteins based on their intrinsic charge, or pI. In the second (vertical) dimension, proteins are separated based on their relative molecular masses using sodium dodecyl sulfate-polyacrylamide gel electrophoresis. After protein staining, the different proteins, as well as the various forms of the same protein (isoforms or PTMs), appear as discrete and, one hopes, unique spots, since they differ in their pI and/or molecular mass. 2-DE is a powerful method for the separation of proteins which, for striated muscle, enables us to resolve over 1,500 protein spots on a single gel (Fig. 19.1). However, alternative methods of protein separation, exploiting hydrophobicity or other intrinsic protein characteristics, are attractive as they overcome limitations associated with 2-DE (e.g., very large molecular weight proteins, hydrophobic proteins, and proteins with extreme pIs). Mass spectrometry (peptide mass fingerprinting and amino acid sequencing), western blotting, N-terminal amino acid sequencing, or some combination, is used to determine the identity of each protein spot and, when applicable, its modification.

Proteomic analysis (separation and identification) of myofilament proteins with high molecular weights (e.g., myosin heavy chain), extreme pIs (e.g., TnI), and strong protein-protein interactions (Tn complex) is especially challenging. In addition, it is necessary to detect and quantify all the myofilament protein alterations, even those present in modest quantities, which can occur in disease. In the rest of this chapter, we discuss the complications of myofilament proteins and their proteomic analysis, and how technological advances are overcoming these limitations.

19.3.1
Creating a Myofilament Proteome

Subproteomic analysis simplifies the daunting task of attempting to simultaneously resolve thousands of cellular proteins. It is particularly important for striated muscle because of the enormous difference in abundances between the myofilament proteins (high) and the vast majority of other cellular proteins (low) (Fig. 19.1). 2-DE gel separations of total muscle homogenates will reveal only a very limited part of the cytosolic and membrane proteome; the shear quantity of the few myofilament proteins makes it impossible to observe properly the rest of the cellular proteome. Fig. 19.1b shows that using low protein loads (whole cell/tissue) allows one to resolve many of the myofilament proteins but the rest of the proteome is virtually undetectable. In other words, there is a range of abundance of the different cellular proteins, with a pronounced disparity between the contractile proteins (high) and the other cellular proteins (low). Theoretically, higher protein loads would reveal the rest of the proteome but the quantity of protein required to visualize it would result in overloading of the myofilament proteins

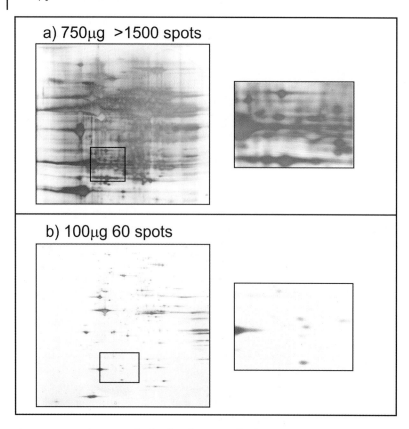

a) 750μg >1500 spots

b) 100μg 60 spots

Fig. 19.1 2-DE silver-stained gels of cardiac tissue with high (a) and low (b) protein loads (pH 3–10, 12% SDS page). Note: higher protein loads permitted more proteins to be visualized although analysis was limited because many spots overlapped and/or were poorly focused. Expanded views of boxed sections of gels shown on right.

which will distort the focusing (Fig. 19.1a) and cause overlap of the spots associated with many of the cellular proteins, thus preventing or limiting their detection and analysis. Therefore, a major goal is to separate the abundant myofilament proteins from the rest of the proteome, while "preserving" the proteome; this should allow more efficient observation and analysis of the less abundant cellular proteins.

Fractionation methods are often used to produce a discrete soluble protein fraction of biological interest; the myofilament proteins constitute, in some regards, a perfect subproteome to study. As previously discussed, they are the final effectors of contraction and have a complex array of protein-protein interactions such that even a (seemingly) modest change can drastically affect efficiency and/or force production. Fractionation or extraction exploits specific protein characteristics such as their inherent chemical properties (biospecificity, hydrophobicity, charge) or differential cellular compartmentalization. Currently, for myofilament proteo-

a)

b)

c)

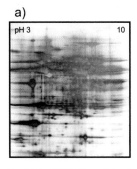
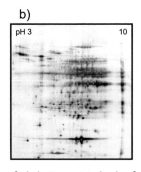
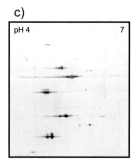

Fig. 19.2 2-DE silver-stained gels of whole tissue (a) and cytoplasmic-enriched (b) and myofilament-enriched (c) extracts from cardiac tissue using IN sequence. Note: high protein loads of cytoplasmic proteins were possible without the high abundance myofilament proteins altering focusing or obscuring the lower abundance proteins.

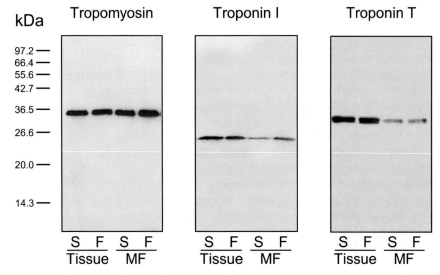

Fig. 19.3 Western blots showing the changes in the mole ratio between tropomyosin, TnI and TnT in tissue and myofibrils from rat extensor digital longus (100% fast – sham (S) and fatigued (F), 12% SDS page). Tissue and MF loadings were normalized for actin. While the Tm:actin ratio was the same between tissue and MFs, the actin:TnI and actin:TnT ratios were altered. Western blotting of the cytosolic and lipid fractions revealed the presence of sTnI and sTnT and, to a lesser extent, tropomyosin and actin.

mic analysis, there are two main approaches to isolation: traditional myofibril preparations or IN Sequence. Myofibril isolation was developed specifically to create a fraction of the contractile/structural proteins for use in *in vitro* functional assays in order to provide insight into the status of the contractile proteins [52]. IN Sequence is an extraction method we recently developed specifically for striated muscle protein that uses buffers and detergents compatible with all protein separation techniques (e.g., 2-DE, HPLC) [34, 53] (Fig. 19.2). The ideal extraction should both "preserve" the proteome during preparation (i.e., no artifacts or loss of PTMs) and use buffers/detergents compatible with downstream protein separation (i.e., HPLC and 2-DE) and identification methods (western blot, MS).

The traditional myofilament (myofibril) preparation is widely used as it provides enzymatically active contractile proteins for *in vitro* functional assays. In our experience, the mole ratios of the various myofilament proteins are altered during purification. The native thin filament is mainly composed of actin : Tm : Tn at a mole ratio of 7 : 1 : 1 (Tm being a dimer). Purification of myofilament proteins relies on the dif-

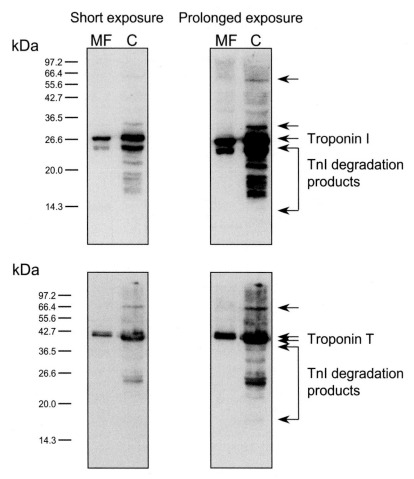

Fig. 19.4 Representative 1-DE western blots of myofibrils purified from cultured rabbit cardiac cells (C) probed with an anti-TnI (top) and anti-TnT (bottom) antibody. Albumin from the cell culture medium affects the migration of intact TnT. Most of the TnI fragments and all of the TnT fragments were lost during purification of the myofilament proteins used for ATPase assays. Also, the higher molecular weight TnT and TnI products were not detected.

ferent solubilities of the various cellular components, allowing the separation of myofibril, cytosolic and lipid (cell membrane) fractions (although sometimes a modified method which only partially removes the membrane proteins is used). Fractionation of myofibrils is imperfect, as various amounts of each myofilament protein are lost to the cytosolic and/or lipid fractions (data not shown). Critically, during purification, unequal amounts of the different myofilament proteins were lost (Fig. 19.3), even when wash steps were drastically shortened (data not shown). The quantity of Tm and a-actinin present in both the tissue and the isolated myofibrils were similar. However, upon isolation of the myofibrils there was a differential loss in cTnT, cTnI, and cTnC compared to that in the tissue; in other words, the mole ratio between the various thin filament proteins was altered following isolation of the myofibrils. Altering the mole ratio of the myofilament proteins can have drastic effects on ATPase activity (e.g., co-operativity, basal and maximal activities, and Ca^{2+} sensitivity; e.g., [54, 55]). Furthermore, PTMs can alter protein solubility and their interactions and affinities with other proteins. Thus, alterations of the myofilament proteome may exacerbate the change in the mole ratio because of how a modified protein behaves during purification. For example, proteolysis of TnI and TnT has been reported to occur in both cardiac and skeletal muscle; however, proteolytic fragments of TnI and TnT do not necessarily remain associated with the purified myofibrils (Fig. 19.4). Thus, the consequences of such proteolysis on *in vivo* function would not be apparent during subsequent biochemical analysis. Ideally, traditional myofilament preparations should allow one to ascribe functional changes to changes in the proteome. Fig. 19.3 and 19.4 suggest that traditional myofibril preparations can misrepresent the diseased proteome. Therefore, when preservation of the complete myofilament proteome is required (depending on one's research question), IN-sequence provides a compelling and suitable alternative.

19.3.2
Protein Separation – 2-DE

The myofilament proteome, despite having relatively few (~ 20) proteins, contains several proteins problematic for 2-DE. These include: a) highly charged proteins (i.e., TnI (pI ~ 9.8) and troponin C (TnC, pI ~ 4.0)); b) large molecular weight proteins (i.e., myosin heavy chain, a-actinin); and c) protein complexes with extremely strong protein-protein interactions (i.e., the Tn complex). Both cardiac and skeletal TnI and TnT are problematic because of the strong interactions between the troponin subunits, as well as the relative insolubilities of the individual proteins themselves and the number of potential isoforms which may be present (some at very low abundances). Most TnI and TnT isoforms possess nearly identical pIs and molecular weights, requiring very low protein loads and the use of zoom gels for their resolution. Use of 1M NaCl in the homogenization buffer presumably helps the destruction of inter- and intra-molecular interactions, thereby increasing the resolution of some of these troublesome proteins as distinct spots without streaking [56]. For example, Fig. 19.5 shows the resolution of cardiac and skeletal TnT with and without the addition of 1M NaCl to the homogenization

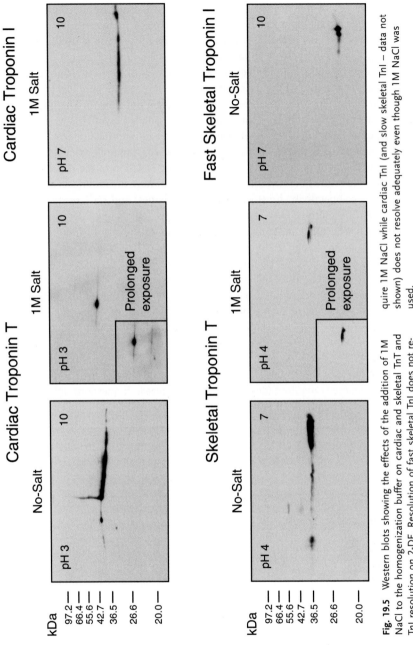

Fig. 19.5 Western blots showing the effects of the addition of 1M NaCl to the homogenization buffer on cardiac and skeletal TnT and TnI resolution on 2-DE. Resolution of fast skeletal TnI does not re- quire 1M NaCl while cardiac TnI (and slow skeletal TnI – data not shown) does not resolve adequately even though 1M NaCl was used.

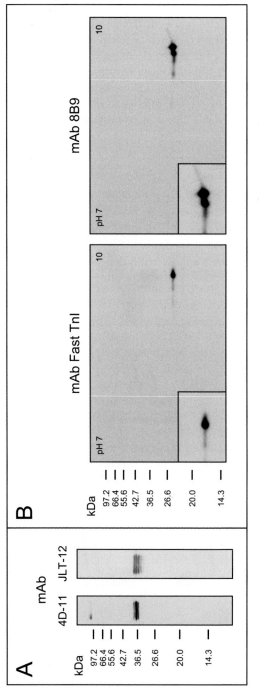

Fig. 19.6 1-DE and 2-DE western blots of skeletal muscle probed with anti-troponin T (A, 4D-11 and JLT-12) and anti-troponin I (B, fast TnI and 8B9) antibodies. The detection of PTMs or isoforms can depend on the mAb. mAb 4D-11 detected three isoforms while mAb JLT-12 detected five isoforms of skeletal TnI. The detection of a PTM of fast skeletal TnI was revealed only by mAb 8B9 (two spots) while mAb Fast TnI detected only the unmodified or native (one spot) form.

buffer (final concentration of NaCl in IEF buffer >25 mM). In the absence of 1M NaCl, substantial streaking of TnT was observed even though other myofilament proteins (e.g., actin, Tm, and MLC-1 and MLC-2) were resolved. Furthermore, basic proteins were difficult to focus during IEF due to electroendoosmotic effects at high pH and the resulting cathodic drift within the gel. Surprisingly, fast skeletal TnI (Fig. 19.6) was suitably resolved while cardiac and slow skeletal TnI were not (Fig. 19.5). Interestingly, these proteins have high amino acid sequence homology but their behaviors differ radically with solubilization and/or separation by 2-DE.

19.3.3
Detection, Quantification and Identification

To repeat, the challenge in protein detection is to use a method that is 1) compatible with downstream protein identification methods, 2) able to detect all the myofilament proteins, and 3) able to detect both the native and modified forms within a linear range, thereby allowing accurate quantification. Protein visualization is most often carried out with silver, fluorescent, or coomassie blue stains. Silver staining is extremely sensitive and is the most common means to detect proteins

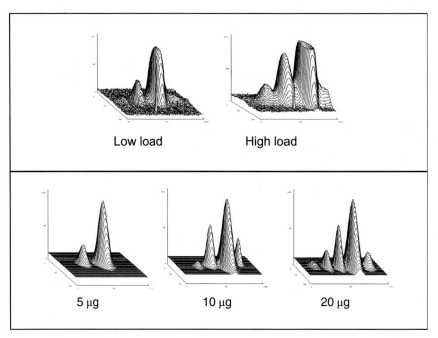

Low load   High load

5 µg   10 µg   20 µg

Fig. 19.7A Computer-generated surface blots of an unidentified protein from silver-stained 2-DE gels. Saturation or non-linearity of protein spots on silver-stained 2-DE gel can occur when the protein is present at a high concentration in the sample, but a high load may be required in order to detect the PTM(s) (top). Therefore, different sample loadings are necessary to optimize each protein present in the extract (bottom).

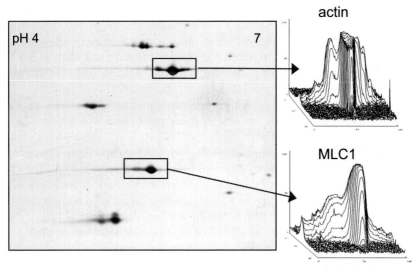

Fig. 19.7 B 2-DE silver-stained gel of myofilament-enriched extract from cardiac tissue. Differential abundance of myofilament proteins requires multiple protein loads to ensure linearity but still permit detection of all PTMs.

for proteomic analysis. The choice of stain represents a tradeoff between sensitivity and linear range. Any disparity in quantity, when present, between the native and modified forms of a protein can lead to problems in the detection and quantification of all forms of the protein. This can result in missing a critical protein alteration, as illustrated in Fig. 19.7, which shows an example of a silver-stained 2-DE gel with 10 µg of a myofilament-enriched extract from cardiac muscle obtained by IN sequence. While MLC-1 is visualized within the linear range, actin is overexposed (main peak), leading to its underestimation. Optimal staining of actin will result in the failure to detect some modified forms of actin, as well as other myofilament proteins including MLC-1. Ideally, the perfect stain should have the sensitivity of silver with a broad linear range equal or better than that of coomassie blue or fluorescent stains. Fluorescent stains have recently gained attention as they offer better sensitivity than coomassie blue and have a larger linear range. Their costs, however, have tempered wide acceptance.

The key regulatory proteins, the Tn complex, all stain poorly and unevenly with either silver (especially with non-glutaraldehyde silver staining which is compatible with subsequent MS) or coomassie blue stain, presumably due to the large clusters of charged amino acid residues throughout their sequences [56]. The myofilament proteins also stain with silver at different rates as a result of both their abundances and abilities to bind silver (data not shown). In cardiac muscle, actin is the first protein observed during silver staining, followed by MLC-1, Tm, MLC-2 and the remaining low abundance proteins. Therefore, it is imperative that each gel be stained to completion to ensure proper visualization and thus accurate quantification of all proteins on each gel with respect to one another. A series of gels with different protein loads may be required for accurate quantification when using silver staining. One interesting aspect of silver staining is that different proteins stain different colors depending on their amino acid composition. For the myofilament proteins, MLC-1 and MLC-2 stain black while Tm has a green tone and actin and desmin stain a golden red; diffractive scattering of light by the silver grains is responsible for this effect [57].

Proteins that are problematic for current staining techniques can be visualized by immunodetection. In our experience, western blot analysis is required for the detection and quantification of the individual Tn subunits. This, however, complicates matters because of inherent problems with antibody specificity. For example, an antibody may not be able to detect all isoforms or a specific PTM (Fig. 19.6). We have characterized over 25 anti-TnI antibodies which are differentially sensitive to either isoform specificity or PTMs, including degradation and phosphorylation. The anti-TnT antibody JLT-12 (Sigma) also cross-reacts to the cytosolic protein, GAPDH. This cross-reactivity caused confusion about whether cTnT is degraded in globally ischemic rat hearts [58]. Therefore, extreme care must be taken when interpreting immunoblots.

19.3.4
Alternative Methods of Protein Separation

Chromatography encompasses several techniques that can be used alone or in combination. It includes ion exchange, size exclusion, reversed-phase and affinity chromatography (conventional and high-performance liquid chromatography (HPLC)), capillary IEF, and capillary zone electrophoresis. Affinity-based separation may be achieved by conventional column chromatography, immunoprecipitation, or by the newly developed "protein chip" method (e.g., SELDI; for review, see [59]). These methods can be used alone or in various combinations and, in some cases, can be carried out under either native or denaturing conditions. The advantage of isolating proteins in their native state is that one can exploit protein-protein interactions and use the isolated proteins in biochemical assays. The caveat is that one has to ensure that no modification(s) is introduced by processing.

We separated the myofilament-enriched fraction (obtained using IN Sequence) using reversed-phase HPLC which separates proteins based on hydrophobicity [53]. In proteomic analysis, reversed-phase HPLC has been used mainly to separate peptides, obtained from proteolytic (in-gel or whole proteome) digestion, prior to their online injection into an electrospray mass spectrometer, although other methods for peptide fragment methods are being developed (see [60] for review). We have used reversed-phase HPLC to separate proteins rather than the digested protein fragments. Optimization of protein peak resolution and peak number was obtained by altering flow, gradient rate, and the organic modifiers, isopropanol and acetronitrile. The subproteome was successfully separated and alterations of cTnT and MHC were observed in hearts in failure. One must remember that the goal of proteomic analysis is to observe as much of the proteome or subproteome as possible. Using a combination of 2-DE analysis and HPLC (or any other protein separation method) will increase one's ability to completely characterize the entire proteome.

19.3.5
The Next Step – Clinical Applications of Proteomics: The Development of Biomarkers

Some of the earliest benefits that will emerge from proteomics of skeletal and/or cardiac muscle will be in the area of biomarkers (also termed diagnostic markers). In particular, myofilament proteins detected in the serum following release from the diseased cell are potential biomarkers. Currently, cTnI is widely used for the diagnosis and management of several myocardial diseases (e.g., myocardial infraction and angina; see [61]). The current method of detection (ELISA) does not specifically differentiate between the different forms of cTnI associated with disease. Recently, we modified 1D SDS-PAGE to optimize the separation of serum proteins (western blot-direct serum analysis, WB-DSA; [49]). Normally, the high abundance proteins (i.e., albumin and IgG) in serum will distort the migration of other proteins. WB-DSA minimizes this distortion, allowing detection by western blot of low quantities of proteins with masses below 60 kDa. Using WB-DSA, we have observed proteolytic fragments of cTnI in serum from patients with AMI [49]. We have also observed that patients with AMI may not necessarily possess the same forms of cTnI [49]. This also applies to patients with stable and unstable angina and who have low levels of cTnI in the serum [62]. These findings, if confirmed, suggest that future biomarkers may involve not just detection of a particular protein but also quantification of the various forms of proteins previously associated with a specific disease process or stage.

We have also applied WB-DSA to the detection of the various isoforms of sTnI in serum, based on the assumption that skeletal muscle will, like the heart, when sufficiently injured, release cellular proteins, including sTnI, into the blood [51]. We have preliminary evidence that the forms (i.e., intact fast vs intact slow vs degraded forms) of sTnI detected in serum vary and therefore, may reflect different degrees or types of injury to fast vs slow skeletal muscle fibers. Thus, the amount of sTnI may not be as important as what forms are present. In addition, whether or not cTnI is present (and in which forms) should provide insights into the relative contributions of the skeletal (e.g., respiratory) and cardiac components to many diseases.

19.4
Conclusion

Our understanding of the molecular basis of muscle dysfunction is just beginning. The interplay between the myofilament proteins is extraordinarily complex and still needs to be defined in health irrespective of the many protein alterations that develop during disease/injury revealed by proteomics. Each protein alteration will have specific effects on contraction, an effect complicated by the many known (and unknown) alterations coexisting in each disease state. Moreover, these changes in the proteome will be further complicated by many factors (e.g., medications, diet, genetics, gender, age, exercise, and hormones) which influence disease progression. This complexity, in turn, will be confounded by the experimental model and the proteomic approach used. This will undoubtedly create a complex array of possibilities which we are currently ill equipped to handle. Even so, we anticipate many diagnostic and pharmacological targets will become evident, some of which should result in improved treatments of many diseases.

19.5
Acknowledgements

We thank Kent Arrell and Michelle Quick; their expertise in 2-DE analysis of skeletal and cardiac muscle was critical to the results depicted in this chapter. We also thank Ralf Labugger for his insights which contributed greatly to the successful analysis of troponin in samples of tissue and serum.

JAS was supported by an Ontario Graduate Student Award (Science and Technology), funds from the Block Term Grant to Queen's University from the Ontario Thoracic Society, and the School of Graduate Studies and Research, Queen's University.

This research was funded by grants to SI (Canadian Institutes of Health Research, the Ontario Thoracic Society, Block Term Grants to Queen's University from the Ontario Thoracic Society, and the William M. Spear Endowment Fund) and to JVE (Canadian Institutes of Health Research, the Ontario Heart and Stroke Foundation, the Canadian Foundation for Innovation, and the National Institutes of Health (USA)). JVE is a Canadian Heart and Stroke Foundation Career Investigator.

19.6
References

1 ALIVERTI, A., P.T. MACKLEM. How and why exercise is impaired in COPD. *Respiration* **2001**, *68*, 229–239.

2 CARUANA, L., M.C. PETRIE, J.J. MCMURRAY, N.G. MACFARLANE. Altered diaphragm position and function in patients with chronic heart failure. *Eur. J. Heart Fail.* **2001**, *3*, 183–187.

3 MACFARLANE, N.G., G.M. DARNLEY, G.L. SMITH. Cellular basis for contractile dysfunction in the diaphragm from a rabbit infarct model of heart failure. *Am. J. Physiol.* **2000**, *278*, C739–C746.

4 STASSIJNS, G., G. GAYAN-RAMIREZ, P. DE LEYN, V. DE BOCK, R. DOM, et al. Effects of dilated cardiomyopathy on the diaphragm in the Syrian hamster. *Eur. Respir. J.* **1999**, *13*, 391–397.

5 MEYER, F.J., M.M. BORST, C. ZUGCK, A. KIRSCHKE, D. SCHELLBERG, et al. Respiratory muscle dysfunction in congestive heart failure: clinical correlation and prognostic significance. *Circ.* **2001**, *103*, 2153–2158.

6 VESCOVO, G., G.B. AMBROSIO, L. DALLA LIBERA. Apoptosis and changes in contractile protein pattern in the skeletal muscle in heart failure. *Acta. Physiol. Scand.* **2001**, *171*, 305–310.

7 AMERICAN THORACIC SOCIETY. Standards for the Diagnosis and Care of Patients with Chronic Obstructive Pulmonary Disease. *Am. J. Respir. Crit. Care. Med.* **1995**, *152*, S77–S120.

8 MADOR, M.J., E. BOZKANAT. Skeletal muscle dysfunction in chronic obstructive

pulmonary disease. *Respir. Res.* **2001**, *2*, 216–224.

9 SIMPSON, J. A., J. E. VAN EYK, S. ISCOE. Hypoxemia-induced modification of troponin I and T in canine diaphragm. *J. Appl. Physiol.* **2000**, *88*, 753–760.

10 GORDON, A. M., E. HOMSHER, M. REGNIER. Regulation of contraction in striated muscle. *Physiol. Rev.* **2000**, *80*, 853–924.

11 RUEGG, J. C. Cardiac contractility: how calcium activates the myofilaments. *Naturwissen* **1998**, *85*, 575–582.

12 CHALOVICH, J. M. Actin mediated regulation of muscle contraction. *Pharmacol. Ther.* **1992**, *55*, 95–148.

13 TOBACMAN, L. S. Thin filament-mediated regulation of cardiac contraction. *Annu. Rev. Physiol.* **1996**, *58*, 447–481.

14 SOLARO, R. J., H. M. RARICK. Troponin and tropomyosin: proteins that switch on and tune in the activity of cardiac myofilaments. *Circ. Res.* **1998**, *83*, 471–480.

15 GERGELY, J. Molecular switches in troponin. *Adv. Exp. Med. Biol.* **1998**, *453*, 169–176.

16 PERRY, S. V. Troponin I. inhibitor or facilitator. *Mol. Cell. Biochem.* **1999**, *190*, 9–32.

17 FILATOV, V. L., A. G. KATRUKHA, T. V. BULARGINA, N. B. GUSEV. Troponin: structure, properties, and mechanism of functioning. *Biochemistry (Moscow)* **1999**, *64*, 969–985.

18 GAGNÉ, S. M., M. X. LI, R. T. MCKAY, B. D. SYKES. The NMR angle on troponin C. *Biochem. Cell. Biol.* **1998**, *76*, 302–312.

19 SQUIRE, J. M., E. P. MORRIS. A new look at thin filament regulation in vertebrate skeletal muscle. *FASEB J.* **1998**, *12*, 761–771.

20 LEHMAN, W., V. HATCH, V. KORMAN, M. ROSOL, L. THOMAS, et al. Tropomyosin and actin isoforms modulate the localization of tropomyosin strands on actin filaments. *J. Mol. Biol.* **2000**, *302*, 593–606.

21 SAGGIN, L., L. GORZA, S. AUSONI, S. SCHIAFFINO. Troponin I switching in the developing heart. *J. Biol. Chem.* **1989**, *264*, 16299–16302.

22 MURPHY, A. M., L. JONES, H. F. SIMS, A. W. STRAUSS. Molecular cloning of rat cardiac troponin I and analysis of troponin I isoform expression in developing rat heart. *Biochemistry* **1991**, *30*, 707–712.

23 COOPER, T. A., C. P. ORDAHL. A single troponin T gene regulated by different programs in cardiac and skeletal muscle development. *Science* **1984**, *226*, 979–982.

24 SAMSON, F., L. MESNARD, M. MIHOVILOVIC, T. G. POTTER, J. J. MERCADIER, et al. A new human slow skeletal troponin T (TnTs) mRNA isoform derived from alternative splicing of a single gene. *Biochem. Biophys. Res. Commun.* **1994**, *199*, 841–847.

25 WU, Q. L., P. K. JHA, M. K. RAYCHOWDHURY, Y. DU, P. C. LEAVIS, et al. Isolation and characterization of human fast skeletal beta troponin T cDNA: comparative sequence analysis of isoforms and insight into the evolution of members of a multigene family. *DNA Cell. Biol.* **1994**, *13*, 217–233.

26 BUCHER, E. A., DE LA BROUSSE F. C, C. P. J. EMERSON. Developmental and muscle-specific regulation of avian fast skeletal troponin T isoform expression by mRNA splicing. *J. Biol. Chem.* **1989**, *264*, 12482–12491.

27 ADAMCOVA, M., V. PELOUCH. Isoforms of troponin in normal and diseased myocardium. *Physiol. Res.* **1999**, *48*, 235–247.

28 THIERFELDER, L., H. WATKINS, C. MACRAE, R. LAMAS, W. MCKENNA, H. P. VOSBERG, et al. Alpha-tropomyosin and cardiac troponin T mutations cause familial hypertrophic cardiomyopathy: a disease of the sarcomere. *Cell* **1994**, *77*, 701–712.

29 ANDERSON, P. A., G. E. MOORE, R. N. NASSAR. Developmental changes in the expression of rabbit left ventricular troponin T. *Circ. Res.* **1988**, *63*, 742–747.

30 SAGGIN, L., L. GORZA, S. AUSONI, SCHIAFFINO. Cardiac troponin T in developing, regenerating and denervated rat skeletal muscle. *Development* **1990**, *110*, 547–554.

31 WILKINSON, J. M. Troponin C from rabbit slow skeletal and cardiac muscle is the product of a single gene. *Eur. J. Biochem.* **1980**, *103*, 179–188.

32 DE TOMBE, P. P., R. J. SOLARO. Integration of cardiac myofilament activity and regulation with pathways signaling hypertrophy and failure. *Ann. Biomed. Eng.* **2000**, *28*, 991–1001.

33 DENG, Y., A. SCHMIDTMANN, A. REDLICH, B. WESTERDORF, K. JAQUET, et al. Effects of

phosphorylation and mutation R145G on human cardiac troponin I function. *Biochemistry* 2001, *40*, 14593–14602.

34 ARRELL, D.K., I. NEVEROVA, H. FRASER, E. MARBAN, J.E. VAN EYK. Proteomic analysis of pharmacologically preconditioned cardiomyocytes reveals novel phosphorylation of myosin light chain 1. *Circ. Res.* 2001, *89*, 480–487.

35 MARSTON, S.B., J.L. HODGKINSON. Cardiac and skeletal myopathies: can genotype explain phenotype? *J. Muscle Res. Cell. Motility* 2001, *22*, 1–4.

36 SOLARO, R.J., J.E. VAN EYK. Altered interactions among thin filament proteins modulate cardiac function. *J. Mol. Cell. Cardiol.* 1996, *28*, 217–230.

37 MITTMANN, C., T. ESCHENHAGEN, H. SCHOLZ. Cellular and molecular aspects of contractile dysfunction in heart failure. *Cardiovasc. Res.* 1998, *39*, 267–275.

38 DALLOZ, F., H. OSINSKA, J. ROBBINS. Manipulating the contractile apparatus: genetically defined animal models of cardiovascular disease. *J. Mol. Cell. Cardiol.* 2001, *33*, 9–25.

39 KIMURA, A., H. HARADA, J.E. PARK, H. NISHI, M. SATOH, et al. Mutations in the cardiac troponin I gene associated with hypertrophic cardiomyopathy. *Nature Genetics* 1997, *16*, 379–382.

40 MORNER, S., P. RICHARD, E. KAZZAM, B. HAINQUE, K. SCHWARTZ, et al. Deletion in the cardiac troponin I gene in a family from northern Sweden with hypertrophic cardiomyopathy. *J. Mol. Cell. Cardiol.* 2000, *32*, 521–525.

41 KOKADO, H., M. SHIMIZU, H. YOSHIO, H. INO, K. OKEIE, et al. Clinical features of hypertrophic cardiomyopathy caused by a Lys183 deletion mutation in the cardiac troponin I gene. *Circ.* 2000, *102*, 663–669.

42 ILKOVSKI, B., S.T. COOPER, K. NOWAK, M.M. RYAN, N. YANG, et al. Nemaline myopathy caused by mutations in the muscle alpha-skeletal-actin gene. *Am. J. Hum. Genet.* 2001, *68*, 1333–1343.

43 MICHELE, D.E., J.M. METZGER. Physiological consequences of tropomyosin mutations associated with cardiac and skeletal myopathies. *J. Mol. Med.* 2000, *78*, 543–553.

44 RICHARDSON, R.S., S.C. NEWCOMER, E.A. NOYSZEWSKI. Skeletal muscle intracellular PO_2 assessed by myoglobin desaturation: response to graded exercise. *J. Appl. Physiol.* 2001, *91*, 2679–2685.

45 TRAHAIR, T., T. YEOH, T. CARTMILL, A. KEOGH, P. SPRATT, et al. Myosin light chain gene expression associated with disease states of the human heart. *J. Mol. Cell. Cardiol.* 1993, *25*, 577–585.

46 SCHAUB, M.C., M.A. HEFTI, R.A. ZUELLIG, I. MORANO. Modulation of contractility in human cardiac hypertrophy by myosin essential light chain isoforms. *Cardiovasc. Res.* 1998, *37*, 381–404.

47 VAN EYK, J.E., A.M. MURPHY. The role of troponin abnormalities as a cause for stunned myocardium. *Coron. Artery. Dis.* 2001, *12*, 343–347.

48 MCDONOUGH, J.L., R. LABUGGER, W. PICKETT, S. MACKENZIE, D. ATAR, et al. Cardiac troponin I is modified in the myocardium of bypass patients. *Circ.* 2001, *103*, 58–64.

49 LABUGGER, R., L. ORGAN, C. COLLIER, D. ATAR, J.E. VAN EYK. Extensive troponin I and T modification detected in serum from patients with acute myocardial infarction. *Circ.* 2000, *102*, 1221–1226.

50 MURPHY, A.M., H. KOGLER, D. GEORGAKOPOULOS, J.L. MCDONOUGH, D.A. KASS, et al. Transgenic mouse model of stunned myocardium. *Science* 2000, *287*, 488–491.

51 SIMPSON, J.A., R. LABUGGER, G.G. HESKETH, C. D'ARSIGNY, D.E. O'DONNELL, et al. Differential detection of skeletal troponin I isoforms in serum of a patient with rhabdomyolysis: markers of muscle injury? *Clin. Chem.* 2002 (in press).

52 SOLARO, R.J., D.C. PANG, F.N. BRIGGS. The purification of cardiac myofibrils with Triton X-100. *Biochim. Biophys. Acta* 1971, *245*, 259–262.

53 NEVEROVA, I., J.E. VAN EYK. Application of reversed phase high performance liquid chromatography for subproteomic analysis of cardiac muscle. *Proteomics* 2002, *2*, 22–31.

54 STRAUSS, J.D., J.E. EYK, Z. BARTH, R.J. WIESNER, et al. Recombinant troponin I substitution and calcium responsiveness in skinned cardiac muscle. *Pflugers Arch.* 1996, *431*, 853–862.

55 Moss, R. L., G. G. Giulian, M. L. Greaser. The effects of partial extraction of TnC upon the tension-pCa relationship in rabbit skinned skeletal muscle fibers. *J. Gen. Physiol.* **1985**, *86*, 585–600.

56 Labugger, R., J. L. McDonough, I. Neverova, J. E. Van Eyk. Proteomic analysis of cardiac myofilament proteins: 2DE focusing and detection of troponin T. *Proteomics* **2002** (in press).

57 Merril, C. R., M. E. Bisher, M. Harrington, A. C. Steven. Coloration of silver-stained protein bands in polyacrylamide gels is caused by light scattering from silver grains of characteristic sizes. *Proc. Nat. Acad. Sci. USA* **1988**, *85*, 453–457.

58 Barbato, R., R. Menabo, P. Dainese, E. Carafoli, S. Schiaffino, F. Di Lisa. Binding of cytosolic proteins to myofibrils in ischemic rat hearts. *Circ. Res.* **1996**, *78*, 821–828.

59 Weinberger, S. R., T. S. Morris, M. Pawlak. Recent trends in protein biochip technology. *Pharmacogenomics* **2000**, *1*, 395–416.

60 Liu, H., D. Lin, J. R. Yates. Multidimensional separations for protein/peptide analysis in the post-genomic era. *Biotechniques* **2002**, *32*, 898–902.

61 Alpert, J. S., K. Thygesen, E. Antman, J. P. Bassand. Myocardial infarction redefined – a consensus document of The Joint European Society of Cardiology/ American College of Cardiology Committee for the redefinition of myocardial infarction. *J. Am. Coll. Cardiol* **2000**, *36*, 959–969.

62 Colantonio, D. A., W. Pickett, R. J. Brison, C. E. Collier, J. E. Van Eyk. Detection of cardiac troponin I early after onset of chest pain in six patients. *Clin. Chem.* **2002**, *48*, 668–671.

Section 3
Future Perspectives

20

Genomics Perspective for Drug Discovery

A. J. Marian and Michael H. Gollob

20.1
Introduction

The history of drug discovery could be divided into three eras. The first era was marked by the empiric use of natural products based on anecdotal reports. At the turn of the 20th century, the paradigm shifted when discoveries of chemistry were joined with the tools of pharmacology to design and develop drugs based on rational biological experimentations. During this phase, which spanned the entire 20th century, clinical utilization of drugs gradually transformed from the empiric use to evidence-based effectiveness. The majority of our current therapeutic armamentarium developed during this era. Advances in the fields of molecular biology and genetics brought forth a new era in drug discovery at the turn of the new millennium. As a result, genomics, proteomics, and bio-informatics are used in complementation with the modern tools of chemistry and pharmacology to design and develop new drugs.

Genomics, defined as information derived from the structure and organization of the genes of an organism, are used to identify targets, validate targets and lead compounds and to individualize therapy. Advances in genomics in conjunction with the modern techniques of combinatorial chemistry and high-throughput screening have provided an unprecedented wealth of information for efficient and less costly drug discovery. Various genomic-based techniques, such as high-throughput DNA and protein sequencing, DNA microarray chips for expression profiling, mass spectrometry, and bio-informatics have been developed and used to identify and validate targets for pharmacological interventions. In parallel with these efforts, a complementary phase of genomic research has already begun with the goal of identification and construction of single nucleotide polymorphism (SNP) and haplotype maps of the human genome. The ultimate goals are to map and identify the susceptibility genes for common complex diseases in order to develop new therapeutic targets and to individualize drug therapy based on genetic profiling (often referred to as pharmacogenetics). Furthermore, efforts are underway to generate 3-D structures of a large number of target molecules in order to design and develop drugs that are targeted against molecular structure and have the least unwanted effects. The potential impact of genomics on drug discovery is evident from the fact that the current pharmacopoeia is based on products of less than 500 molecules [1], while the potential exists for the expansion of our pharmacological armamentarium to specifically target products of approximately 35,000 genes that comprise the human genome [2]. Furthermore, genetic profiling – SNP and haplotype mapping – could also lead to more effective treatment of the responders and identification of those who are at risk of developing side effects. Collectively, utilization of genome-based technology has the potential to introduce a large number of "biotech" drugs into the market. In the year 2001, 6 "biotech" drugs were introduced, while the number is expected to double in 2002. Genomic-based drug discovery and treatment could also reduce the cost and the time it takes to bring a product to the market. At the present time it is estimated that it takes about US $500 million and 15 years to develop and market a drug [3]. In 1998, the pharmaceutical industry spent approximately $20 billion in developing new drugs [3]. Genomics, by providing for a more efficient target identification and validation as well as more intelligent drug design and testing, could reduce the cost of developing new drugs. Altogether, genome-based drug discovery has the potential to revolutionize the practice of medicine in the coming years.

20.2
Human Genome

Miescher was the first to isolate DNA in 1869 [4]. However, the biological significance of this discovery was unrecognized until 1944, when Avery, McLeod and McCarty [5] and Hershey and Chase later on in 1952 [6], showed DNA, rather than protein, was responsible for inheritance. The works of Nobel Laureates Wilkins, Watson and Crick [7, 8] established the double-stranded structure of DNA and laid the foundation of modern molecular biology and genetics. The birth of recombinant DNA technology occurred in 1970 with the discovery of a type II restriction enzyme in *Hemophilus influenza* [9], that along with identification of DNA ligase three years earlier [10], made it possible to cut DNA at specific sites and rejoin the DNA fragments. Subsequently, a DNA fragment was subcloned into a self-replicating plasmid *in vitro* [11]. These advances, along with isolation of DNA polymerases and reverse transcriptase [12], the development of techniques of DNA sequencing [13], and the polymerase chain reaction [14] provided the milieu necessary for initiation of the Human Genome Project (HGP). The HGP was started in 1990 with the initial goal of developing genetic and physical maps of the human genome and ultimate goal of sequencing the entire genome. The project was successfully completed in 2001 several years ahead of schedule and the first draft of the human genome was published [2].

The human genome is comprised of approximately 3.2 billion base pairs organized in approximately 35,000 genes and inter-genic regions [2]. Genes encoding proteins, the major interest of pharmacogenomics, comprise about 1% of the genome and the remainder is comprised of repeat sequences and segments with unknown function. The most direct utility of genomics with regard to drug discovery is to identify and characterize the encoded proteins from all genes in the human genome (proteomics), which provide for the drug targets. Another powerful feature of the human genome is the presence of sequence variation among individuals, referred to as polymorphism. While >99% of the genome sequence is identical between individuals, subtle variation in the sequence exists that are used to map and identify the susceptibility genes for complex diseases and thus, identify new drug targets. The most common type of polymorphism is the single nucleotide polymorphism (SNP), which is estimated to occur 1 in every 300 bases in the human genome. Each gene is expected to contain several SNPs. Systematic screening of a 5.5 kb fragment encompassing the gene encoding apolipoprotein E shows 21 SNPs [15]. Similarly, at least 13 SNPs have been identified in a 27 kb fragment encompassing the angiotensin-1 converting enzyme 1 gene [16]. As of December 13, 2001, the SNP database (dbSNP) for the human genome contains more than 4.1 million SNPs (*http://www.ncbi.nlm.nih.gov/SNP*). The majority of SNPs are located in the non-coding regions and are useful as genetic markers. A smaller number of SNPs are located in the coding (cSNPs) and regulatory regions of genes (rSNP) and thus, could affect structure or expression level of the encoded proteins. Therefore, SNPs are powerful tools not only for gene mapping and target identification, but also for delineating the genetic basis for the inter-individual variation in response to drug therapy (pharmacogenetics). The emphasis by the public and private consortiums is to develop SNP and haplotype maps of the human genome in order to perform a genome-wide search for the susceptibility genes for common diseases, such as atherosclerosis and hypertension, and to perform genetic profiling in order to individualize medical therapy.

20.3
Genomics and the Process of Drug Discovery

The current process of drug discovery could be classified into four stages of target identification, target validation, lead identification and lead validation. The next phase following drug discovery is to individualize therapy based on genetic information. We briefly discuss the potential impacts of genomics and the related sciences on each phase of drug discovery as well as individualization of drug therapy.

20.3.1
Genomics and Target Identification

Perhaps, the most important contribution of genomics to drug discovery is toward understanding the molecular pathogenesis of human diseases. One direct consequence of such contribution is the identification of drug targets. Prior to the beginning of the modern era of molecular genetics, target identification was restricted by the ability to find and isolate new proteins. Drug development also required *a priori* knowledge of involvement of the isolated proteins in the pathogenesis of disease of interest. Angiotensin-1 converting enzyme 1 (ACE-1) inhibitors and β-blockers were developed when experimental data suggested their involvement in the pathogenesis of heart failure and hypertension. The approach of isolating and identifying a protein as the initial step for drug discovery, while tedious, has led to identification of the vast majority of the current cardiovascular drugs, such as ACE-1 inhibitors, angiotensin II receptor 1 blockers, β-blockers, thrombolytics, IIb/IIIa inhibitors, and endothelin-1 receptor blockers. The limited number of known proteins, however, hinders the utility of this approach in drug discovery. As a result, our current pharmacopoeia is aimed at approximately 500 molecules, the majority of which are targeted to a few receptors and enzymes [1]. As such, over 50% of our current drugs directly affect G protein coupled receptors [1]. The genomic approach, which initially was referred to as "the reverse genetics", does not require *a priori* isolation of the encoded proteins or *a priori* knowledge of the candidate gene in the pathogenesis of the disease. Instead, potential drug targets are identified through the use of genomic tools. Accordingly, the number of proteins and their isoforms encoded from the approximately 35,000 genes in the human genome determines the number of potential drug targets. Potential drug targets are identified through genome-wide chromosomal mapping, profiling of mRNA expression (also referred to as transcriptome), profiling of protein expression (proteomics), genotype-phenotype correlation studies and large-scale SNP/haplotype association studies. Consequently, a genomic approach could provide for identification of novel molecular pathways as targets for drug therapy. In addition, it could lead to identification of known pathways that were not previously implicated in the pathogenesis of disease of interest and thus, were not considered targets for drug therapy. Given the number of genes in the human genome, estimated at approximately 35,000, and the number of encoded proteins and their isoforms, it is quite conceivable that the potential targets for drug development may be amplified at least by an order of magnitude or even greater.

Genomics has had the greatest impact on identification of the causal genes for single-gene disorders. As of 02/09/2002 over 30,000 mutations in more than 1,000 human genes have been identified (*http://archive.uwcm.ac.uk/uwcm/mg/docs/haha-ha.html*). During the past 15 years, the causal genes for more than 100 different diseases involving the cardiovascular system have been identified. The direct consequence of identification of the causal gene is to design and develop drugs that could specifically target the mutant protein. For a recessive disorder, whereby the disease results from the lack of a specific protein, identification of the causal gene could lead to replacement of the deficient protein. Enzyme replacement therapy has been tested in human patients with Fabry disease; an X-linked recessive lysosomal storage disorder caused by a deficiency of α-galactosidase A [17]. The phenotype manifests as painful neuropathy, progressive renal, cardiovascular, and cerebrovascular dysfunction and premature death [17]. Preliminary studies show safety and efficacy of single infusions of α-galactosidase A prepared from transfected human fibroblasts [18]. For an autosomal dominant disorder, specific drugs could be designed to manipulate expression or function of the mutant protein. This is illustrated in the case of familial hypercholesterolemia, caused by mutations in the low-density lipoprotein receptor (*LDLR*) gene [19]. The phenotype is characterized by the deficiency of LDL receptors in hepatocytes, severe hyperlipidemia and premature atherosclerosis. While the primary therapeutic target protein for reduction of LDL-cholesterol is HMG CoA reductase, insight into the regulation of the *LDLR* gene has provided additional targets. It is shown that the expression of LDL receptors is regulated by the sterol regulatory binding protein (SREBP) transcription factors and SREBP cleavage protein (SCAP) [20]. New compounds have been developed that bind to SCAP and activate cleavage of SREBPs [21]. Increased expression level of the active SREBPs in the cell nucleus activates transcription of LDL receptors in the hepatocytes [21], resulting in increased LDL receptor density, increased hepatic uptake of LDL and decreased levels of plasma LDL and triglycerides [21]. Thus, SCAP ligands have emerged as potential new drugs for treatment of patients with dyslipidemia that could be used in conjunction with HMG CoA reductase inhibitors to maximize the benefits.

Genetic and molecular biology studies also could provide significant insight into the pathogenesis of the disease of interest and thus identification of potential targets. One example is the case of hypertrophic cardiomyopathy, an autosomal dominant disease for which currently no specific therapy is available. During the past 10 years the molecular genetic basis of HCM has been elucidated and significant insight has been gained into the pathogenesis of its phenotype [22] . In addition, expression profiling has led to identification of genes that are differentially expressed in the heart of patients with HCM [23]. The collective results of these experiments have led to the notion that hypertrophy and fibrosis are secondary phenotypes due to activation of stress-responsive signaling molecules and kinases, and thus, potentially reversible [24]. Insight into the pathogenesis of HCM – derived from genetic studies, has led to the new application of existing drugs to reverse and attenuate cardiac phenotype [25, 26]. We have shown blockade of signaling kinases implicated in cardiac hypertrophy with simvastatin could reverse evolving cardiac hypertrophy and fibrosis and improve cardiac function in a transgenic rabbit model of HCM [25]. Similarly, we have shown that losartan, an angiotensin II receptor 1 blocker, could reverse and normalize interstitial fibrosis in a transgenic mouse model of HCM [26]. Thus, genomics, by providing insight into the pathogenesis of disease, not only could provide for the opportunity to develop new drugs, but also could provide for new applications of existing drugs.

While thus far, genomics has had the greatest impact on identification of the causal genes for monogenic disorders, construction of SNP and haplotype maps of the human genome is now shifting the paradigm toward mapping and identification of the susceptibility genes for common complex disorders. Similar to monogenic disorders, identification of genes involved in susceptibility to complex phenotypes, such as atherosclerosis, hypertension, and dyslipidemia, will identify a variety of new targets for drug development. This point is illustrated for identification and cloning of human angiotensin-1 converting enzyme 2 (ACE2) [27], which was identified through 5′ sequencing of a human heart failure ventricular cDNA library. Unlike ACE1, ACE2 is expressed only in the heart, kidney, and testis and is localized predominantly to the endothelium of coronary and intra-renal vessels and to renal tubular epithelium [27]. ACE2 hydrolyzes angiotensin 1 to angiotensin 1–9, but does not cleave bradykinin [27]. Therefore, given the organ- and cell-specific expression of ACE2 and its unique cleavage of key vasoactive peptides, it is an attractive drug target for blocking expression of the local renin-angiotensin system in the heart and kidney in patients with heart failure and hypertension.

Genomic Tools for Drug Target Identification. A cadre of genomic tools is currently available for identification of potential drug targets and the list is expanding very rapidly. It encompasses tools used to profile gene expression – at mRNA or protein level – in a pathological state, bio-informatics and structural biology, and many others. The utility of these techniques is not restricted to identification of potential drug targets but also extends to target validation as well. The majority of the currently used techniques, such as expression profiling using micro-array DNA chips and proteomics, have been discussed in separate chapters. Therefore, our discussion will be limited to their utility on drug discovery.

Expression profiling is performed to identify genes that are differentially expressed, at the mRNA or protein level, in a disease state. Identification of the differentially expressed genes not only could provide insight into the pathogenesis of the disease under investigation but also could provide for new targets for drugs. Currently, several techniques are available for detection of differentially expressed genes at the mRNA level, including subtraction hybridization, DNA micro-array chips, and serial analysis of gene expression (SAGE). High-throughput sequencing of Expressed Sequence Tags (ESTs) has led to development of a compendium of differentially expressed genes in the cardiovascular system [28]. Similarly, subtraction hybridization has led to identification of differentially expressed genes – in part – in human hypertrophic cardiomyopathy [23]. The advantage of techniques, such as subtraction hybridization, EST sequencing and SAGE, is their ability to detect expression of the known and novel genes without *a priori* knowledge of their involvement in the pathogenesis of the disease. However, these techniques are somewhat cumbersome and may require several sets of experiments for comprehensive analysis. In contrast, micro-array DNA chips, while restricted to known sequences, provide for screening of a large number of genes in a single hybridization assay. Until recently, microarray chips provided for screening of a relatively limited number of known genes as potential drug targets and thus precluded discovery of unknown genes. The completion of The Human Genome Project and identification of expressed sequences have diminished this problem significantly. The recent microarray DNA chips comprise approximately 39,000 human transcripts from more than 33,000 well-characterized human genes (*http://www.affymetrix.com*). Micro-array technology has been utilized to detect changes in gene expression in a variety of cardiovascular pathobiologies including myocardial ischemia, heart failure and reperfusion injury [29–31]. It also provides for the opportunity to monitor gene expression following pharmacological interventions [32, 33]. DNA micro-array chips have been used to detect gene expression following myocardial infarction before and after treatment with an ACE-1 inhibitor [33]. Profiling gene expression before and after pharmacological intervention could identify additional previously unidentified targets.

Proteomics provide global analysis of gene expression at the protein level and could be used to analyze protein content of a given organ or cells at a given time. Several methods also have been developed to identify the differentially expressed proteins, the ultimate drug targets, in disease states. The conventional approach of using 2D gel electrophoresis has been conjoined with the power technique of mass spectrometry to identify the proteins based on their mass. Recently, a high-throughput protein analyzer based on Matrix Assisted Laser Desorption Ionization (MALDI) source and Time of Flight optics has been developed, which could accelerate discovery and identification of the differentially expressed proteins by a magnitude of order. Data derived from expression profiling along with advances in computational techniques and bioinformatics have provided a wealth of information that could be used as potential drug targets.

20.3.2
Genomics and Target Validation

Genomic based techniques commonly lead to identification of a large number of potential drug targets. To identify the most relevant candidates, it is necessary to validate the role of each specific candidate in the pathogenesis of the disease of interest. The initial approach is largely based on selection of targets that are biologically plausible candidates as suggested by the existing biological data. Subsequent experiments, to validate individual genes and proteins as the potential candidates for drug development, may involve *in vitro* and *in vivo* experiments using techniques of functional genomics. Transgenic animal models are commonly used to determine the impact of organ-specific over-expression of the candidate gene. Inversely, knock out and gene-targeting experiments are used to determine the impact of absence or reduced expression levels of the candidate genes in the pathogenesis of the disease of the interest. Gene-targeting experiments are sometimes confounded by the functional redundancy of the candidate genes and embryonic lethality. Therefore, several methods of conditional gene expression have been developed whereby expression of the gene of interest could be induced and regulated in a specific organ as desired. In addition, gene transfer experiments and antisense oligonucleotides are used to validate the function of the candidate gene in the pathogenesis of the disease of interest.

Molecular genetics and biology studies also have led to identification of several attractive drug targets for treatment of dyslipidemia including targets involved in "reverse-cholesterol transport". A variety of experiments have explored the utility of ATP binding cassette 1 (ABCA1) protein, scavenger receptor class B, type 1 (SR-B1), Lecithin:cholesterol acyltransferase (LCAT), and cholesteryl ester transfer protein (CETP) as targets for lipid lowering drugs. The process usually includes identification of the causal gene for a familial phenotype characterized by dyslipidemia. With regard to ABCA1, molecular genetic studies of Tangier disease, a phenotype characterized by very low plasma levels of HDL-C and ApoA1 and an increased risk of premature coronary atherosclerosis, led to identification of mutations in the *ABCA1* gene [34, 35]. In addition, SNPs in ABCA1 have been associated with the severity of coronary atherosclerosis in population studies [36, 37]. Therefore, ABCA1 protein has emerged as an attractive drug target for manipulating plasma levels of HDL and treatment of atherosclerosis. Subsequent generation of transgenic and knock out mouse models has provided additional credence to ABCA1 as a potential drug target for regulating plasma levels of lipids and reducing the risk atherosclerosis [38–43]. Accordingly, over-expression of ABCA1 results in increased plasma levels of HDL-C and apoA1 and confers protection against atherosclerosis [39–41]. In contrast, knock out deletion of ABCA1 leads to virtual absence of HDL-C [28, 43]. Collectively, the results of these experiments along with the result of many others validate ABCA1 as a potential drug target for treatment of dyslipidemia in humans.

Genetic and molecular biology studies have provided credence to the utility of CETP, LCAT and Acyl coenzyme A-cholesterol acyltransferase (ACAT) as potential drug targets for lipid lowering effects. With regard to CETP, the first direct clue to the potential involvement of CETP in reverse cholesterol transport emerged from the identification of 2 Japanese families who had CETP deficiency and extremely high levels of HDL-C [44, 45]. Subsequent cloning of the CETP cDNA [46], SNPs association studies [47], and transgenesis [48], provided further evidence as for the validity of CETP as a potential drug target to raise plasma levels of HDL-C. An example of a lead molecule to inhibit CETP is JTT-705, designed based on amino acid sequence of CETP. It forms a disulfide bond with CETP, inhibits its activity, increases HDL cholesterol, decreases non-HDL cholesterol and slows the progression of atherosclerosis in animal models [49]. Phase 1 clinical trials are ongoing to determine safety and efficacy of the inhibition of CETP in humans. Another attractive target for drug inhibition is ACAT, since partial inhibition of ACAT improves dyslipidemia and induces regression of atherosclerosis in experimental models [50]. An alternative approach to inhibition is to promote expression of the target gene. This approach has been tested for two targets in dyslipidemia. One example is upregulation of expression of LCAT, which provide salutary effects on lipid profile [51]. Similarly, promotion of expression of SCAP, which cleaves and activates SREBP, is an attractive modality for treatment of high plasma levels of LDL-C, as discussed earlier [21]. Collectively, these examples illustrate the role of genomics in target identification and validation and the potential utility of such targets in treatment of dyslipidemia in the general population.

Genomics also have been used to validate the role of renin-angiotensin-aldosterone system (RAAS) and the kallikrein-kinin system as effective targets in the treatment of systemic cardiovascular diseases such as hypertension, heart failure, and renal failure. Overexpression of human tissue kallikrein in transgenic mice lowers the blood pressure [52]. Similarly, gene transfer and expression of human tissue kallikrein lowers blood pressure that persists for longer than 6 weeks, attenuates cardiac hypertrophy, and improves renal function [53, 54]. Several sets of experiments also have validated components of the renin-angiotensin-aldosterone system (RAAS) as potential drug targets, despite the well-established role of inhibition of the RAAS in treatment of hypertension and heart failure. Transgenic mice expressing angiotensinogen exhibit a dose-dependent rise in blood pressure [55]. In contrast, anti-sense oligonucleotides targeted against angiotensinogen mRNA lowers the blood pressure and attenuate cardiac hypertrophy in spontaneously hypertensive rats [56–58]. Similar experiments using antisense technology and gene targeting experiments [59, 60] also validate other components of the RAAS as drug targets for chronic long-term treatment of hypertension.

The adrenergic system plays a major role in a variety of cardiovascular pathologies, in particular heart failure. Experimental studies implicate activation of β-adrenergic kinase 1 (βARK1) in phosphorylation and desensitization of the β-adrenergic receptors in heart failure [61]. Over-expression of βARK1 in the heart attenuates myocardial contractile response to adrenergic stimulation [62]. In contrast, cardiac-restricted expression of βARK1 inhibitor leads to an enhanced cardiac contractile performance [62]. In addition, blockade of β-receptors is associated with reduction of βARK1 activity [63]. Collectively, these data suggest βARK1 is a suitable target for drug targeting in heart failure. To validate the potential significance of inhibition of βARK1 in heart failure, transgenic mice have been generated that over-express an inhibitor of βARK1 in the heart and crossed to mouse models of heart failure. Over-expression of inhibitor of βARK1 rescued the phenotype of heart failure in transgenic mice that overexpress the sarcoplasmic reticulum Ca^{2+}-binding protein, calsequestrin [63]. The results of target validation studies have paved the way toward development and refinement of lead compounds to inhibit function of βARK1 or down-regulate its expression in the heart.

20.3.3
Genomics and Lead Identification and Refinement

Identification of new drug targets and their subsequent validation shifts the emphasis toward search for a lead compound that interacts with the target protein and affects its expression or function. The lead compound is subsequently refined to design and develop a specific drug with the desirable therapeutic index (median toxic dose/median effective dose). The conventional approach is to perform high-throughput screening of combinatorial libraries, comprised of large collections of all possible combinations of a set of smaller chemical structures, against the potential targets. Then the large pool of combinatorial libraries is fractionated into smaller pools optimized for an activity of interest until compounds with the highest levels of activity are identified (deconvlution). Advances in combinatorial chemistry and high-throughput screening techniques have made it possible to test the effect of thousands of compounds at several concentrations on a target molecule in a short period of time. However, the random testing of a large number of compounds has had limited success in accelerating the rate of drug discovery and the results have been compounded by the limitation of cell-based assays to reflect the complexity of the molecular biology of living organisms. Thus, the focus is toward the utilization of tools of structural genomics to simplify the screening process and support the results with biology evidence.

The process of drug design and sampling of the compounds against the target protein has been significantly accelerated by advances in structural genomics. Tools of structural biology, such as X-ray crystallography and magnetic resonance spectroscopy in conjunction with bio-informatics have led to accelerated characterization of the three dimensional structure of proteins, which could provide major clues for appropriate selection and design of drugs. The three dimensional structure of a variety of proteins expressed in the cardiovascular system have been determined, including cardiac troponin C [64], myosin heavy chain [65], FKBP12.6, an isoform of FKBP12 that selectively binds to the cardiac ryanodine receptor (RyR2) [66], HERG potassium channel N terminus [67], calsequestrin [68], cytochrome c oxidase [69], integrin-binding fragment of human VCAM-1 [70], human recombinant cystathionine beta-synthase [71], hormone-binding domain of a guanylyl-cyclase-coupled receptor [72], N-terminal two domains of intercellular adhesion molecule-1 [73], VEGF in complex with domain 2 of the Flt-1 receptor [74], among many others.

The initial lead compound usually does not have the desirable therapeutic index, necessitating further refinement of the lead molecule. Typically, structural modifications are made to the lead molecule and the modified molecules are tested in cell cultures and animal models until a compound with a desirable therapeutic index is identified and selected for testing in humans. Delineating the crystal structure of the target and lead molecules could accelerate the process of lead refinement. This is illustrated for HMG CoA reductase and its inhibitors (statins). Despite HMG CoA reductase inhibitors being the cornerstone of treatment of dyslipidemia, there is substantial inter-individual variability in the response of plasma lipids to treatment with statins. In an attempt to further refine and develop new HMG Co A reductase inhibitors, the structural basis for binding of statins to HMG CoA reductase has been delineated [75]. Accordingly, statins occupy the enzymatic active site of HMG CoA reductase by forming multiple polar and van der Waals interactions with the enzyme, blocking its access to the substrate [75]. However, statins do not occupy the nicotinamide-binding site of the enzyme. This information provides the opportunity to design and develop statins with the nicotinamide-like moiety. Such drugs, through covalent attachment to the nicotinamide-binding site, in addition to binding to the active enzymatic site, will be expected to confer higher potency.

Genomic tools also have been utilized in designing and developing more clot-specific thrombolytic agents, such as the mutant forms of tissue type plasminogen (tPA) [76, 77] and streptokinase [78]. Molecular biology studies suggest a catalytic domain in the N terminus of streptokinase can activate plasminogen through both fibrin-dependent and fibrin-independent mechanisms. Deletion of 59 amino acids from the N terminus catalytic domain decreases the fibrin-independent plasminogen activity by >600-fold [78]. As a result, *in vitro* assays show the mutant streptokinase confers total clot lysis with minimal fibrinogen degradation [78]. These findings raise the possibility of developing safer and more effective mutant forms of streptokinase. Similarly, several variants of tPA have been developed and tested with the goal of establishing safer, more effective and simple to use agents.

An example of mutant tPA is the TNK-tPA, which has a longer plasma half-life, enhanced fibrin-specificity, and increased resistance to the plasminogen activator inhibitor-1 [76, 77].

While structural genomics provides major clues for the design, selection, and refinement of the lead compounds, protein-protein interactions also could lead to additional target and lead identification. Molecular biology tools, such as yeast two-hybrid screening, could identify proteins that interact with the original target as the potential secondary targets for drug design and validation. Yeast, a unicellular organism, has mammalian homologue genes and could be used for screening and identification of lead compounds [79]. Thus, lead compounds are identified based on the structural and biological function of the target molecule.

20.3.4
Genomics and Lead Validation

Tools of genomics are utilized in conjunction with pharmacological experiments to validate leads as therapeutic agents, to establish their safety, efficacy and potency and define their side effects. In general, lead validation encompasses three phases of *in vitro* and cell culture studies, testing the compounds in animal models of human disease, and ultimately through performing clinical trials in humans. Cell culture and *in vitro* experiments determine whether the candidate lead compound confers the expected phenotype, such as blocking the target receptor or enzyme, at desirable specificity and concentration. The lead compounds are tested in animal models to determine their efficacy, safety, and side effects. The general goal is to determine whether administration of the candidate compound produces the desirable effect on the phenotype of interest without causing major side effects. Ultimately, clinical trials are performed to determine the efficacy, potency and toxicity of the candidate compound in humans (different phases of clinical trials).

In the first and second stage of lead validation a variety of molecular biology experiments and genetic manipulations are used. Gene transfer studies and functional genomics in animal models could provide crucial efficacy and safety data. Examples are abound and includes a variety of studies in animal models including transgenic animal models of human diseases, such as validation of a CETP inhibitor in a rabbit model of dyslipidemia, as discussed earlier [49]. Similarly, validation studies of TNK-tPA were initially performed in a rabbit model of thrombosis, which showed significantly higher potency and clot selectivity of TNK-tPA than the wild type tPA. Initial validation studies of mutant streptokinase Δ59 were performed in *in vitro* assays [78]. Furthermore, validation studies of inhibition of βARK1 have been performed through transgenesis and crosses of transgenic mice [63]. Finally, knock out models, such as apoE-/- [80, 81] and LDLR-/- [82] mice provide excellent opportunities to test the effects of lead compounds on the phenotype of interest, such as dyslipidemia and atherosclerosis.

20.3.5
Genomics and Individualization of Drug Therapy

Inter-individual variation in response to drugs, both beneficial and harmful, has been recognized since the dawning of medicine. One of the first examples is the variability in response to isoniazid and individuals were considered either as "fast" or "slow" metabolizers. It was subsequently recognized that differences in the rate of acetylation of isoniazid, the major route of its elimination, was responsible for the inter-individual variability. Another well-recognized example is response to warfarin therapy, which is encountered commonly in the daily practice of cardiology. While the familial basis of variability in response to drug therapy has been recognized for many years, its molecular genetic basis remain largely unknown. The contribution of genomics is to identify the genetic basis of variability in drug responses and thus, to establish individualized drug therapy.

Completion of The Human Genome Project and development of SNP and haplotype maps have provided the opportunity to map and identify genes and their variants involved in response to drug therapy. As a result, the field of identification of genetic determinants of response to therapy, often referred to as "pharmacogenetics", has become the focus of extensive research by the pharmaceutical and biotechnology industry. The ultimate goal is to tailor drug therapy according to the genetic profile in order to maximize the beneficial effects while minimizing the side effects. Extensive review of this topic is beyond the scope of this Chapter and the reader is referred to an excellent recent review [83].

The existing knowledge is largely based on SNP-association studies, which are subject to a high rate of spurious results [84]. Large-scale studies are ongoing to identify the genetic determinants of response to drugs. Nevertheless, despite the provisional results of association studies, it has become evident that SNPs are major determinants of variability to drug response in each stage of drug effects. SNPs and haplotypes could contribute to variability in response to therapy by affecting the pharmacokinetics, namely absorption, distribution, metabolism and elimination of the drug. This is exemplified in cloning and identification of SNPs in cytochrome p450 genes, which have led to identification of several variants that affect drug metabolism. SNPs in *CYP2C9* have been found to affect the anticoagulant effect of warfarin [85]. Similarly, SNPs in *CYP2D6*, encoding debrisoquine hydroxylase, which catalyze the oxidation of drugs containing a basic nitrogen, affect blood levels of β blockers [86]. Subjects with SNPs in *CYP2D6* that results in loss of function are the "poor metabolizers" and subjects with gain of function SNPs are considered "rapid metabolizers". Indeed, inter-ethnic variation in response to drugs is partly due to differences in the prevalence of loss of function and gain of function SNPs in *CYP2D6* among different ethnic populations. In addition, SNPs in the drug metabolizing genes could affect drug-drug interactions as exemplified by the inter-individual variability in interactions between erythromycin and antiarrhythmic drugs. Accordingly, not only the differences in the magnitude of beneficial response to therapy but also the frequency and severity of side effects and drug-drug interactions are partly determined by SNPs that affect drug metabolism, such as those located in cytochrome p450 genes.

SNPs in target proteins, by affecting the sensitivity of the target to drugs (pharmacodynamics), are also major determinants of response to drug therapy. SNPs in CETP have been implicated in response to pravastatin [87] and an insertion/deletion polymorphism in the ACE-1 gene has been associated with response to treatment with ACE-1 inhibitors [88]. SNPs and haplotypes in human ether a go go-related (*HERG*) and *KCNH2* (formerly known as Mirp1) have been associated with drug-induced long QT syndrome and torsade de pointes [89], those in β2-adrenergic receptors with progression of heart failure [90], vascular responsiveness [91] and response to bronchodilators [92], and those in ryanodine receptor 2 with anesthesia induced malignant hyperthermia [93, 94].

The ultimate goal of drug therapy is to develop drugs that are targeted to a specific phenotype and provide maximum beneficial effects with minimal side effects. Pharmacogenomic and pharmacogenetic research could provide the opportunity to identify new drug targets, design drugs more selective to the phenotype and to individualize drug therapy. Major strides have already been made and advances in genomics will accelerate the pace of our progress toward achieving these goals.

Tab. 20.1 Genomics-based approach to drug discovery and therapy

---

Target identification (gene mapping and expression profiling)
  Identifying the causal gene (single gene disorders)
  Identifying the susceptibility genes (complex traits)
  Identifying the modifier genes
  Identification of differentially expressed mRNA
  Identification of differentially expressed proteins

Target validation (functional genomics)
  Cell culture studies
  Transgenesis
  Gene targeting studies
  Gene transfer

Lead identification (Structural genomics)
  Determining 3-dimensional structure of proteins
  Identification of protein-protein interaction

Lead validation (Gene based therapy)
  Cell culture studies
  Transgenesis
  Gene targeting
  Gene transfer and other gene therapy approaches

Genomics and individualized drug therapy
  SNP and haplotype association studies

20.4
Acknowledgements

This work is supported in part by grants from the National Heart, Lung, and Blood Institute, Specialized Centers of Research P50-HL42267-01 and 1R01HL/DK68884-01.

20.5
References

1 Drews, J. 2000. Drug discovery: a historical perspective. *Science* 287:1960–1964. LANDER, E.S., *et al.* 2001. Initial sequencing and analysis of the human genome. *Nature* 409:860–921.

3 MICHELSON, S., JOHO, K. 2000. Drug discovery, drug development and the emerging world of pharmacogenomics: prospecting for information in a data-rich landscape. *Curr. Opin. Mol. Ther.* 2:651–654.

4 JAMES,J. 1970. Miescher's discoveries of 1869. A centenary of nuclear chemistry. *J. Histochem. Cytochem.* 18:217–219.

5 AVERY, O.T., MACLEOD, C.M., MCCARTY, M. 1944. Studies on the chemical transformation of penumococcal type. *J. Exp. Med.* 79:137–158.

6 HERSHEY, A.D., CHASE, M. 1952. Independent function of viral protein and nucleic acid in growth of bacteriophage. *J. Gen. Physiol* 36:39–56.

7 WATSON, J.D., CRICK, F.H. 1953. Molecular structure of nucleic acids: a structure for deoxyribose nucleic acid. *Nature* 171:737–738.

8 WILKINS, M.H.F., STOKES, A.R., WILSON, H.R. 1953. Molecular structure of deoxypentose nucleic acids. *Nature* 171:748–750.

9 SMITH, H.O., WILCOX, K.W. 1970. A restriction enzyme from Hemophilus influenzae. I. Purification and general properties. *J. Mol. Biol.* 51:379–391.

10 OLIVERA, B.M., LEHMAN, I.R. 1967. Linkage of polynucleotides through phosphodiester bonds by an enzyme from Escherichia coli. *Proc. Natl. Acad. Sci. USA* 57:1426–1433.

11 COHEN, S.N., CHANG, A.C., BOYER, H.W., HELLING, R.B. 1973. Construction of biologically functional bacterial plasmids in vitro. *Proc. Natl. Acad. Sci. U.S.A* 70:3240–3244.

12 BALTIMORE, D. 1970. RNA-dependent DNA polymerase in virions of RNA tumour viruses. *Nature* 226:1209–1211.

13 SANGER, F., NICKLEN, S., COULSON, A.R. 1977. DNA sequencing with chain-terminating inhibitors. *Proc. Natl. Acad. Sci. USA* 74:5463–5467.

14 MULLIS, K., et al. 1986. Specific enzymatic amplification of DNA in vitro: the polymerase chain reaction. *Cold Spring Harb. Symp. Quant. Biol. 51* Pt 1:263–273.

15 NICKERSON, D.A., et al. 2000. Sequence diversity and large-scale typing of SNPs in the human apolipoprotein E gene. *Genome Res* 10:1532–1545.

16 ZHU, X., et al. 2001. Linkage and association analysis of angiotensin I-converting enzyme (ACE)-gene polymorphisms with ACE concentration and blood pressure. *Am. J. Hum. Genet.* 68:1139–1148.

17 BRADY, R.O., SCHIFFMANN, R. 2000. Clinical features of and recent advances in therapy for Fabry disease. *JAMA* 284:2771–2775.

18 SCHIFFMANN, R., *et al.* 2000. Infusion of alpha-galactosidase A reduces tissue globotriaosylceramide storage in patients with Fabry disease. *Proc. Natl. Acad. Sci. USA* 97:365–370.

19 HOBBS, H.H., BROWN, M.S., GOLDSTEIN, J.L. 1992. Molecular genetics of the LDL receptor gene in familial hypercholesterolemia. *Hum. Mutat.* 1:445–466.

20 BROWN, M.S., GOLDSTEIN, J.L. 1999. A proteolytic pathway that controls the cholesterol content of membranes, cells, and blood. *Proc. Natl. Acad. Sci. USA* 96:11041–11048.

21 GRAND-PERRET, T., et al. **2001**. SCAP ligands are potent new lipid-lowering drugs. *Nat. Med. 7*:1332–1338.

22 MARIAN, A.J., SALEK, L., LUTUCUTA, S. **2001**. Molecular genetics and pathogenesis of hypertrophic cardiomyopathy. *Minerva Med. 92*:435–451.

23 LIM, D.S., ROBERTS, R., MARIAN, A.J. **2001**. Expression profiling of cardiac genes in human hypertrophic cardiomyopathy: insight into the pathogenesis of phenotypes. *J. Am. Coll. Cardiol 38*:1175–1180.

24 MARIAN, A.J. **2000**. Pathogenesis of diverse clinical and pathological phenotypes in hypertrophic cardiomyopathy. *Lancet 355*:58–60.

25 PATEL, R., *et al.* **2001**. Simvastatin induces regression of cardiac hypertrophy and fibrosis and improves cardiac function in a transgenic rabbit model of human hypertrophic cardiomyopathy. *Circulation 104*:317–324.

26 LIM, D.S., et al. **2001**. Angiotensin II blockade reverses myocardial fibrosis in a transgenic mouse model of human hypertrophic cardiomyopathy. *Circulation 103*:789–791.

27 DONOGHUE, M., et al. **2000**. A novel angiotensin-converting enzyme-related carboxypeptidase (ACE2) converts angiotensin I to angiotensin 1–9. *Circ. Res. 87*:E1–E9.

28 LIEW, C.C., et al. **1994**. A catalogue of genes in the cardiovascular system as identified by expressed sequence tags. *Proc. Natl. Acad. Sci. USA 91*:10645–10649.

29 STANTON, L.W., et al. **2000**. Altered patterns of gene expression in response to myocardial infarction. *Circ. Res. 86*:939–945.

30 RAZEGHI, P., et al. **2001**. Metabolic gene expression in fetal and failing human heart. *Circulation 104*:2923–2931.

31 BARRANS, J.D., STAMATIOU, D., LIEW, C. **2001**. Construction of a human cardiovascular cDNA microarray: portrait of the failing heart. *Biochem. Biophys. Res. Commun. 280*:964–969.

32 FRIDDLE, C.J., KOGA, T., RUBIN, E.M., BRISTOW, J. **2000**. Expression profiling reveals distinct sets of genes altered during induction and regression of cardiac hypertrophy. *Proc. Natl. Acad. Sci. U.S.A 97*:6745–6750.

33 JIN, H., et al. **2001**. Effects of early angiotensin-converting enzyme inhibition on cardiac gene expression after acute myocardial infarction. *Circulation 103*:736–742.

34 BODZIOCH, M., et al. **1999**. The gene encoding ATP-binding cassette transporter 1 is mutated in Tangier disease. *Nat. Genet. 22*:347–351.

35 BROOKS-WILSON, A., et al. **1999**. Mutations in ABC1 in Tangier disease and familial high-density lipoprotein deficiency. *Nat. Genet. 22*:336–345.

36 LUTUCUTA, S., BALLANTYNE, C.M., EL-GHANNAM, H., GOTTO, A.M., JR., MARIAN, A.J. **2001**. Novel polymorphisms in promoter region of atp binding cassette transporter gene and plasma lipids, severity, progression, and regression of coronary atherosclerosis and response to therapy. *Circ. Res. 88*:969–973.

37 CLEE, S.M., et al. **2000**. Age and residual cholesterol efflux affect HDL cholesterol levels and coronary artery disease in ABCA1 heterozygotes. *J. Clin. Invest. 106*:1263–1270.

38 CAVELIER, L.B., et al. **2001**. Regulation and activity of the human ABCA1 gene in transgenic mice. *J. Biol. Chem. 276*:18046–18051.

39 JOYCE, C.W., et al. **2002**. The ATP binding cassette transporter A1 (ABCA1) modulates the development of aortic atherosclerosis in C57BL/6 and apoE-knockout mice. *Proc. Natl. Acad. Sci. USA 99*:407–412.

40 VAISMAN, B.L., et al. **2001**. ABCA1 overexpression leads to hyperalphalipoproteinemia and increased biliary cholesterol excretion in transgenic mice. *J. Clin. Invest. 108*:303–309.

41 SINGARAJA, R.R., et al. **2001**. Human ABCA1 BAC transgenic mice show increased high density lipoprotein cholesterol and ApoAI-dependent efflux stimulated by an internal promoter containing liver X receptor response elements in intron 1. *J. Biol. Chem. 276*:33969–33979.

42 ORSO, E., et al. **2000**. Transport of lipids from golgi to plasma membrane is defec-

tive in tangier disease patients and Abc1-deficient mice. *Nat. Genet.* 24:192–196.

43 McNEISH, J., et al. **2000**. High density lipoprotein deficiency and foam cell accumulation in mice with targeted disruption of ATP-binding cassette transporter-1. *Proc. Natl. Acad. Sci. USA* 97:4245–4250.

44 YOKOYAMA, S., KURASAWA, T., NISHIKAWA, O., YAMAMOTO, A. **1986**. High density lipoproteins with poor reactivity to cholesteryl ester transfer reaction observed in a homozygote of familial hyperalphalipoproteinemia. *Artery* 14:43–51.

45 KURASAWA, T., YOKOYAMA, S., MIYAKE, Y., YAMAMURA, T., YAMAMOTO, A. **1985**. Rate of cholesteryl ester transfer between high and low density lipoproteins in human serum and a case with decreased transfer rate in association with hyperalphalipoproteinemia. *J. Biochem. (Tokyo)* 98:1499–1508.

46 DRAYNA, D., et al. **1987**. Cloning and sequencing of human cholesteryl ester transfer protein cDNA. *Nature* 327:632–634.

47 HERRERA, V.L., et al. **1999**. Spontaneous combined hyperlipidemia, coronary heart disease and decreased survival in Dahl salt-sensitive hypertensive rats transgenic for human cholesteryl ester transfer protein. *Nat. Med.* 5:1383–1389.

48 INAZU, A., et al. **1994**. Genetic cholesteryl ester transfer protein deficiency caused by two prevalent mutations as a major determinant of increased levels of high density lipoprotein cholesterol. *J. Clin. Invest* 94:1872–1882.

49 OKAMOTO, H., et al. **2000**. A cholesteryl ester transfer protein inhibitor attenuates atherosclerosis in rabbits. *Nature* 406:203–207.

50 KOGUSHI, M., et al. **1996**. Anti-atherosclerotic effect of E5324, an inhibitor of acyl-CoA:cholesterol acyltransferase, in Watanabe heritable hyperlipidemic rabbits. *Atherosclerosis* 124:203–210.

51 HOEG, J.M., et al. **1996**. Overexpression of lecithin:cholesterol acyltransferase in transgenic rabbits prevents diet-induced atherosclerosis. *Proc. Natl. Acad. Sci. USA* 93:11448–11453.

52 CHAO, J., CHAO, L. **1996**. Functional analysis of human tissue kallikrein in trans-

genic mouse models. *Hypertension* 27:491–494.

53 WANG, C., CHAO, L., CHAO, J. **1995**. Direct gene delivery of human tissue kallikrein reduces blood pressure in spontaneously hypertensive rats. *J. Clin. Invest.* 95:1710–1716.

54 YAYAMA, K., WANG, C., CHAO, L., CHAO, J. **1998**. Kallikrein gene delivery attenuates hypertension and cardiac hypertrophy and enhances renal function in Goldblatt hypertensive rats. *Hypertension* 31:1104–1110.

55 KIM, H.S., et al. **1995**. Genetic control of blood pressure and the angiotensinogen locus. *Proc. Natl. Acad. Sci. USA* 92:2735–2739.

56 MAKINO, N., SUGANO, M., OHTSUKA, S., SAWADA, S., HATA, T. **1999**. Chronic antisense therapy for angiotensinogen on cardiac hypertrophy in spontaneously hypertensive rats. *Cardiovasc. Res* 44:543–548.

57 TANG, X., et al. **1999**. Intravenous angiotensinogen antisense in AAV-based vector decreases hypertension. *Am.J. Physiol.* 277:H2392–H2399.

58 TOMITA, N., et al. **1995**. Transient decrease in high blood pressure by in vivo transfer of antisense oligodeoxynucleotides against rat angiotensinogen. *Hypertension* 26:131–136.

59 KREGE, J.H., et al. **1995**. Male-female differences in fertility and blood pressure in ACE-deficient mice. *Nature* 375:146–148.

60 ESTHER, C.R., JR., et al. **1996**. Mice lacking angiotensin-converting enzyme have low blood pressure, renal pathology, and reduced male fertility. *Lab. Invest.* 74:953–965.

61 FREEDMAN, N.J., et al. **1995**. Phosphorylation and desensitization of the human beta 1-adrenergic receptor. Involvement of G protein-coupled receptor kinases and cAMP-dependent protein kinase. *J. Biol. Chem.* 270:17953–17961.

62 KOCH, W.J., et al. **1995**. Cardiac function in mice overexpressing the beta-adrenergic receptor kinase or a beta ARK inhibitor. *Science* 268:1350–1353.

63 HARDING, V.B., JONES, L.R., LEFKOWITZ, R.J., KOCH, W.J., ROCKMAN, H.A. **2001**. Cardiac beta ARK1 inhibition prolongs

survival and augments beta blocker therapy in a mouse model of severe heart failure. *Proc. Natl. Acad. Sci. USA* 98:5809–5814.

64 LI, Y., LOVE, M. L., PUTKEY, J. A., COHEN, C. 2000. Bepridil opens the regulatory N-terminal lobe of cardiac troponin C. *Proc. Natl. Acad. Sci. USA* 97:5140–5145.

65 RAYMENT, I., HOLDEN, H. M., SELLERS, J. R., FANANAPAZIR, L., EPSTEIN, N. D. 1995. Structural interpretation of the mutations in the beta-cardiac myosin that have been implicated in familial hypertrophic cardiomyopathy. *Proc. Natl. Acad. Sci. USA* 92:3864–3868.

66 DEIVANAYAGAM, C. C., CARSON, M., THOTAKURA, A., NARAYANA, S. V., CHODAVARAPU, R. S. 2000. Structure of FKBP12.6 in complex with rapamycin. *Acta. Crystallogr. D. Biol. Crystallogr.* 56 (Pt 3):266–271.

67 MORAIS CABRAL, J. H., et al. 1998. Crystal structure and functional analysis of the HERG potassium channel N terminus: a eukaryotic PAS domain. *Cell.* 95:649–655.

68 HAYAKAWA, K., et al. 1994. Crystallization of canine cardiac calsequestrin. *J. Mol. Biol.* 235:357–360.

69 TOMIZAKI, T., et al. 1999. Structure analysis of bovine heart cytochrome c oxidase at 2.8 A resolution. *Acta. Crystallogr. D. Biol. Crystallogr.* 55 (Pt 1):31–45.

70 TAYLOR, P., BILSLAND, M., WALKINSHAW, M. D. 2001. A new conformation of the integrin-binding fragment of human VCAM-1 crystallizes in a highly hydrated packing arrangement. *Acta. Crystallogr. D. Biol. Crystallogr.* 57:1579–1583.

71 JANOSIK, M., et al. 2001. Crystallization and preliminary X-ray diffraction analysis of the active core of human recombinant cystathionine beta-synthase: an enzyme involved in vascular disease. *Acta. Crystallogr. D. Biol. Crystallogr.* 57:289–291.

72 VAN DEN, A. F., et al. 2000. Structure of the dimerized hormone-binding domain of a guanylyl-cyclase-coupled receptor. *Nature* 406:101–104.

73 CASASNOVAS, J. M., STEHLE, T., LIU, J. H., WANG, J. H., SPRINGER, T. A. 1998. A dimeric crystal structure for the N-terminal two domains of intercellular adhesion molecule-1. *Proc. Natl. Acad. Sci. USA* 95:4134–4139.

74 WIESMANN, C., et al. 1997. Crystal structure at 1.7 A resolution of VEGF in complex with domain 2 of the Flt-1 receptor. *Cell.* 91:695–704.

75 ISTVAN, E. S., DEISENHOFER, J. 2001. Structural mechanism for statin inhibition of HMG-CoA reductase. *Science* 292:1160–1164.

76 LANGER-SAFER, P. R., et al. 1991. Replacement of finger and growth factor domains of tissue plasminogen activator with plasminogen kringle 1. Biochemical and pharmacological characterization of a novel chimera containing a high affinity fibrin-binding domain linked to a heterologous protein. *J. Biol. Chem.* 266:3715–3723.

77 KEYT, B. A., et al. 1994. A faster-acting and more potent form of tissue plasminogen activator. *Proc. Natl. Acad. Sci. USA* 91:3670–3674.

78 REED, G. L., et al. 1999. A catalytic switch and the conversion of streptokinase to a fibrin-targeted plasminogen activator. *Proc. Natl. Acad. Sci. USA* 96:8879–8883.

79 HUGHES, R. E., et al. 2001. Altered transcription in yeast expressing expanded polyglutamine. *Proc. Natl. Acad. Sci. USA* 98:13 201–13 206.

80 PLUMP, A. S., et al. 1992. Severe hypercholesterolemia and atherosclerosis in apolipoprotein E-deficient mice created by homologous recombination in ES cells. *Cell.* 71:343–353.

81 ZHANG, S. H., REDDICK, R. L., PIEDRAHITA, J. A., MAEDA, N. 1992. Spontaneous hypercholesterolemia and arterial lesions in mice lacking apolipoprotein E. *Science* 258:468–471.

82 MERKEL, M., VELEZ-CARRASCO, W., HUDGINS, L. C., BRESLOW, J. L. 2001. Compared with saturated fatty acids, dietary monounsaturated fatty acids and carbohydrates increase atherosclerosis and VLDL cholesterol levels in LDL receptor-deficient, but not apolipoprotein E-deficient, mice. *Proc. Natl. Acad. Sci. USA* 98:13 294–13 299.

83 RODEN, D. M., GEORGE, A. L. JR. 2002. The genetic basis of variability in drug responses. *Nature Reviews* 1:37–44.

84 MARIAN, A. J. 2001. On genetics, inflammation, and abdominal aortic aneurysm:

can single nucleotide polymorphisms predict the outcome? *Circulation* 103:2222–2224.

85 AITHAL, G.P., DAY, C.P., KESTEVEN, P.J., DALY, A.K. 1999. Association of polymorphisms in the cytochrome P450 CYP2C9 with warfarin dose requirement and risk of bleeding complications. *Lancet* 353:717–719.

86 KOYTCHEV, R., et al. 1998. Influence of the cytochrome P4502D6*4 allele on the pharmacokinetics of controlled-release metoprolol. *Eur. J. Clin. Pharmacol.* 54:469–474.

87 KUIVENHOVEN, J.A., et al. 1998. The role of a common variant of the cholesteryl ester transfer protein gene in the progression of coronary atherosclerosis. The Regression Growth Evaluation Statin Study Group. *N. Engl. J. Med. 338*:86–93.

88 UEDA, S., MEREDITH, P.A., MORTON, J.J., CONNELL, J.M., ELLIOTT, H.L. 1998. ACE (I/D) genotype as a predictor of the magnitude and duration of the response to an ACE inhibitor drug (enalaprilat) in humans. *Circulation* 98:2148–2153.

89 MITCHESON, J.S., CHEN, J., LIN, M., CULBERSON, C., SANGUINETTI, M.C. 2000. A structural basis for drug-induced long QT syndrome. *Proc. Natl. Acad. Sci. USA* 97:12329–12333.

90 BRODDE, O.E., et al. 2001. Blunted cardiac responses to receptor activation in subjects with Thr164Ile beta(2)-adrenoceptors. *Circulation 103*:1048–1050.

91 DISHY, V., et al. 2001. The effect of common polymorphisms of the beta2-adrenergic receptor on agonist-mediated vascular desensitization. *N. Engl. J. Med.* 345:1030–1035.

92 MARTINEZ, F.D., GRAVES, P.E., BALDINI, M., SOLOMON, S., ERICKSON, R. 1997. Association between genetic polymorphisms of the beta2-adrenoceptor and response to albuterol in children with and without a history of wheezing. *J. Clin. Invest. 100*:3184–3188.

93 QUANE, K.A., et al. 1994. Detection of a novel common mutation in the ryanodine receptor gene in malignant hyperthermia: implications for diagnosis and heterogeneity studies. *Hum. Mol. Genet.* 3:471–476.

94 MacLENNAN, D.H., et al. 1990. Ryanodine receptor gene is a candidate for predisposition to malignant hyperthermia. *Nature 343*:559–561.

21

Proteomics: A Post-Genomic Platform
for Drug Discovery and Development

STEPHEN T. RAPUNDALO

The recent completion of the sequencing of the human genome [1, 2] has marked the beginning of a new era in modern biology and medicine. It has spawned a variety of technologies and techniques for genomic research, gene discovery, and in particular functional genomics. Most importantly it already has provided an enormous information load that needs to be converted into a comprehensive understanding of gene function and regulation, and to new approaches in the diagnosis, prevention and treatment of diseases.

Despite the power of genomic technologies and the information derived from their use, it is clear that gene function, and in reality cell function, is manifested by the activity of its protein product(s). The poor correlation between mRNA and protein levels (i.e., coefficients of only 0.5) [3] clearly demonstrates that genes may be present, mutated, but not necessarily transcribed, or if so, then not translated. Proteomics, or the analysis of global patterns of gene expression at the protein level, has the potential to yield information about a system of interest that cannot be obtained by gene profiling alone, namely relative protein abundance, post-translational modifications (e.g., phosphorylation, glycosylation, acetylation, demethylation), turnover/activity, localization, and protein-protein interactions, all of which undergo differential regulation as a result of various physiological, pharmacological or disease stimuli. Accordingly, the true value of the genome sequence information will only be realized after a function has been assigned to all of the encoded proteins. Proteomics, and its many variant and encompassing technological approaches, is destined to provide that functional information for all proteins, and bridge the gap between genomics and systems' phenotypes.

The fact that proteins constitute the vast majority of pharmaceutical targets underscores a further need to comprehend their function, location, and interaction with other proteins in order to design new drug therapies. In turn, there is an urgency to apply high-throughput technologies for protein expression analysis, analogous to those in use for RNA profiling. Since most proteins don't act alone, there is a need to focus on mapping protein-protein interactions and protein complexes to better understand disease pathways and the mode of action of existing drugs. One would anticipate the identification of many more drug targets by selecting alternative points in cellular protein networks that would ultimately lead to the development of agents with greater efficacy, specificity and safety. This much is true – whether as drug targets, or even as new drugs themselves, proteins are key to the future of therapeutics as never before. The challenge for pharmaceutical and biotech research is to unlock the secrets of the biological systems of choice through modern proteomic analysis and to mine the colossal amount of information that can be unraveled from them. This chapter describes how companies primarily, but academic investigators too, are developing their proteomic strategies and applications in the pursuit of novel disease modifying/preventing agents and diagnostic/prognostic markers.

21.1
Perspectives on the Drug Discovery and Development Process

The pharmaceutical industry has been exceedingly attentive to the rapid strides in genomic research and of its significant implications for identifying potentially useful molecular targets suitable for drug discovery and development. Elucidation of the approximately 30,000 protein-encoding human gene sequences [1, 2] has in turn provided estimates of 3,000 to 5,000 potentially interesting protein drug targets [4]. This vastly outnumbers by an order of magnitude the number of molecular targets, i.e. about 500 gene products, accounting for all marketed drugs [5, 6]. Although a significant number of proteins will belong to well-characterized families with predicted biological function, the biomedical community and pharmaceutical industry are anticipating many proteins that will be entirely novel with unknown structure and function [7]. The challenge for pharmaceutical research now is to identify those targets amenable to drug intervention through the elucidation of the genetic and molecular pathophysiology of human diseases and development of novel biomedicine [8–10].

Pharmaceutical research organizations have historically favored disease-oriented strategies seeking to identify and validate the role of "novel" candidate targets that are hypothesis-driven. Typically companies were aligned by therapeutic area and performed drug discovery and development in sequential and disciplined steps, as reviewed comprehensively by Williams [11]. This involved the identification of putative targets using low-throughput strategies. Targets were then validated thoroughly using molecular, biochemical, cellular or *in vivo* means before initiation of a drug discovery effort. Thus, their relationship with a disease state was often well established before chemical screening was initiated. Any identified lead compounds were optimized upon completion of chemical library profiling. At this stage candidate selection relied on further *in vitro* studies, whole animal investigations of efficacy, metabolism, pharmacokinetics and toxicokinetics, and perhaps the development of second-generation or "backup" compounds with improved properties. A broad goal of the pre-clinical evaluation was the integration of knowledge gained from this phase into the decision-making process related to the design and implementation of early clinical studies. Phase 1 clinical studies, conducted in healthy subjects or, in some cases, patients, primarily would provide information on acute tolerability and safety, drug plasma levels, routes of metabolism and elimination, and initial estimates of therapeutic variability. In turn, the data could be used for selection of drug formulation, dosing regimens and routes of administration in the target patient population as part of the Phase II studies evaluating efficacy. Finally, it is in Phase III that large clinical trials are developed to provide evidence of efficacy and safety in the broader target patient population, and that adverse reaction profiles are scrutinized. This information then guides product labeling and patient dosing regimens.

In recent years the pharmaceutical industry has relied on technological innovations in an attempt to maximize efficiency in the drug discovery and development process. Advances in biology such as mapping of the human genome were supposed to help maintain full drug discovery pipelines. Yet the integration of novel biotechnologies with traditional methods of drug discovery, such as screening rational design and/or combinatorial chemical libraries, has proven difficult. Many compounds created this way lacked characteristics that would have made them suitable for use as safe and efficacious clinical therapies. Thus, there has been a clear fall in drug discovery and development productivity. The industry's output of new and approved drug entities has only seen a modest increase despite an enormous growth of research and development funding allocations. Instead of narrowing the list of potential novel compounds that may be useful as therapies, technological advances, especially automation, have broadened it – greatly increasing the number of compounds validated without yet delivering commensurate growth in safe and effective drugs. It is apparent that new science and techniques are developing faster than can be utilized practically by big pharma.

The last several years has been marked by a noticeable paradigm shift in the approaches that pharmaceutical companies are taking towards the discovery and development of new therapeutic agents. As illustrated in Fig. 21.1, the process is now one that has necessitated the acquisition and application of new skill sets, including a variety of high-throughput technologies, bioinformatics, genomics, proteomics, and metabonomics, among others. In short, this evolving paradigm aims to integrate different information sets that ultimately will lead to a systems biology approach to the study of biological function. The hope is that a seamless integration of target discovery, validation, drug design, safety, and clinical validation, coupled with the use of automation technologies, will result in a more efficient R&D process, and a higher quality of drug candidates entering the market.

A first critical step in the drug discovery process is how to identify key targets in disease pathways. The key to formulating the next generation of therapeutics is the context of gene and protein expression, both in relationship to disease pathophysiology. Simplistically, any alteration in gene and protein expression can be viewed as a potential molecular target ripe for drug discovery. Traditional biochemical, pharmacology, and molecular biology will still need to be applied to the identification of targets. However, the drug discovery process can now be augmented significantly with genomic information at the level of chromosomal DNA (human genetics), disease gene associations (via analysis of single nucleotide polymorphisms or SNPs), mRNA profiling (via gene expression or "gene chip" array analysis), genetic animal models of disease (with transgenic knockouts, overex-

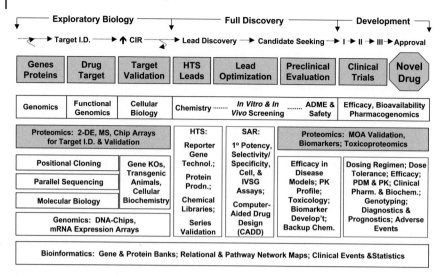

Fig. 21.1 The modern drug discovery and development process and the role of proteomics. Depicted, from top to bottom, are the broad discovery and development stages that are comprised of critical phases. The specific steps from genes to marketing of a new drug are listed in the third tier, and followed by their disciplinary focus. Finally, the various technologies are listed relative to their application in the overall process. The role of proteomics is noted in the shaded boxes. While the technologies are illustrated in discrete timeframes, in many cases much overlap occurs that renders the overall drug discovery and development process as quite dynamic, and certainly adaptive to the needs of any one particular project. CIR = confidence in rationale; I, II, III = Clinical Phases 1, II, III; HTS = high throughput screening; ADME = absorption, distribution, metabolism and elimination/excretion; SAR = structure-activity relationship; MOA = mechanism of action; KO = knockout; IVSG = in vitro; PK = pharmacokinetic; PDM = pharmacodynamics and metabolism.

pression, phenotypic screening), and bioinformatic/computational searching (for paralogues of known drug targets – see the International Human Genome Consortium description of identifying eighteen new targets including dopamine and insulin-like growth factor receptors [2]). Additionally, proteomics and functional genomics can provide a critical insight into the functional role of specific proteins. Advances in analytical proteomic techniques will, one hopes, allow for a greater level of sensitivity in detection and a quantitative assessment of protein changes in disease states. This, coupled with bioinformatic analysis, should allow the development of protein network maps that can provide meaningful and novel targets for drug discovery. Application of high-throughput automation with all the aforementioned approaches should make target identification a far more rapid process. Clearly, the knowledge of the drug target and its function will permit a better understanding of the mechanism of drug action that will guide clinical trials and facilitate drug development.

Target validation has become an increasingly important component in the drug discovery process as the number of potential new drug targets grows. Already there are the emerging signs that large numbers and/or poorly characterized genomic targets can create a bottleneck in drug discovery pipelines. In contrast to previous R&D paradigms where target validation was in many instances a definable stage with limited means for confirmation (i.e., low-throughput animal studies and cellular assays), the process of validating a molecular target can now be viewed as being continuous, though high-throughput approaches have not yet been applied. Indeed, a target can only be defined as truly validated when the drug candidate is approved in the clinic – thus meeting its characterization as being an effector of a therapeutic agent that has a desired clinical use and benefit when modulated in humans. In addition, target validation can now be achieved through multiple approaches, performed independently or collectively at molecular, cellular, animal or patient levels. New technologies are improving the pace of target validation by assisting in the establishment of a genotype-phenotype relationship – these include SNP analysis, gene and protein expression assessment using microarrays and proteomics, gene- and protein-specific "knockouts" either in cells or animals (as performed on a high-throughput level by Deltagen, Lexicon Genetics and Paradigm Therapeutics), antisense and RNAi technologies, structural biology for elucidating drug-target interactions, and chemical genomics (where members of combinatorial libraries are linked with special functionalities and used in *in vitro* screens [e.g., as developed by Novalon, Scriptgen, Chiron, Neo-Genesis] to identify specific ligand-target interactions). Target validation now can take 9–12 months generally, sometimes less, as compared to 18 months in the past. However, the sheer numbers of targets and compounds being generated now have neither provided the high-quality leads anticipated nor even increased the rate at which such leads are validated. The pharmaceutical industry has only begun to grapple with this issue, and it is likely that target validation will continue to receive considerable attention in an attempt to improve its utility in the drug discovery and development process.

The lead optimization step in drug discovery is where early chemical leads (generated from high-throughput screening (HTS) of large chemical libraries) can be optimized for biological activity, pharmacodynamic and pharmokinetic profiles, safety, and clinical predictiveness. Most companies have now incorporated many of these components as early as possible, and placed them on parallel, rather than sequential tracks, to more quickly ascertain the quality of lead compounds. Typically, this comprises an iterative process between chemical modifications and biological activities, with each cycle improving the compound until results against all clinical parameters (i.e., efficacy, selectivity/specificity, toxicology, potency, metabolic stability, and delivery) are optimized. Pharmaceutical companies have launched major initiatives to integrate predictive screening for absorption, metabolism, distribution, and elimination/excretion (ADME), and physiochemical optimization (i.e., permeability, solubility, logD, pKa, and stability) early in the testing schemes, sometimes even prior to *in vivo* efficacy studies. Broadly speaking, an acute and unmet need is the application of automated high-throughput technologies to lead optimization. The intent is to replace traditional low-throughput testing, much of it animal-based, and still impeded by considerable regulatory guidelines that have not yet embraced the validity of all new technologies. Novel technologies for lead optimization now include genetically modified animals or cells that can provide predictive *in vivo* results using "humanized" genes/proteins. Technologies such as toxicoproteomics and toxicogenomics, which deal primarily with effects of compounds on protein and gene expression patterns in target cells or tissues, are emerging as key approaches in the pharmaceutical industry's current initiative to introduce high-quality, chemical entities more rapidly, and with less attrition and cost. For instance, the development of toxicology-relevant genomic and proteomic databases, as noted in sections below, is an emerging activity within several biotech and big pharma companies. It will now be possible to help predict toxicity of new compounds by comparing their profiles to those of compounds with known mechanisms of action.

Another major challenge in accelerating drug development is the clinical evaluation stage. Here the pharmaceutical industry has begun to place more emphasis on overlapping phases of clinical development. This includes providing more focus in a given phase, simplifying clinical studies in order to shorten timelines, and providing a greater effort towards the discovery, and then utility, of biomarkers and surrogate endpoints as indicators of therapeutic benefit and clinical endpoints. Proteomic and genomic technologies applied to target identification can simultaneously identify genes and proteins that are co-regulated with drug targets. Both targets and co-regulated genes/proteins can be potential surrogate biomarkers for use in pre-clinical and clinical studies. This is an example of the growing integration of the early and late stages of drug discovery and development. Similarly, both of these technologies are being incorporated into clinical trial designs and are beginning to generate databases of individualized and population data that should prove useful in understanding a disease state or the effects of specific drug classes. One current problem with biomarkers is that it takes just as long and is just as expensive to discover the biomarker, as it is to do the clinical trials, thus potentially negating the utility of the biomarker in the first place. On the other hand, pharmacogenomics is now beginning to provide an objective measure of a drug's biological efficacy, including its potential adverse effects. However, it will take a few more years and considerable clinical trial data to reveal the true impact of pharmacogenomics on general medical and healthcare practices, and on target discovery and lead compound design. Thus it seems apparent that of all the stages in the research and development (R & D) process, it is the clinical development phase that requires further evolution in its approaches to insure that attrition rates for new drug entities are reduced.

Pharmaceutical research companies are thus at a crossroads in terms of how their drug discovery and development processes should be adjusted to account for the current realities of clinical market needs, business forces, socio-economic determinants, and health- and managed-care policies. Challenges to selecting the most pharmacologically accessible or "druggable" targets, with specific and optimal physicochemical, safety and bioavailability properties, are multi-faceted. Solving them will require the integration of technology platforms across the R & D process in such a manner as to ensure predictive paradigms for drug action and disease modification, whether at the pre-clinical or clinical stages. Proteomics, as an ever-evolving set of technologies, is certainly poised to play a significant role in making the drug discovery and development process more efficient and robust, and examples of its emerging utility in target identification and validation, drug safety and biomarker discovery, are described below.

21.2
Emerging Proteomic Technologies in R & D

The enabling technologies that contributed to the success of the genome initiatives now provide new tools to examine complex biological systems and processes. To achieve the latter will require an automated, quantitative and global measurement of gene expression at the protein level [9, 12].

Pharmaceutical companies have actively embraced the acquisition, development and application of proteomic technologies to drug discovery and the study of disease pathophysiologies. The most widespread strategy for studying global protein expression in biological systems has of course employed two-dimensional polyacrylamide gel electrophoresis (2D-PAGE or 2-DE) followed by enzymatic degradation of isolated protein spots, analytical peptide mapping, and bioinformatics searches for identity matching. Using this method, thousands of proteins can be resolved in a gel and their expression quantified [13, 14]. However, certain types of proteins possessing important cellular functions are not easily analyzed using this strategy. These proteins include membrane, low copy number, highly basic, and very large (>150 kDa) and small (<10 kDa) proteins. To this end, the application of sample fractionation, narrow pH range immobiline strips, novel and/or serial solubilization approaches, and enhanced detection methods including fluorescence staining, has addressed partially the aforementioned deficiencies in 2-DE technology [15]. Fluoroscence 2D difference gel electrophoresis (2D-DIGE) is a variant of 2-DE that uses molecular weight- and pI-matched, spectrally resolvable Cy dyes to label protein samples prior to 2-DE [16]. This approach allows multiplexing (multiple co-separation of samples on the same gel) that has been suggested to significantly increase the throughput, accuracy, ease and speed of comparison, while demonstrating an apparent 10^5 dynamic range in sensitivity. Nonetheless, both conventional 2-DE and 2D-DIGE have shortcomings that have necessitated the search for alternative approaches to assessing system proteomes on the large-scale required for drug research.

New separation strategies have emerged that are amenable to mass spectrometric techniques as a way to meet the growing need for the higher-throughput and simultaneous monitoring of all types of proteins in a biological system. For a more detailed understanding of these technologies, many of which are analytical based, the reader is directed to other chapters within this book or to recent comprehensive reviews [17, 18]. However, all the techniques mentioned below must be viewed as emerging only, and their true utility has yet to be fully evaluated. There is a dire need for advances in sample fractionation, both at the protein and peptide level, and significant improvements in the informatics and software tools necessary to support the analysis and management of the massive amounts of data generated in the process. Thus, until such time as these analytical technologies are fully developed and validated, any proteomics initiative would be best served by still relying on complementary 2-DE methodology.

An exciting and recent alternative to the traditional 2-DE approach for a comparative and quantitative global protein assessment is the isotope-coded affinity tag (ICAT) method combined with direct MS/MS analysis (as developed by Ruedi Aebersold and his group at the Institute for Systems Biology) [17, 19, 20]. This technology involves the cysteine-specific, covalent labeling of proteins with isotopically normal or heavy ICAT reagents, proteolysis of the combined, labeled protein mixture, followed by the isolation and MS analysis of the labeled peptides [19]. This approach offers the potential for high sensitivity, coverage and throughput, and the ability to be quantitative in its ability to assess differential expression. The ICAT technology has now been applied to the proteomic studies of yeast [19], human myeloid leukemia or HL-60 cells [21], and normal and metastatic human prostate epithelial cells [20]. A number of improvements have been made to the technology since its initial implementation including optimized labeling conditions [22], enhanced software able to differentiate peptides based on their abundance [20], and the application of ESI-TOF MS [20] and MALDI Q-TOF [23] to augment identification and quantification. Most recently, Aebersold's group has described a new strategy based on the 2-DE separation of proteins labeled with ICAT reagents, and their identification and quantification by MALDI-TOF [24]. This ICAT variant seems particularly useful for the study of differentially processed or post-translationally modified proteins. Another variation of the ICAT approach involves a phosphoprotein ICAT approach or PhIAT to quantify measurements of differences in phosphorylation state of proteins [25, 26]. The Aebersold group has also devised an ICAT variant using a solid-phase (i.e., beads) isotope tagging or SPAT approach that involves labeling post-proteolysis [27]. A side-by-side comparison with the ICAT method revealed that the SPAT approach is simpler (i.e., a single step for labeling and isolation of peptides), more efficient (i.e., almost exclusive recovery of cysteinyl peptides) and efficient (i.e., unaffected by proteolytic enzymes, denaturants and detergents that would require their removal). Lastly, the commercialization of ICAT reagents by Applied Biosystems has enabled the establishment of this technology in numerous labs, particularly that of pharmaceutical and biotech (e.g., Celera and Oxford Glycosciences) companies, and the beginning of its wider application to quantitative proteome projects.

The extension of ICAT beyond the choice of cysteine as a labeling site to other features has formed the basis for the global internal standard technology or GIST developed by Regnier *et al.* [28]. The GIST technology involves tryptic digestion of samples prior to their differential isotopic labeling of the resulting peptides, mixing of the samples, fractionation of the peptide mixture by RP-LC, and isotope ratio analysis by MALDI-MS and ESI-MS. This technology has the potential advantage of uniformity in the labeling of all peptides, accuracy in quantifying differential protein expression, and flexibility in MS analysis. All this should lend itself particularly well to the assessment of post-translational modifications of proteins following regulated changes in cell function.

Stults and colleagues at Genentech have developed an analytical procedure, referred to as the mass western experiment, which is based on ICAT methodology, but does not require electrophoresis or other initial purification steps prior to analysis by LC-MS [29]. The underlying concept of this technology is that the specificity of the CID fragmentation pattern of a peptide is analogous to using an antibody for the identification of a protein as in Western blotting. This approach differs from true Western blots in that protein size cannot be obtained. The mass western experiment apparently maximizes sensitivity and reduces the possibility that some peptides go otherwise undetected as demonstrated by the application of this approach to the quantitative identification of the cell surface proteins prostate stem cell antigen and ErbB2 in prostate and breast tumor cell lines.

An automated method named multidimensional protein identification technology or MudPIT, has been developed for the large-scale analysis of complex protein/peptide mixtures by Yates and co-researchers [30, 31]. This technology has been impressive in its ability to identify large numbers of proteins, many infrequently seen in previous proteomic studies, including low-abundance proteins like transmembrane proteins, transcription factors and protein kinases [31]. In subsequent studies, improvements to the MudPIT system have demonstrated a dynamic range of 10,000:1 between the most and least abundant proteins in a mixture, i.e. cell lysate, which translates to a level of detection at low-femtomole levels or 100 copies/cell [32]. The MudPIT technology potentially provides a robust alternative in proteome analysis though it is limited by its qualitative properties, low sample throughput and constraints in data management.

Recently, an experimental strategy was described for systematically sequencing and quantifying proteins in complex mixtures based on differential guanidination of C-terminal lysine residues on tryptic peptides followed by capillary LC/ESI-MS [33]. This approach, termed mass-coded abundance tagging or MCAT, is simple, economic, sensitive in the pmol to fmol range, and can be applied to large-scale studies of relative protein abundance present in distinct cell states under the effects of mutation, disease, or pharmacological treatment. Furthermore, the MCAT approach can be performed in tandem with multidimensional separation techniques, such as MudPIT, to achieve an even more detailed and robust proteomic analysis.

Another emerging labeled tag technology for the rapid comparative analysis of proteomes is capillary isoelectric focusing (CIEF) coupled with Fourier transform ion cyclotron resonance mass spectrometry (FTICR) being developed by Richard Smith's group at Pacific Northwest National Labs [34, 35]. This approach claims to provide a means of characterizing large numbers of proteins/peptides (i.e., ~1,000 proteins from >100,000 peptides detected) with exceptional resolution, higher mass measurement accuracy (i.e., ~1 ppm), and greater sensitivity than possible with conventional MS-based strategies described above [36, 37]. The ESI-FTICR-MS approach has also been applied recently to the identification of phosphopeptides [38].

Ficarro *et al.*, working in conjunction with MDS Proteomics, described an interesting approach to "phosphoprofiling" in yeast [39]. Proteins derived from whole cell lysates were digested and resulting peptides converted to methyl esters, enriched for phosphopeptides by immobilized metal-affinity chromatography or IMAC, and analyzed by nanoflow HPLC/ESI-MS. This method has the apparent advantage of eliminating nonspecific binding, thereby allowing for the identification of hundreds of phosphorylation sites in a single analysis with a sensitivity of 5 fmol. Indeed, the method achieves a level of detection sufficient to identify rare protein phosphorylation, such as tyrosine phosphorylation and of proteins with low codon bias or abundance.

Researchers at MDS Proteomics, as well as at Cellzome, have independently developed technologies with the ability to characterize hundreds of distinct multiprotein complexes as described in two new large-scale studies [40, 41]. These approaches involved the tagging of individual proteins in *Saccharomyces cerevisiae* that in turn were used to pull down associated proteins for identification by MS. Ho *et al.* detected over 3,600 associated proteins covering 25% of the yeast proteome with high-throughput mass spectrometric protein complex identification or HMS-PCI, an approach that only used 10% of the predicted yeast proteins as bait [41]. The HMS-PCI data set revealed a high degree of cellular connectivity and an average 3-fold higher success rate in detection of known complexes compared with large-scale two-hybrid studies. Gavin *et al.* identified over 1,400 distinct proteins within 232 multi-protein complexes using tandem affinity purification or TAP [40]. In doing so, they predicted the function of many previously unknown proteins, principally through homology, and more importantly, reconstructed a proteome network of functional units. While these two new approaches are clearly powerful, they do suffer from significant number of false-positive interactions (as high as 30% according to Gavin *et al.* [40]), and a failure to identify many known associations. Nonetheless, the HMS-PCI and TAP approaches, in conjunction with others hold great promise and should make it feasible to characterize all cellular proteins interactions, though this will surely be labor intensive.

A promising technology for rapid protein pattern analysis developed by Ciphergen Biosystems is surface-enhanced laser desorption ionization TOF (SELDI-TOF) mass spectrometry [42, 43]. Protein samples are first prepared, directly applied and separated on solid supports, or ProteinChips™, engineered with hydrophobic, normal-phase, metal-affinity, and cationic or anionic bait surfaces. Mass signature "peaks" derived from the MS are displayed as a standard chromatograph but routine identification of the protein peaks is not yet achievable. Nonetheless, with this technique, peptide and protein profiles are easily obtained within minutes and hundreds of proteins can be profiled simultaneously from a small number of cells or complex mixtures (tissue of body fluids). It also has the advantage of having greater sensitivity for protein <25 kDa, but on average resolves approximately only half the number of proteins compared to 2-DE. The SELDI technology may be an important tool for the molecular fingerprinting of disease samples, especially for providing insights into protein expression changes as potential diagnostic and prognostic markers.

Lastly, the use of CALI (chromophore-assisted laser inactivation) in protein target validation should be mentioned here. This approach involves the conversion of a nonfunction-blocking antibody into a neutralizing molecule through photochemical modifications [44, 45]. A temporal and locally restricted protein "knockout" is achieved in cell-based assays that results in systematic inhibition of function. This is followed by subsequent identification of the target proteins by MS and other techniques. Xerion Pharmaceuticals has adapted the CALI approach on an industrial scale with a technology termed XCALIbur that combines CALI with automation (i.e., for 96- and 384-well platforms), combinatorial binder generation and cell-based assays. Most recently, this approach has been modified to apply fluoroscein-labeled probes for inactivation referred to as fluorophore-assisted light inactivation (FALI) [46]. The use of CALI in a disease-based setting precludes prior knowledge of a target, takes into account post-translational modifications, and can be applied to both cell-free as well as cell-based assays, demonstrating its versatility for validating disease relevant targets.

Clearly, it is anticipated that the aforementioned quantitative analytical technologies will become fully integrated and adopted for the high-throughput, automated, global analysis of protein expression profiling. Additional applications and refinements to bioanalytical separations for protein profiling are certainly expected in the near future and will no doubt assist in revolutionizing the proteomics field.

21.3
Target Identification

A proteomics approach offers great advantages in identifying potential molecular targets. Strategies for target-driven drug discovery and subsequent rational drug design require identifying key cellular proteins that are causally related to disease processes and the validation of such molecular mechanisms as targets for therapeutic intervention. As noted previously, the functions of gene products are frequently unknown and many proteins undergo post-translational modifications that greatly influence their biological properties. The value of proteomics is in its ability to provide a global way of understanding molecular mechanisms involved in a defined system as well as their interactions. Well-conceived biological questions and carefully designed experimental paradigms that have measurable physiological and molecular changes will allow for the elucidation of the interrelationships between expressed proteins. The development of a functional protein network database begins to establish the means to identify therapeutic targets of consequence.

21.3.1
Cardiovascular Diseases

The complex nature of cardiovascular (CV) diseases offers great potential for discoveries of novel therapeutic targets by applying proteome-wide analysis to human and animal models of CV pathophysiology. However, it is just such complexity that underscores the many challenges, such as cellular heterogeneity, awaiting proteomic studies in CV biology. The assessment of various models of CV disease and dysfunction have revealed changes in cardiac protein expression patterns, and identified a significant number that were previously unknown. For comprehensive and recent reviews the reader is directed to Dunn [47], Arrell *et al.* [48], Van Eyk [49], and Macri and Rapundalo [50]. A few notable CV-related proteomic studies are highlighted in the sections below.

One area of CV disease research that has received significant attention through proteomics is cardiomyopathy or progressive heart failure. A number of groups including those led by Michael Dunn (London, UK), Peter Jungblutt, and Joachim Klose (the latter two in Berlin) have been pioneers in characterizing cardiac protein profiles from human, rat, canine and bovine models of cardiomyopathy and establishing public databases of identified cardiac proteomes.

A number of other studies have utilized 2-DE protein profiling of myocardial tissue and cells in an effort to elucidate the cellular and molecular mechanisms associated with cardiac hypertrophy and failure. Arnott *et al.* at Genentech observed that phenylephrine-induced hypertrophy in neonatal rat cardiomyocytes was associated with substantial protein expression changes including molecular chaperones, ubiquitin-related and cytoskeletal proteins, and enzymes associated with cellular metabolism [51]. Recent studies in our laboratory have also characterized novel changes in protein expression in rat neonatal cardiomyocytes made hypertrophic following endothelin treatment [52]. In rats with coronary ligation-induced heart failure we observed alterations in protein profiles that were significant, time-dependent (i.e., 1 day–16 weeks of ligation) and correlated with cardiac dysfunction as measured by echocardiography [53]. Proteome maps of rat cardiac sarcoplasmic reticulum and sarcolemma were also developed, given the importance that these membranes play in mediating beat-to-beat cardiac function through Ca^{+2} regulatory proteins like phospholamban [54]. These studies have been extended to the assessment of protein expression changes in mouse hearts made hyperdynamic through transgenic knockout of phospholamban, the key regulator of sarcoplasmic reticulum Ca^{2+}-ATPase activity [55].

Hemostasis or thrombotic disorders are of particular clinical interest and a major focus for several of the large pharmaceutical companies. A primary goal has been to better understand the exact molecular mechanisms in blood platelet activation and aggregation that lead to the formation of thrombi and vascular plaques. While there have been many proteomic studies performed using platelets, it is only recently that the techniques have been robust and sensitive enough to provide meaningful identification and functionality of proteins [56–58]. Researchers at Bayer established a non-activated human blood platelet protein map, and in particular identified tyrosine-phosphorylated proteins in the cytosolic fraction, typically a major challenge for protein analysis [56]. A more comprehensive analysis of the platelet proteome was developed by Watson's group that revealed at least 5 new proteins (along with a host of transcription factors and signaling molecules) and has helped build a basis for future identification of new drug targets and therapeutic approaches [58]. Another study that focused on protein changes following activation of platelets with thrombin was associated with the translocation of novel candidate proteins to the cytoskeletal actin scaffold [57]. Our own proteomic efforts towards characterizing the platelet proteome±pharmacological treatment (i.e., activators like thrombin, ADP, collagen, and inhibitors such a glycoprotein IIb/IIIa antagonists, the P2Y12 receptor antagonist clopidogrel, aspirin and thromboxane inhibitors) utilizing 2-DE and ICAT have confirmed and extended the aforementioned studies (unpublished results). Thus, the application of proteomics to platelets is increasingly leading to a more useful molecular understanding of the role this blood cell type may play in health and disease.

The utility of proteomic analysis in identifying differentially expressed proteins under conditions related to hypertension has been demonstrated in only a few instances [59–63]. An early study by Kohane *et al.* noted an absence of HSP induction using an *in vivo* deoxycorticosterone-salt-treated rat model of hypertension or following treatment of aortic smooth muscle cells with either norepinephrine or angiotensin II [60]. Abnormal vascular smooth muscle cell (VSMC) growth following treatment with either hypertrophic (angiontensin II) or hyperplastic (platelet-derived growth factor) agents was accompanied by differential expression of a group of proteins involved in protein synthesis and folding [59]. These had not previously been recognized as being regulated by growth factors in VSMCs nor identified by RNA expression profiling. The investigators speculated that the transition from hyperplasia to hypertrophy may be a critical event regulated by growth factor signaling events as identified by proteomic analysis, and differentially-regulated proteins may serve as putative sites for therapeutic intervention. Pleibner *et al.* demonstrated that despite significant myocardial hypertrophy, the expression pattern of the most abundant myocardial proteins was not altered in hypertrophic, non-failing hearts in renovascular hypertension [61]. Berk's group studied the secretion of stress-induced oxidative factors (SOXF) from VSMC in response to LY83583, a generator of reactive oxygen species, and showed that the secreted proteins stimulated ERK1/2 signaling [62]. The functional role of SOXF candidate proteins remains to be elucidated, but these proteins are speculated to be potential therapeutic targets under pathophysiological conditions of oxidative stress such as hypertension, ischemia/reperfusion or atherosclerosis. Lastly, Shusta *et al.* utilized a novel technique that combines a tissue-specific polyclonal antiserum with a cDNA library expression cloning system in an attempt to analyze differential protein expression in the brain microvasculature [63]. This subtractive expression cloning strategy revealed changes in a number of tissue-specific membrane proteins that could either be tracked as marker proteins or be useful in identifying molecular transporters for non-invasive drug delivery to the brain. Thus, hypertension as a disease is by and large lacking in a suitable understanding of its etiology as assessed by protein profiling. By the same token, it is prime area for proteomic study, especially by pharmaceutical researchers, as there is a paucity of viable and novel molecular targets suitable for drug discovery. This is particularly important given the many patients who are non-responders to the stable of available anti-hypertensive therapies.

21.3.2
Cancer

The direct comparison of proteomic profiles of normal vs malignant tissues is rapidly becoming a fertile area of drug research and has revealed a number of potential therapeutic targets. Petricoin's group has found that the known negative regulator of Ras associated small GTP-binding proteins, RhoGDI, over-expresses in invasive ovarian cancer [64]. This protein has also been associated with chemoresistance in several cancer cell lines [65], suggesting that it is potentially a useful therapeutic target. A separate study revealed decreased expression of the molecular chaperone 14-3-3 sigma in breast cancer cells [66]. The researchers suggested that restoring higher levels of the protein in those cells might lead to their decreased proliferation since this protein directly associates with cyclin-dependent kinases to negatively regulate cell growth. Pearl *et al.* investigated the signaling network of fibroblast growth factor FGF-2, a pleiotropic polypeptide known to be involved in many forms of cancer cell growth and metastasis [67]. Differential protein expression was noted in MCF-7 cells by studying both the rapid changes in intracellular signaling and modifications in protein synthesis induced by FGF-2. In addition to the expected induction of tyrosine phosphorylation, expression of four proteins was upregulated within the first 12 hours of FGF-2 stimulation and these included proliferating cell nuclear antigen (PCNA), HSP90 and HSP70, and the transcriptionally-controlled tumor protein. Geldanamycin, an inhibitor of HSP90 activity, completely blocked the FGF-2 induced proliferation and growth of breast cancer cells. The data suggested that expression of this protein is causally linked to breast epithelial cell tumorigenesis, and thus constitutes a potential drug discovery target. In the largest published study of its kind, researchers at Large Scale Biology, in concert with collaborators at the NCI, tested over 60,000 compounds for their ability to inhibit growth of 60 different cancer cell lines representing different organs of origin [68]. The patterns of activity or "signatures" observed provided incisive information on possible molecular targets and modulators of activity within the cancer cells. In addition, it demonstrated an effective approach to profile, or fingerprint, candidate therapeutic agents. Pharmaceutical companies in a variety of therapeutic areas and disease states are now actively following this approach.

21.3.3
Infectious Diseases

Microbial proteomes are particularly amenable to protein profiling since their genomes are small and mostly sequenced, and they can be mutated and cultured easily to monitor diverse cellular conditions. Many groups have now described proteomes of various microbes and/or the characterization of protein changes under certain physiological states as reviewed in Washburn and Yates [69], and Cash [70].

Several reviews have reported on research efforts towards the identification of novel and heterogeneous classes of microbial targets for drug discovery and devel-

opment [71, 72]. In one such study, approximately 40 proteins were down-regulated in sera from patients infected with *H. pylori* as determined by 2-DE [73].
Over one-third of these were membrane or membrane-associated proteins regarded as important putative targets for the development of vaccines designed to
interfere with bacterial colonization of host tissues. Langen *et al.* [74] at Roche
have described the most comprehensive proteomic analysis of *H. influenzae*, and
Jungblut and co-workers have done the same for *H. pylori* [75] and *M. tuberculosis*
[76], which represent excellent models for the identification of new anti-infective
drug targets. VanBogelen's group at Pfizer has pioneered the use of "protein signatures" or "protein phenotypes" as unique and coordinated reflections of microbial physiology following pharmacological manipulation or genetic circumstance
[77]. This approach has enabled the elucidation of the modes of action for various
physical and chemical agents, and by extension putative molecular targets for
drug discovery.

21.3.4
Inflammation and Immune Function

Several studies have recently characterized proteins in neutrophils [78], polymorphonuclear leukocytes (PMNs) [79], and blood monocytes [80], all cells involved in
inflammatory processes and host-defence mechanisms. Proteins secreted from human neutrophil granules activated by calcium revealed many previously observed
soluble components and several novel and unknown proteins [78]. These granule
proteins were identified as being involved in anti-microbial-/-fungal functions and
have been investigated as targets for novel classes of pharmaceutical drugs. Approximately 4% of the over 900 distinct proteins identified from monocyte-derived
dendritic cells exhibited quantitative changes during cell differentiation and maturation [79]. This represented proteins with Ca^{2+} binding, fatty acid binding, chaperone activities, as well as proteins involved in cell motility, and offered a number of potential targets for drug intervention of antigen processing and presentation. A group from GlaxoSmithKline has reported on the development of the first
2-DE map of rat PMN proteins representative of different cellular compartments
[80]. The authors predicted that such a proteomic map would facilitate the study
of protein regulation in preclinical models of inflammation and a search for new
disease targets and markers.

21.4
Target/Drug Validation

Target and/or drug validation is an important step in the drug discovery process.
A validated target is a molecular effector of a therapeutic agent that when modulated *in vivo* has a desirable physiological effect and/or clinical use. It is anticipated that genomic and proteomic approaches will greatly enhance the number of
viable new drug targets, thereby eliminating an important impediment in the

pharmaceutical drug discovery pipelines. However, a new bottleneck is created that relates to the assessment of those new targets, since the availability of substantive information on the physiological role of a putative target is necessary before target validation can occur. Proteomics aims to play a critical role in contributing meaningful assessments of molecular targets in the context of their global cellular and functional environment(s).

Classic examples of target validation via proteomics are the studies demonstrating the actions of various cholesterol-lowering agents (fluvastatin, lovastatin, atorvastatin, gemfibrozil, probucol, and niacin) on hepatic proteins involved in cholesterol metabolism [81–83]. Rats were treated for 7 days with statin levels comparable to recommended daily doses for humans. Major changes in protein patterns were evident in the cholesterol biosynthesis pathway including the induction of enzymes upstream (i.e., hydroxymethylglutaryl coenzyme A synthase, or HMG-CoA synthase) and downstream (i.e., isopentenyl-diphosphate δ-isomerase or IPP-isomerase), of the target protein HMG-CoA reductase. In so doing the mechanism of action for these class of agents was validated. The investigators suggested that both HMG-CoA synthase and IPP-isomerase are alternative drug targets for the regulatation of cholesterol biosynthesis.

Several groups have applied proteomics to investigate drug efficacy based on the presence of distinct protein signature patterns related to the agents' mechanisms of action [84–91]. An excellent example is a series of studies that characterized a broad range of xenobiotics associated with increased proliferation of peroxisomes [i.e., agonists of the nuclear peroxisome proliferator activator receptor *a* (PPAR*a*)] and liver tumors [84–88]. These experiments collectively assessed various potent agonists in several species and models, and validated the proposed PPAR*a* receptor-based mechanism of action. By comparison, treatment with a non-proliferator compound revealed changes that were independent of the target. A subsequent proteomic study by Dahllöf's group examined the effects of the PPAR*a* agonist WY16463 on hepatic proteins from obese (*ob/ob*) mice. The identification of specific proteins in the peroxisomal fatty acid β-oxidation pathway that were up-regulated clearly demonstrated the role of this pathway in the drug's therapeutic efficacy.

Properties of compounds can be evaluated against a set of known responses under the premise that two compounds with the same chemical properties induce matching effects in protein expression. Evers *et al.* characterized Ro-64-1874, a 2,4-diaminopyrimidine derivative similar to trimethoprin by proteomic analysis using *H. influenzae* [92]. The drug's effects were then compared to that of other antibiotics with known cellular mechanisms of action (including inhibitors of transcription, translation, DNA gyrase, tRNA synthesis, etc.) and a correlation was demonstrated only with those of trimethoprin. The mechanism of action of Ro-64-1874 was thus validated. In a separate study, a novel antimicrobial agent with *in vitro* gyrase inhibitory activity induced a secondary *in vivo* response indicative of protein synthesis inhibition based on the 2-DE protein expression profiles. In turn, this *in vivo* activity was validated in an *in vitro* transcription/translation assay.

The effects of non-steroidal anti-inflammatory drugs (NSAIDs) on serum proteins in rodent inflammation models has been demonstrated by two groups [89, 90]. In both cases distinct signature patterns revealed expression changes common to all NSAIDs characterized. However, differentiation among the agents was clearly evident. In turn such protein profiles could be used to screen new NSAIDs and predict their efficacy.

21.5
Toxicoproteomics

An increasingly important application of proteomics is its utility in profiling the safety of new drug entities [93]. Toxicology groups within pharmaceutical companies are contending with the pressure to handle potentially large numbers of compounds that are emerging from HTS and combinatorial chemistry efforts. Proteomics has the advantage of being able to characterize hundreds to thousands of protein changes in a single experiment as compared to conventional protocols that typically focus on specific proteins of interest. Thus proteomic analyses can be used robustly in drug candidate selection on both qualitative and quantitative levels before undertaking lengthier and costly development studies. This is a real asset for interpreting adverse drug effects and assessing risks for humans accurately. Drug candidate attrition rates should be reduced as well, thereby increasing the probability of successful development in the clinical phases.

The Large Scale Proteomics Corp. was an early pioneer in applying 2-DE in toxicology [91]. They established a comprehensive Molecular Effects of Drugs™ database that described the toxicity of a wide spectrum of established drugs at a protein expression level. In turn this could be used as a reference for comparative purposes when investigating the side effects of new chemotypes.

Several studies using proteomics to investigate the molecular effects of drugs have been described in recent years. Proteomics played a critical role in elucidating novel molecular mechanisms involved in cyclosporine A nephrotoxicity [94]. The significant downregulation of the calcium binding protein, calbindin D28, in the kidneys of cyclosporin A (CsA)-treated rats provided an explanation for the observed accumulation of calcium in tubules and consequent tubular toxicity. Subsequent studies helped develop a structure-activity relationship (SAR) around CsA derivatives and other immunosuppressive compounds, like FK506 and rapamycin [95], and demonstrated that a similar phenomenon occurred in humans with CsA-related nephrotoxicity [96].

Proteomics has been used to study other hepatotoxicities induced by a variety of drugs either established or in clinical development. This has included characterizing adverse effects of acetaminophen [97, 98], substituted pyrimidine derivatives leading to hepatomegaly [99], the dithiolethione compounds oltipraz and its analogs under development for aflatoxin-induced liver cancer [100], etomoxir, an inhibitor of carnitine palmitoyl transferase developed as a potential anti-diabetic agent [101], and lastly, the hypoglycemic agent SDZ PGU 693 known to induce hepatocellular hypertrophy [102]. Researchers in the Proteomics Group at AstraZeneca have routinely used fluorescence 2-D DIGE technology in their projects which has allowed for simplified imaging, better confidence in spot matching, and thus differential expression, as compared to conventional 2-DE methods. One such study profiled effects of paracetamol treatment on liver damage in mice [16].

A number of studies and applications have focused on the toxicity of CV agents, most notably the lipid-lowering statins [81–83]. Anderson *et al.* at Large Scale Proteomics Corp., demonstrated that treatment of rats for 7 days with high doses of either lovastatin [81] or fluvastatin [82] resulted in hepatotoxicity as assessed by 2-DE. Alterations in a heterogeneous set of cellular stress proteins involved in cytoskeletal structure (i.e., cytokeratin 18, major vault protein), calcium homeostasis (i.e., senescence marker protein-30 or SMP-30) and protease activity (i.e., the serine protease inhibitor 2 or SPI-2) were noted as general indicators of toxicity. Low dose treatment with the same drugs showed little or no changes in protein indicators of toxicity. In a patent application submitted by the same group, the entire range of marketed statins, as well as other lipid-lowering agents such as gemfibrozil, niacin and probucol, were characterized in a similar fashion [83]. Differentiation was observed among the various agents based on their protein expression profiles.

Finally, with the advent of new proteomic technologies such as SELDI-TOF, the use of protein patterns or fingerprints from cells or sera of patients may become an essential component in assessing drug treatments and adverse effects. Ardekani *et al.* demonstrated that serum samples from Sprague-Dawley rats treated with toxic doses of isoproterenol had distinctive protein patterns compared to those from sera of saline-treated rats [103]. The investigators concluded that SELDI-TOF offers a fast and reliable method for predicting outcomes with different drug therapies, and thus could be useful in a clinical setting.

It is apparent that great strides are being made towards identifying protein indicators of toxicity in preclinical trial. These are emerging from a focus on molecular evolution of defense mechanisms/systems, which can be induced by damage or xenobiotics via an effect on associated genes which regulate activity. Other opportunities for new and/or improved toxicology surrogates will center around damage-specific responses, host defense cell signaling, markers of cell death, homeostasis and/or integrity, tissue/organ-specific proteins, and secreted biomolecules. It will also be important to "proteo"-type individual differences when assessing humans. This will require a much more integrated approach to biomarker development and the use of novel techniques. Clinical pharmacologists need to become involved at the preclinical stages in the choice of markers and discovery, and pharmaceutical development scientists must be involved in the laboratory techniques used in clinical trials. Thus, it will be essential that the discovery and use of protein safety markers be built into the development plan of each new drug entity.

21.6
Biomarker Discovery

An organism's phenotype is its overall biochemical and physiological state at any given time, and can be characterized by biological markers, or biomarkers, of its function. Biomarkers are objectively measurable phenotypic parameters that characterize a system's state of health or disease, or response to a particular therapeutic intervention.

Pharmaceutical companies have placed a priority on the identification, validation and utility of circulating and soluble protein markers of disease and drug efficacy in their effort to speed up the drug development process and reduce attrition of clinical drug candidates [104]. This is partly due to recent regulatory requirements, referred to as the "fast track" provision, put in place as part of the Federal Drug Administration (FDA) Modernization Act of 1997. The "fast track" recognizes that a single pivotal clinical trial, with other confirmatory evidence, may be adequate to demonstrate safety and efficacy. It is generally accepted that the "other confirmatory evidence" could be based on pharmacokinetic/pharmacodynamic studies that rely on appropriate biomarkers. One of the major barriers to the successful "fast track"-ing of many drug candidates has been the paucity of validated and clinically accepted surrogate endpoints. Pharmaceutical companies and clinicians alike are very much interested in identifying suitable biomarkers useful for a variety of applications, including stratifying patients by disease type and/or stage, response to therapy, and as diagnostic targets. Consequently there appears to a growing collaboration between government, academia and industry to translate basic pathological and biochemical observations into accepted biomarkers.

Proteomics has the potential to increase the yield of novel biomarkers through its "systems biology" or global approach. Differentially expressed proteins in disease vs normal state or pre- and post-treatment, i.e., proteins either up- or down-regulated, may serve as possible biomarkers and merit further validation. The usefulness of a biomarker lies in its ability to provide an early diagnosis of disease (both in terms of detection and of potential risk), a key challenge when it comes to controlling many progressive disorders such as cancer, atherosclerosis, congestive heart failure, diabetes, and neurodegenerative disorders, among others.

21.6.1
Cardiovascular Diseases

CV disease is a systemic multi-factorial pathology involving many different etiologies and it remains the leading cause of mortality in the United States. Currently, there exists a rather poor repertoire of relevant, validated and clinically accepted markers of heart disease.

In a recent study by Skehel et al., serum from hyperlipoproteinemic Apolipoprotein E (ApoE)*3-Leiden transgenic mice was phenotyped using 2-DE and MS identification [105]. Mice with this variant typically have a reduced affinity for the low density lipoprotein receptor and the mutated ApoE was indeed identified on the 2-DE gels. A protein charge "train" or "ladder" (i.e., linear series of spots differing in charge but representing the same protein) of β-haptoglobin and attributable to post-translational modifications was observed in ApoE*3 transgenic mice fed a high-cholesterol diet. These protein modifications would not ordinarily have been found using genomic or transcriptomic approaches, thus reinforcing the utility of proteomics to identify specific changes as hallmarks of disease or variant gene products.

Several companies, including Millenium Pharmaceuticals, Incyte Genomics, Myriad Genetics, GlaxoSmithKline, Novartis (and their partner Geneva Proteomics), Proteome Sciences, SYNX Pharma, Bristol Myers Squibb, and Pfizer, have launched efforts towards identifying biomarkers of CV disease. In a proof-of-concept biomarker study (unpublished observations) our lab examined urinary protein expression from spontaneously hypertensive, heart failure rats following treatment with a novel matrix metalloprotease inhibitor known to attenuate cardiac remodeling. Differential expression was noted in the drug-treated group. Clearly, the ability to measure primarily disease-specific, as well as treatment-specific, markers will be of enormous value in the diagnostic/prognostic assessment of CV disease and for the development of clinical drug candidates.

21.6.2
Cancer

Proteomic approaches to biomarker discovery have been most extensively applied in recent years to the study of various types of cancer in an attempt to elucidate disease predictive and progression indicators [42, 106, 107]. Only a brief survey is given below of oncology-related proteomic approaches in biomarker discovery due to the many recent advancements in this area – the reader instead is guided to several excellent reviews [108–111].

A number of groups have studied body fluids in the search for biomarkers of various cancers [108, 112–115]. Celis' laboratory has used 2-DE to identify putative urinary protein biomarkers, particularly that of psoriasin or S100A7, for the follow-up of squamous cell carcinoma bearing patients [108, 114]. Oxford Glycoscience investigators have profiled serum from patients with hepatocellular carcinoma by 2-DE proteomic methods [115]. Furthermore, they characterized protein-associated oligosaccharides as potential biomarkers since modifications in the latter are one of many alterations that accompany malignant cellular changes. Using SELDI-TOF technology, Petricoin *et al.* identified proteomic patterns in serum that correctly distinguished all neoplastic from non-neoplastic ovarian disease cases. The investigators concluded that protein pattern profiling by SELDI was an accurate, easy, rapid, and prospective screening tool for all stages of ovarian cancer in high-risk and general populations [113]. Wright's group has used the same technology to discover several potential serum and urinary biomarkers, including a 33-kD and an 18-kD protein found to be up-regulated in patients with prostate and bladder cancers [112]. In another study, the same group identified several protein differences, including five potential novel biomarkers and seven unique protein clusters in urine from normal patients and those with transitional cell carcinoma and benign urogenital diseases [116, 117].

Many groups have undertaken a comprehensive proteomics approach to identify potential biomarkers in a variety of cancers using tumor tissue and cell lines [109, 110, 118, 119]. The cellular heterogeneity of cancerous tissue has precluded many substantive and direct proteomic comparisons in the past. New techniques such as laser capture microdissection have allowed the characterization of protein expression changes in histopathologically-defined, homogenous cell populations in human tissue specimens [120]. The National Cancer Institute-FDA Tissue Proteomics Initiative, involving the Liotta and Petricoin labs respectively, has applied this pioneering approach to the study of protein expression in a variety of cancer types [64, 121–123]. These analyses revealed that proteins such as the 52 kDa FK506 binding protein and Rho G-protein dissociation inhibitor (RhoGDI) are uniquely over-expressed in ovarian cancer [64], whereas 14.3.3 protein and proliferating cell nuclear antigen are up-regulated in prostate cancer [121]. Petricoin's group has also used laser capture microdissection in conjunction with SELDI-based proteomics and observed the downregulation of a 28 kDa protein in prostate cancer samples [124, 125], as well as extending these studies to esophageal cancer [122]. These are just a few of the many proteomic profiling studies undertaken to identify new diagnostic and prognostic biomarkers of cancers.

An important corollary of large-scale protein expression profiling should be the ability to meaningfully integrate such data with clinical outcomes. In one recent study, Voss *et al.* compared protein expression patterns obtained by 2-DE with clinical features from 24 patients with B-cell chronic lymphocytic leukemia, a disease characterized by broad clinical variability [126]. Proteins were identified and then analyzed using stringent statistical methods that clearly discriminated between karyotype or with patient survival (the latter using Kaplan-Meier survival plots). Patients with deletions of the tumor suppressor p53 with defined chromosomal characteristics displayed a down-regulation of glutathione-S-transferase π and thioredoxin peroxidase I. The latter protein, along with protein disulfide isomerase, was also down-regulated in patients with shorter survival times.

21.6.3
Neurological Diseases

An excellent review by Rohlff clearly demonstrates that proteomics has begun to facilitate the comprehensive assessment of various central nervous system (CNS) diseases for the identification of specific diagnostic and prognostic biomarkers [127]. Some groups have performed their proteomic analyses of CNS disorders by examining the more easily obtained cerebrospinal fluid (CSF) from patients [128, 129]. These studies demonstrated differential protein expression of many synaptic proteins in schizophrenia and Alzheimer's disease using SELDI-TOF MS technology. Others have extended protein expression findings in CSF through proteomic analysis of specific brain areas [130, 131]. Synaptosomal-associated protein 25 previously identified in CSF [129] was also found to be a disease-specific protein in several cortical lobes and cerebellum of patients with Alzheimer's disease [131]. Schonberger *et al.* used quantitative proteome analysis to compare levels of differentially expressed proteins in six brain regions from patients with Alzheimer's disease, and those from tissues in age-matched, non-demented controls [130]. A separate study by the same group revealed a significant decrease in diazepam-binding inhibitor, a regulator of the γ-aminobutyric acid(A) or GABA(A) receptor, in hippocampal tissue from patients with Alzheimer's disease or schizophrenia [132]. The authors concluded from both studies that proteome analysis is a viable means by which to assess brain disorders and identify putative biomarkers that underlie psychopathology. The disclosure of the first 2-DE map of human brain proteins should greatly facilitate the search of potential biomarkers through the characterization of protein expression changes in various CNS disorders.

A number of proteomics technology companies have now filed patent applications on the identification of protein markers for the diagnosis and treatment of various neurological disorders. These include schizophrenia [133–136], Alzheimer's [133, 137], epilepsy [138], Parkinson's [133], bipolar affective disorder (BAD) [136, 139, 140], and multiple sclerosis [133]. Indeed, two of the noted patents make claims to specific protein markers identified either as a family of related proteins (i.e., BAD-associated protein isoforms) [139] or as a single biomolecule (i.e., DPI-6) [140], the latter proposed as a putative drug target too.

21.6.4
Inflammatory Diseases

Two reports from pharmaceutical groups have described the application of 2-DE to the analysis of plasma and urinary proteins from patients under inflammatory conditions [89, 141]. The relative abundance of 19 plasma proteins was examined under acute (i.e., parenteral typhoid vaccination) and chronic (i.e., patients with rheumatoid arthritis) conditions. In both cases, levels of serum amyloid A, haptoglobin isoforms, and Apo E were up-regulated, while that of Apo A-I, Apo A-IV, and various α-glycoproteins were downregulated, compared to normals [89]. The Bristol-Myers Squibb group conducted a thorough study of urinary proteins following acute inflammation in volunteer patients, and for the first time, compared the techniques of 2-DE, HPLC/MS, and multidimensional LC in their ability to identify putative biomarkers [141]. In all three cases, a set of proteins was identified representing possible markers of disease (particularly orosomucoid) and which had been previously linked to inflammation. This study was able to demonstrate that a preclinical research component of the drug discovery process was not needed to identify potential biomarkers, thereby saving significant drug development time, effort and costs. Lastly, Oxford Glycosciences has received a patent on the identification of protein markers of rheumatoid arthritis using 2-DE approaches [142].

21.6.5
Infectious Diseases

Several groups have identified differentially expressed proteins following infectious microbial and viral infections. Greco *et al.* characterized host proteins whose expression was up-regulated following herpes simplex virus type 1 infection [143]. Nilsson used MALDI-TOF to fingerprint lysates and extracts from six strains of *H. pylori* [144]. A set of *H. pylori* specific, and a probable set of strain-specific, biomarkers could be identified and used for bacterial typing. Chow and co-workers at Genelabs Technologies Inc. describe in a patent the characterization and isolation of proteins recovered from *H. pylori* useful in the detection of active status infection via diagnostic kits employing one or more of the described proteins [145]. Hendrickson *et al.* at Corixa Corp. used proteome analysis in conjunction with LC/MS/MS to identify a novel protein, Mtb81, which could be useful for the serodiagnosis of tuberculosis, especially for patients coinfected with human immunodeficiency virus, HIV [146].

21.6.6
Miscellaneous Diseases

Proteomic analysis has been successfully applied on a more limited scale to search for lung disease markers [147], and non-insulin-dependent diabetes [148].

The use of proteomics in biomarker discovery has clearly increased in recent years. Nonetheless, numerous technical problems need to be addressed before global proteomic strategies can demonstrate a clear clinical utility. These include the heterogeneity of biopsy materials, problems associated with the separation and focusing of some types of proteins (e.g., membrane, basic, high-molecular weight, etc.), contending with the large dynamic range of protein expression, and a need for better image analysis systems and bioinformatic tools. Some of these problems may be dealt with by newer and quantitative analytical approaches, the use of laser dissection capture technology to isolate specific cell types within biopsies, and the development of specialized protein microarrays to interrogate targets relevant to disease progression.

21.7
The Changing Pharmaceutical R & D Landscape

Proteomics is emerging as a key new application in pharmaceutical research in this post-genome world. By using protein expression and function studies, biotech and pharmaceutical companies aim to more rapidly select functional gene and protein products as potential drug targets from among thousands identified through gene sequencing, and thereby seek a competitive advantage in the race for useful new therapies and diagnostics. With the development of new proteomic approaches to analyzing drug action described in this review, proteomics will begin to influence a much broader range of drug development operations, from drug metabolism and safety assessment within preclinical research, to direct assessment of a drug's performance in clinical trials. Marketed drugs may also become a focus of attention for applying proteomics, both in mode of action assessment and in population studies designed to examine differentiation of a drug's clinical performance as compared to competitor therapies. Thus proteomics is increasingly viewed as having a direct impact on the entire drug discovery and development process (Fig. 21.1).

Proteomics has not yet reached its peak practicality or high throughput capability, but the platform landscape is changing rapidly. The technology is attracting substantial investment from pharmaceutical and biotechnology industry leaders such as Amgen, Incyte, GlaxoWellcome, Pfizer, AstraZeneca, Novartis, Pharmacia, Celera and Millenium. Many others have more focused, or exploratory proteomic programs. The success of commercial proteomic company pioneers such as Oxford GlycoSciences, Large Scale Proteomics, Proteome and MDS-Proteomics is evidence of the interest proteomics has generated. This is in addition to the many academic proteomic initiatives being developed, and alongside those established at the NCI and centers of excellence programs sponsored by the National Heart Lung & Blood Institute. Lastly, one should not forget the myriad of technology companies whose very existence has allowed the development and availability of the various analytical, biocomputational and imaging systems needed to perform proteomic analyses. However, not all approaches are equal – diverse instrumentation and automation platforms exist, and incompatibilities amongst systems are rampant. Thus companies that are able to select the right tools and apply them well strategically should see a direct benefit to their discovery efforts.

It is clear that proteomics has a huge potential application in drug discovery and development. However, it will have far greater, and more immediate, impacts in some therapeutic areas (i.e., oncology) than in others. Much is predicated on asking the right biological and clinical questions, finding the right preclinical or clinical paradigms in which to pose that question, and formulating a study design which allows valid interpretation of the results. Nonetheless, proteomics as an enabling technology, seems poised to augment the quality of drug development pipelines, particularly at a time when most big pharmaceutical companies are experiencing diminishing numbers of clinical lead candidate, patent expirations, and strong generic drug competition.

21.8
References

1 VENTER, J.C., ADAMS, M.D.; MYERS, E., et al. The sequence of the human genome. *Science* **2001**, *291* (5507), 1304–1351.

2 CONSORTIUM, I.H.G.S. Initial sequencing and analysis of the human genome. *Nature* **2001**, *409*, 860–921.

3 GYGI, S., ROCHON, Y., FRANZA, B., AEBERSOLD, R. Correlation between protein and mRNA abundance in yeast. *Mol. Cell. Biol.* **1999**, *19*, 1720–1730.

4 KUMAR, G., et al. Post-map workplan: the hunt for utility. *BioCentury* **2001**, A1–A7.

5 DREWS, J. Genomic sciences and the medicine of tomorrow. *Nature Biotech.* **1996**, *14*, 1516–1518.

6 DREWS, J. Drug discovery: a historical perspective. *Science* **2000**, *287*, 1960–1964.

7 EISENBERG, D., MARCOTTE, E., XENARIOS, I., YEATES, T. Protein function in the post-genome era. *Nature* **2000**, *405*, 823–826.

8 BRENT, R. Genomic biology. *Cell* **2000**, *100*, 169–183.

9 AEBERSOLD, R., HOOD, L., WATTS, J. Equipping scientists for the new biology. *Nature Biotech.* **2000**, *18*, 359.

10 REISS, T. Drug discovery of the future: the implications of the human genome project. *Trends Biotech.* **2001**, *19* (12), 496–499.

11 WILLIAMS, M. Strategies for drug discovery. *NIDA Res. Monogr.* **1993**, *134*, 1–36.

12 KITANO, H. Systems biology: a brief overview. *Science* **2002**, *295*, 1662–1664.

13 PANISKO, E., CONRADS, T., GOSHE, M., VEENSTRA, T. The postgenome age: characterization of proteomes. *Exp. Hematol.* **2002**, *2002*, 97–107.

14 NAABY-HANSEN, S., WATERFIELD, M., CRAMER, R. Proteomics – post-genomic cartography to understand gene function. *Trends Pharmacol. Sci.* **2001**, *22*, 376–384.

15 MOLLOY, M., WITZMANN, F. Proteomics: technologies and applications. *Brief. Funct. Genomics Proteomics* **2002**, *1*, 29–39.

16 TONGE, R., SHAW, J., MIDDLETON, B., ROWLINSON, R., et al. Validation and development of fluorescence two-dimensional differential gel electrophoresis proteomics technology. *Proteomics* **2001**, *1*, 377–396.

17 MOSELEY, M.A. Current trends in differential expression proteomics: isotopically coded tags. *Trends Biotech.* **2001**, *19*, S10–S16.

18 LIU, H., LIN, D., YATES III, J. Multidimensional separations for protein/peptide analysis in the post-genomic era. *BioTechniques* **2002**, *32*, 898–911.

19 GYGI, S., RIST, B., GERBER, S., TURECEK, F., et al. Quantitative analysis of complex protein mixtures using isotope coded affinity tags. *Nature Biotech.* **1999**, *10*, 994–999.

20 GRIFFIN, T., HAN, D., GYGI, S., RIST, B., et al. Toward a high-throughput approach to quantitative proteomic analysis: expression-dependent protein identification by mass spectrometry. *J. Am. Soc. Mass Spectr.* **2001**, *12*, 1238–1246.

21 HAN, D., ENG, J., ZHOU, H., H. AEBERSOLD, H. Quantitative profiling of differentiation-induced microsomal proteins using isotope-coded affinity tags and mass spectrometry. *Nature Biotech.* **2001**, *19*, 946–951.

22 SMOLKA, M., ZHOU, H., PURKAYASTHA, S., AEBERSOLD, R.. Optimization of the isotope-coded affinity tag-labeling procedure for quantitative proteome analysis. *Anal. Biochem.* **2001**, *297*, 25–31.

23 GRIFFIN, T., GYGI, S., RIST, G., AEBERSOLD, R., LOBODA, A., JILKINE, A., ENS, W., STANDING, J.K. Quantitative proteomic analysis using a MALDI quadrupole time-of-flight spectrometer. *Anal. Chem.* **2001**, *73*, 978–986.

24 SMOLKA, M., ZHOU, H., AEBERSOLD, R. Quantitative protein profiling using two-dimensional gel electrophoresis, isotope-coded affinity tag labeling, and mass spectrometry. *Molec. Cell. Proteomics* **2002**, *1*, 19–29.

25 GOSHE, M., CONRADS, T., PANISKO, E., ANGELL, N., et al. Phosphoprotein isotope-coded affinity tag approach for isolating and quantitating phosphopeptides in proteome-wide analysis. *Anal. Chem.* **2001**, *73*, 2578–2586.

26 GOSHE, M., VEENSTRA, T., PANISKO, E., CONRADS, T., et al. Phosphoprotein isotope-coded affinity tags: application to the enrichment and identification of low-abundance phosphoproteins. *Anal. Chem.* **2002**, *74*, 607–616.

27 ZHOU, H., RANISH, J., WATTS, J., AEBERSOLD, R. Quantitative proteome analysis by solid-phase isotope tagging and mass spectrometry. *Nature Biotech.* **2002**, *19*, 512–515.

28 CHAKRABORTY, A., REGNIER, F. Global internal standard technology for comparative proteomics. *J. Chromatogr.* **2002**, *949*, 173–184.

29 ARNOTT, D., KISHIYAMA, A., LUIS, E., LUDLUM, S., MARSTERS JR., J., STULTS, J. Selective detection of membrane proteins without antibodies. *Mol. Cell. Proteomics* **2002**, *1*, 148–156.

30 LINK, A., ENG, J., SCHIELTZ, D., CARMACK, E., et al. Direct analysis of protein complexes using mass spectrometry. *Nature Biotech.* **1999**, *17*, 676–682.

31 WASHBURN, M., WOLTERS, D., YATES III, J. Large-scale analysis of the yeast proteome by multidimensional protein identification technology. *Nature Biotech.* **2001**, *19*, 242–247.

32 WOLTERS, D., WASHBURN, M., YATES III, J. An automated multidimensional protein identification technology for shotgun proteomics. *Anal. Chem.* **2001**, *73*, 5683–5690.

33 CAGNEY, G., EMILI, A. De novo peptide sequencing and quantitative profiling of complex protein mixtures using mass-coded abundance tagging. *Nature Biotech.* **2002**, *20*, 163–170.

34 JENSEN, P., PASA-TOLIC, L., ANDERSON, G., HORNER, J., et al. Probing proteomics using capillary isoelectric focusing-electrospray ionization Fourier transform ion cyclotron resonance mass spectrometry. *Anal. Chem.* **1999**, *71*, 2076–2084.

35 SMITH, R., PASA-TOLIC, L., LIPTON, M., JENSEN, P., et al. Rapid quantitative measurements of proteomes by Fourier transform ion cyclotron resonance mass spectrometry. *Electrophoresis* **2001**, *22*, 1652–1668.

36 SHEN, Y., TOLIC, N., ZHAO, R., PASA-TOLIC, L., et al. High-throughput proteomics using high-efficiency multiple-capillary liquid chromatography with on-line high-performance ESI FTICR mass spectrometry. *Anal. Chem.* **2001**, *73*, 3011–3021.

37 BELOV, M., ANDERSON, G., ANGELL, N., SHEN, Y., et al. Dynamic range expansion applied to mass spectrometry based on data-dependent selective ion ejection in capillary liquid chromatography Fourier transform ion cyclotron resonance for enhanced proteome characterization. *Anal. Chem.* **2001**, *73*, 5052–5060.

38 FLORA, J., MUDDIMAN, D. Selective, sensitive, and rapid phosphopeptide identification in enzymatic digests using ESI-FTICR-MS with infrared multiphoton dissociation. *Anal. Chem.* **2001**, *73*, 3305–3311.

39 FICARRO, S., MCCLELAND, M., STUKENBERG, P.T., BURKE, D., et al. Phosphoproteome analysis by mass spectrometry and its application to Saccharomyces cerevisiae. *Nature Biotech.* **2002**, *20*, 301–305.

40 GAVIN, A.-C., BOSCHE, M., KRAUSE, R., GRANDI, P., et al. Functional organization of the yeast proteome by systematic analysis of protein complexes. *Nature* **2002**, *415*, 141–147.

41 HO, Y., GRUHLER, A., HEILBUT, A., BADER, G., et al. Systematic identification of protein complexes in Saccharomyces cerevisiae by mass spectrometry. *Nature* **2002**, *415*, 180–183.

42 FUNG, E., WRIGHT, G. Jr., Dalmasso, E. Proteomic strategies for biomarker identification: progress and challenges. *Curr. Opinion Molec. Therap.* **2000**, *2*(6), 643–650.

43 VON EGGELING, F., JUNKER, K., FIEDLER, W., WOLLSCHEID, V., et al. Mass spectrometry meets chip technology: a new proteomic tool in cancer research? *Electrophoresis* **2001**, *22*, 2898–2902.

44 ILAG, L., NG, J., JAY, D. Chromophore-assisted laser inactivation (CALI) to validate drug targets and pharmacogenomic markers. *Drug Dev. Res.* **2000**, *49*, 65–73.

45 RUBENWOLF, S., NIEWOHNER, J., MEYER, E., PETIT-FRERE, C., et al. Functional proteomics using chromophore-assisted laser inactivation. *Proteomics* **2002**, *2*, 241–246.

46 BECK, S., SAKURAI, T., EUSTACE, G., BESTE, G., et al. Fluorophore-assisted light inactivation: a high-throughput tool for direct target validation of proteins. *Proteomics* **2002**, *2*, 247–255.

47 DUNN, M. Studying heart disease using the proteomic approach. *Drug Disc. Today* **2000**, *5*, 76–84.

48 ARRELL, D.K., NEVEROVA, I., VAN EYK, J. Cardiovascular proteomics: evolution and potential. *Circ. Res.* **2001**, *88*, 763–773.

49 VAN EYK, J. Proteomics: unraveling the complexity of heart disease and striving to change cardiology. *Curr. Opin. Molec. Therap.* **2001**, *3*, 546–553.

50 MACRI, J., RAPUNDALO, S. Application of proteomics to the study of cardiovascular biology. *Trends Cardiovasc. Med.* **2001**, *11*, 66–75.

51 ARNOTT, D., O'CONNELL, K., KING, K., STULTS, J. An integrated approach to proteome analysis: identification of proteins associated with cardiac hypertrophy. *Anal. Biochem.* **1998**, *258*, 1–18.

52 MACRI, J., DUBAY, T., MATTESON, D., et al. Characterization of the protein profile associated with endothelin-induced hypertrophy in neonatal rat cardiomyocytes. *J. Mol. Cell. Cardiol.* **2000**, *32*, A60.

53 MACRI, J., WELCH, K., DUBAY, T., DU, P., et al. Changes in protein expression in a rat model of coronary ligation-induced heart failure: a proteomic analysis. *Circ.* **2000**, *102*, II–28.

54 MACRI, J., MCGEE, B., THOMAS, J., DU, P., et al. Cardiac sarcoplasmic reticulum and sarcolemmal proteins separated by two-dimensional electrophoresis: surfactant effects on membrane solubilization. *Electrophoresis* **2000**, *21*, 1685–1693.

55 MACRI, J., CHU, G., DUBAY, T., DU, P., et al. Proteomic analysis of phospholamban knockout mouse hearts: changes in protein expression. *Circ.* **2000**, *102*, II–73.

56 MARCUS, K., IMMLER, D., STERNBERGER, J., MEYER, H. Identification of platelet proteins separated by two-dimensional gel electrophoresis and analyzed by matrix assisted laser desorption/ionization-time of flight-mass spectrometry and detection of tyrosine-phosphorylated proteins. *Electrophoresis* **2000**, *21*, 2622–2636.

57 GEVAERT, K., EGGERMONT, L., DEMOL, H., VANDEKERCKHOVE, J. A fast and convenient MALDI-MS based proteomic approach: identification of components scaffolded by the actin cytoskeleton of activated human thrombocytes. *J. Biotech.* **2000**, *78*, 259–269.

58 O'NEILL, E., BROCK, C., VON KRIEGS-HEIM, A., PEARCE, A., et al. Towards complete analysis of the platelet proteome. *Proteomics* **2002**, *2*, 288–305.

59 PATTON, W., ERDJUMENT-BROMAGE, H., MARKS, A., TEMPST, P., TAUBMAN, M. Components of the protein synthesis and folding machinery are induced in vascular smooth muscle cells by hypertrophic and hyperplastic agents. *J. Biol. Chem.* **1995**, *270*, 21404–21410.

60 KOHANE, D., SARZANI, R., SCHWARTZ, J., CHOBANIAN, A., BRECHER, P. Stress-induced proteins in aortic smooth muscle cells and aorta of hypertensive rats. *Am. J. Physiol.* **1990**, *258*, H1699–H1705.

61 PLEIBNER, K.-P., REGITZ-ZAGROSEK, V., TRENKNER, J., HOCHER, B., FLECK, E. Effects of renovascular hypertension on myocardial protein patterns: analysis by computer-assisted two-dimensional electrophoresis. *Electrophoresis* **1998**, *19*, 2043–2050.

62 LIAO, D.-F., JIN, Z.-G., BAAS, A., DAUM, G., et al. Purification and identification of secreted oxidative stress-induced factors from vascular smooth muscle cells. *J. Biol. Chem.* **2000**, *275*, 189–196.

63 SHUSTA, E., BOADO, R., PARDRIDGE, W. Vascular proteomics and subtractive antibody expression cloning. *Molec. Cell. Proteomics* **2002**, *1*, 75–82.

64 BROWN JONES, M., KRUTZSCH, H., SHU, H., ZHAO, Y., et al. Proteomic analysis and identification of new biomarkers and therapeutic targets for invasive ovarian cancer. *Proteomics* **2002**, *2*, 76–84.

65 SINHA, P., KOHL, S., FISCHER, J., HULTER, G., et al. Identification of novel proteins associated with the development of chemoresistance in malignant melanoma using two-dimensional electrophoresis. *Electrophoresis* **2000**, *21*, 3048–3057.

66 VERCOUTTER-EDOUART, A., LEMOINE, J., LEBOURHIS, X., LOUIS, H., et al. Proteomic analysis reveals that 14-3-3 sigma is down-regulated in human breast cancer cells. *Cancer Res.* **2001**, *61*, 76–80.

67 PEARL, L., PRODROMOU, C. Structure, function, and mechanism of the Hsp90 molecular chaperone. *Curr. Opin. Struct. Biol.* **2000**, *10*, 46–51.

68 MYERS, T., ANDERSON, N.L., WALTHAM, M., LI, G., et al. A protein expression database for the molecular pharmacology of cancer. *Electrophoresis* **1997**, *18*, 647–653.

69 WASHBURN, M., YATES, J. Analysis of the microbial proteome. *Curr. Opin. Microbiol.* **2000**, *3*, 292–297.

70 CASH, P. Proteomics in medical microbiology. *Electrophoresis* **2000**, *21*, 1187–1201.

71 SHEA, J., SANTANGELO, J., FELDMAN, R. Combating gram-positive pathogens: emerging techniques to identify relevant virulence targets. *Emerging Therap. Targets.* **2001**, *5*, 155–164.

72 GRANDI, G. Antibacterial vaccine design using genomics and proteomics. *Trends Biotech.* **2001**, *19*, 181–188.

73 NILSSON, C., LARSSON, T., GUSTAFSSON, E., KARLSSON, K., DAVIDSSON, P. Identification of protein vaccine candidates from Helicobacter pylori using a preparative two-dimensional electrophoretic procedure and mass spectrometry. *Anal. Chem.* **2000**, *72*, 2148–2153.

74 LANGEN, H., TAKACS, B., EVERS, S., BERDNT, P., et al. Two-dimensional map of the proteome of Haemophilus influenzae. *Electrophoresis* **2000**, *21*, 411–429.

75 JUNGBLUT, P., BURMANN, D., HAAS, G., ZIMNY-ARDNT, U., et al. Comparative proteome analysis of Helicobacter pylori. *Mol. Microbiol.* **2000**, *36*, 710–725.

76 JUNGBLUT, P., SCHAIBLE, U., MOLLENKGEF, H.-J., ZIMNY-ARNDT, U., et al. Comparative proteome analysis of Mycobacterium tuberculosis and Mycobacterium bovis BCG strains: towards functional genomics of microbial pathogens. *Mol. Microbiol.* **1999**, *33*, 1103–1117.

77 VAN BOGELEN, R., SCHILLER, E., THOMAS, J., NEIDHARDT, F. Diagnosis of cellular states of microbial organisms using proteomics. *Electrophoresis* **1999**, *20*, 2149–2159.

78 BOUSSAC, M., GARIN, J.. Calcium-dependent secretion in human neutrophils: a proteomic approach. *Electrophoresis* **2000**, *21*, 665–672.

79 LE NAOUR, F., HOHENKIRK, L., GROLLEAU, A., MISEK, D., et al. Profiling changes in gene expression during differentiation and maturation of monocyte-derived dendritic cells using both oligonucleotide microarrays and proteomics. *J. Biol. Chem.* **2001**, *276*, 17920–17931.

80 PIUBELLI, C., GALVANI, M., HAMDAN, M., DOMENICI, E., RIGHETTI, P. Proteome analysis of rat polymorphonuclear leukocytes: a two-dimensional electrophoresis/ mass spectrometry approach. *Electrophoresis* **2002**, *23*, 298–310.

81 STEINER, S., GATLIN, C., LENNON, J., McGRATH, A., et al. Proteomics to display lovastatin-induced protein and pathway regulation in rat liver. *Electrophoresis* **2000**, *21*, 2129–2137.

82 STEINER, S., GATLIN, C., LENNON, J., McGRATH, A., et al. Cholesterol biosynthesis regulation and protein changes in rat liver following treatment with fluvastatin. *Toxicol. Lett.* **2001**, *120*, 369–377.

83 ANDERSON, N.L., STEINER, S. Protein markers for pharmaceutical and related toxicity. In: *PCT Intl. Appl.* **2001**, Large Scale Proteomics Inc., p. 103.

84 CHEVALIER, S., MACDONALD, N., TONGE, R., et al. Proteomic analysis of differential protein expression in primary hepatocytes induced by EGF, tumour necrosis factor alpha or the peroxisome proliferator nafenopin. *Eur. J. Biochem.* **2000**, *267*, 4624–4634.

85 WATANABE, T., LALWANI, N., REDDY, J. Specific changes in the protein composition of rat liver in response to the peroxisome proliferators ciprofibrate, Wy-14,643 and di-(2-ethylhexyl)phthalate. *Biochem. J.* **1985**, *227*, 767–775.

86 WITZMANN, F., JARNOT, B., PARKER, D., CLACK, J. Modification of hepatic immunoglobulin heavy chain binding protein (BiP/Grp78) following exposure to structurally diverse peroxisome proliferators. *Fundamental Applied Toxicology* **1994**, *23*, 1–8.

87 ANDERSON, N.L., ESQUER-BLASCO, R., RICHARDSON, F., FOXWORTHY, P., EACHO, P.

The effects of peroxisome proliferators on protein abundances in mouse liver. *Toxicol. Appl. Pharmacol.* **1996**, *137*, 75–89.

88 GIOMETTI, C., TOLLAKSEN, S., LIANG, X., CUNNINGHAM, M. A comparison of liver protein changes in mice and hamsters treated with the peroxisome proliferator Wy-14,643. *Electrophoresis* **1998**, *19*, 2498–2505.

89 DOHERTY, N., LITTMAN, B., REILLY, K., SWINDELL, A., et al. Analysis of changes in acute phase plasma proteins in an acute inflammatory response and in rheumatoid arthritis using two-dimensional gel electrophoresis. *Electrophoresis* **1998**, *19* (2), 355–363.

90 EBERINI, I., MILLER, I., ZANCAN, V., BOLEGO, C., et al. Proteins in rat serum IV. Time-course of acute-phase protein expression and its modulation by indomethacine. *Electrophoresis* **1999**, *20* (4–5), 846–853.

91 ANDERSON, N.L., ANDERSON, N.G. Proteome and proteomics: new technologies, new concepts, and new words. *Electrophoresis* **1998**, *19*, 1853–1861.

92 EVERS, S., DIPADOVA, K., MEYER, M., FOUNTOULAKIS, M., et al. Strategies towards a better understanding of antibiotic action: folate pathway inhibition in Haemophilus influenzae as an example. *Electrophoresis* **1998**, *19*, 1980–1988.

93 EVERS, S., GRAY, C. Application of proteome analysis to drug development and toxicology. In: *Proteomics*, PENNINGTON, S.R., DUNN, M.J. (Editor), **2001**, BIOS Scientific Publishers, Oxford, p. 225–236.

94 STEINER, S., AICHER, L., RAYMACKERS, J., MEHEUS, L., et al. Cyclosporine A mediated decrease in the rat renal calcium binding protein calbindin-D 28 kDa. *Biochem. Pharmacol.* **1996**, *51*, 253–258.

95 AICHER, L., MEIER, G., NORCROSS, A., JAKUBOWSKI, J., et al. Decrease in kidney calbindin-D as a possible mechanism mediating CsA and FK-506-induced calciura and tubular mineralization. *Biochem. Pharmacol.* **1997**, *53*, 723–731.

96 AICHER, L., WAHL, D., ARCE, A., GRENET, O., STEINER, S. New insights into cyclosporine A nephrotoxicity by proteome analysis. *Electrophoresis* **1998**, *19*, 1998–2003.

97 FOUNTOULAKIS, M., BERNDT, P., BOELSTERLI, U., CRAMERI, F., et al. Two-dimensional database of mouse liver proteins: changes in hepatic protein levels following treatment with acetominophen or its nontoxic regioisomer 3-acetamidophenol. *Electrophoresis* **2000**, *21*, 2148–2161.

98 RUEPP, S., TONGE, R., SHAW, J., WALLIS, N., POGNAN, F. Genomics and proteomics analysis of acetaminophen toxicity in mouse liver. *Toxicol. Sci.* **2002**, *65*, 135–150.

99 NEWSHOLME, S., MALEEFF, B., STEINER, S., ANDERSON, N.L., SCHWARTZ, L. Two-dimensional electrophoresis of liver proteins: characterization of a drug-induced hepatomegaly in rats. *Electrophoresis* **2000**, *21*, 2122–2128.

100 ANDERSON, N.L., STEELE, V., KELLOFF, G., SHARMA, S. Effects of oltipraz and related chemoprevention compounds on gene expression in rat liver. *J. Cell. Biochem.* **1995**, *22*, 108–116.

101 STEINER, S., WAHL, D., MANGOLD, B., ROBISON, R., et al. Induction of the adipose differentiation-related protein in liver of etomoxir-treated rats. *Biochem. Biophys. Res. Comm.* **1996**, *218*, 777–782.

102 ARCE, A., AICHER, L., WAHL, D., ANDERSON, N.L., et al. Changes in the liver protein pattern of female Wistar rats treated with the hypoglycemic agent SDZ PGU 693. *Life Sci.* **1998**, *63* (25), 2243–2250.

103 ARDEKANI, A., HERMAN, E., SISTARE, F., LIOTTA, L., PETRICOIN, E. Molecular profiling of cancer and drug-induced toxicity using new proteomic technologies. *Curr. Therap. Res.* **2001**, *62* (11), 803–819.

104 LESKO, L., ATKINSON Jr, A. Use of biomarkers and surrogate endpoints in drug development and regulatory decision making: criteria, validation, strategies. *Ann. Rev. Pharmacol. Toxicol.* **2001**, *41*, 347–366.

105 SKEHEL, J., SCHNEIDER, K., MURPHY, N., GRAHAM, A., et al. Phenotyping apolipoprotein E*3-Leiden transgenic mice by two-dimensional polyacrylamide gel electrophoresis and mass spectrometric iden-

tification. *Electrophoresis* **2000**, *21*, 2540–2545.

106 SRINIVAS, P., KRAMER, B., SRIVASTAVA, S. Trends in biomarker research for cancer detection. *Lancet Oncol.* **2001**, *2*, 698–704.

107 BISCHEL, V., LIOTTA, L., PETRICOIN III, E. Cancer proteomics: from biomarker discovery to signal pathway profiling. *Cancer J.* **2001**, *7* (1), 69–78.

108 CELIS, J., WOLF, H., OSTERGARD, M. Bladder squamous cell carcinoma biomarkers derived from proteomics. *Electrophoresis.* **2000**, *21*, 2115–2121.

109 ALAIYA, A., FRANZEN, B., AUER, G., LINDER, S. Cancer proteomics: from identifiction of novel markers to creation of artificial learning models of tumor classification. *Electrophoresis* **2000**, *21*, 1210–1217.

110 HONDERMARCK, H., VERCOUTTER-EDOUART, A.-S., REVILLION, F., LEMOINE, J., et al. Proteomics of breast cancer for marker discovery and signal pathway profiling. *Proteomics* **2001**, *1*, 1216–1232.

111 ADAM, B.-L., VLAHOU, A., SEMMES, O., WRIGHT Jr, G. Proteomic approaches to biomarker discovery in prostate and bladder cancers. *Proteomics.* **2001**, *1*, 1264–1270.

112 WRIGHT, G., CAZARES, L., LEUNG, S.-M., NASIM, S., et al. ProteinChip surface enhanced laser desorption/ionization (SELDI). *Prostate Cancer Prostat. Dis.* **2000**, *2*, 264–276.

113 PETRICOIN III, E., ARDEKANI, A., HITT, B., LEVINE, P., et al. Use of proteomic patterns in serum to identify ovarian cancer. *Lancet* **2002**, *359*, 572–577.

114 OSTERGAARD, M., WOLF, H., ORNTOFT, T., CELIS, J. Psoriasin (S100A7): a putative urinary marker for the follow-up of patients with bladder squamous cell carcinomas. *Electrophoresis* **1999**, *20*, 349–354.

115 STEEL, L., MATTU, T., MEHTA, A., HEBESTREIT, H., et al. A proteomic approach for the discovery of early detection markers of hepatocellular carcinoma. *Dis. Markers.* **2001**, *17*, 179–189.

116 VLAHOU, A., MENDRINOS, S., KONDYLIS, F., SCHELLHAMMER, P., et al. Identification of protein changes in bladder cancer patient urine by ProteinChip SELDI affinity mass spectrometry. *Proc. Am. Assoc. Cancer Res.* **2000**, *41*, 852–853.

117 VLAHOU, A., SCHELLHAMMER, P., MENDRINOS, S., PATEL, K. Development of a novel proteomics approach for the detection of transitional cell carcinoma of the bladder in urine. *Amer. J. Pathol.* **2001**, *158*, 1491–1502.

118 HANASH, S., BRICHORY, F., BEER, D. A proteomic approach to the identification of lung cancer markers. *Dis. Markers* **2001**, *17*, 295–300.

119 OSTERGAARD, M., RASMUSSEN, H., NIELSEN, H., VORUM, H., et al. Proteome profiling of bladder squamous cell carcinomas: identification of markers that define their degree of differentiation. *Cancer Res.* **1997**, *57*, 4111–4117.

120 BONNER, R., EMMERT-BUCK, M., COLE, K. Laser capture microdissection: molecular analysis of tissue. *Science* **1997**, *278*, 1481–1483.

121 ORNSTEIN, D., GILLESPIE, J., PAWELETZ, C., DURAY, P., et al. Proteomic analysis of laser capture microdissected human prostate cancer and in vitro prostate cell lines. *Electrophoresis* **2000**, *21*, 2235–2242.

122 EMMERT-BUCK, M., GILLESPIE, J., PAWELETZ, C., ORNSTEIN, D., et al. An approach to proteomic analysis of human tumors. *Mol. Carcinog.* **2000**, *27*, 158–165.

123 WULFKUHLE, J., McLEAN, K., PAWELETZ, C., SGROI, D. New approaches to proteomic analysis of breast cancer. *Proteomics* **2001**, *1*, 1205–1215.

124 PAWELETZ, C., GILLESPIE, J., ORNSTEIN, D., SIMONE, N., et al. Rapid protein display profiling of cancer progression directly from human tissue using a protein biochip. *Drug Dev. Res.* **2000**, *49*, 34–42.

125 ORNSTEIN, D., ENGLERT, C., GILLESPIE, J., PAWELETZ, C., LINEHAN, W., EMMERT-BUCK, M., PETRICOIN III, E. Characterization of intracellular prostate-specific antigen from laser capture microdissected benign and malignant prostatic epithelium. *Clin. Cancer Res.* **2000**, *6*, 353–356.

126 VOSS, T., AHORN, H., HABERL, P., DOHNER, H., WILGENBUS, K. Correlation of clinical data with proteomic profiles in 24 patients with B-cell chronic lymphocytic leukemia. *Intl. J. Cancer* **2001**, *91*, 180–186.

127 ROHLFF, C. Proteomics in molecular medicine: applications in central nervous systems disorders. *Electrophoresis* **2000**, *21*, 1227–1234.

128 PASINETTI, G., Ho, L. From cDNA microarrays to high-throughput proteomics. Implications in the search for preventive initiatives to slow the clinical progression of Alzheimer's disease dementia. *Rest. Neurol. Neurosci.* **2001**, *18*, 137–142.

129 DAVIDSSON, P., NILSSON, C. Peptide mapping of proteins in cerebrospinal fluid utilizing a rapid preparative two-dimensional electrophoretic procedure and matrix-assisted laser desorption/ionization mass spectrometry. *Biochim. Biophys. Acta* **1999**, *1473*, 391–399.

130 SCHONBERGER, S., EDGAR, P., KYDD, R., FAULI, R., COOPER, G. Proteomic analysis of the brain in Alzheimer's disease: molecular phenotype of a complex disease process. *Proteomics* **2001**, *1* (12), 1519–1528.

131 GERBER, S., LUBEC, G., CAIRNS, N., FOUNTOULAKIS, M. Decreased levels of synaptosomal associated protein 25 in the brain of patients with Down Syndrome and Alzheimer's disease. *Electrophoresis* **1999**, *20*, 928–934.

132 EDGAR, P., SCHONBERGER, S., DEAN, B., FAULL, R., et al. A comparative proteome analysis of hippocampal tissue from schizophrenic and Alzheimer's disease individuals. *Mol. Psychiat.* **1999**, *4*, 173–178.

133 HARRINGTON, M. Diagnosing neurologic disorders. In: *PCT Intl. Appl.* **1998**, Neuromark, USA, p. 51.

134 HERATH, H., ATHULA, C., PAREKH, R., ROHLFF, C. Proteins, genes and their use for diagnosis and treatment of schizophrenia. In: *PCT Intl. Appl.* **2001**, Oxford Glycosciences (UK) Ltd., p. 161.

135 HERATH, H., CHANDRASIRI, A., PAREKH, R., ROHLFF, C., et al. Protein and genetic markers for diagnosis and treatment of schizophrenia. In: *PCT International Appl.* **2001**, Oxford Glycosciences (UK) Ltd., Pfizer Inc p. 148.

136 JOHNSTON-WILSON, N., SIMS, C., HOFMANN, J.-P., ANDERSON, N.L., et al. Brain protein markers for diagnosing brain disorders. In: *PCT Intl. Appl.* **2001**, Large Scale Proteomics Corp., p. 43.

137 DURHAM, K., FRIEDMAN, D., HERATH, H., ATHULA, C., et al. Nucleic acid molecules, polypeptides, and uses including diagnosis and treatment of Alzheimer's diseae. In: *PCT Intl. Appl.* **2001**, Oxford Glycosciences (UK) Ltd., Pfizer Inc., p. 162.

138 SCHROTZ-KING, P., KING, A., MANN, M., ANDERSEN, J., KUESTER, B. Human seizure-related proteins and their encoded cDNA sequences. In: *PCT Intl. Appl.* **2001**, Proteome Inc., p. 150.

139 HERATH, H., ATHULA, C., PAREKH, R., ROHLFF, C. Protein, genes and their use for diagnosis and treatment of bipolar affective disorder (BAD) and unipolar depression. In: *PCT Intl. Appl.* **2001**, Oxford Glycosciences (UK) Ltd., p. 163.

140 HERATH, H., ATHULA, C., PAREKH, R., ROHLFF, C., PATEL, T. DPI-6, a putative therapeutic target and biomarker in neuropsychiatric and neurological disorders. In: *PCT Intl. Appl.* **2001**, Oxford Glycosciences (UK) Ltd., p. 91.

141 PANG, J., GINANNI, N., DONGRE, A., HEFTA, S., OPITECK, G. Biomarker discovery in urine by proteomics. *J. Proteome Res.* **2002**, *1*, 161–169.

142 PAREKH, R., PATEL, T., TOWNSEND, R. Two-dimensional electrophoresis in the diagnosis of rheumatoid arthritis. In: *PCT Intl. Appl.* **1999**, Oxford Glycosciences (UK) Ltd., p. 157.

143 GRECO, A., BAUSCH, N., COUTE, Y., DIAZ, J. Characterization by two-dimensional gel electrophoresis of host proteins whose synthesis is sustained or stimulated during the course of herpes simplex virus type 1 infection. *Electrophoresis* **2000**, *21*, 2522–2530.

144 NILSSON, C. Fingerprinting of Helicobacter pylori strains by matrix-assisted laser desorption/ionization mass spectrometric analysis. *Rapid Comm. Mass Spectrom.* **1999**, *13*, 1067–1071.

145 CHOW, T., FRY, K., LIM, M., MCATEE, C. Antigenic composition and method of detection for Helicobacter pylori. In: *PCT Intl. Appl.* **1998**, Genelabs Technologies Inc., p. 402.

146 HENDRICKSON, R., DOUGLASS, J., REY-
NOLDS, L., MCNEILL, P., et al. Mass spectrometric identification of Mtb81, a novel
serological marker for tuberculosis. *J.
Clin. Microbiol.* **2000**, *38* (6), 2354–2361.

147 NOEL-GEORIS, I., BERNARD, A., FAL-
MAGNE, P., WATTIEZ, R. Proteomics as
the tool to search for lung disease markers in bronchoalveolar lavage. *Dis. Markers* **2001**, *17*, 271–284.

148 CAWTHORNE, M., SANCHEZ, J.-C. Protein
expression-based screening and diagnostic methods and compositions relating to
pancreatic islet and *β*-cell dysfunction.
In: *PCT Intl. Appl.* **2001**, Proteome
Sciences PLC, p. 121.

Subject Index